PHYSIOLOGY AND
BEHAVIOUR OF PLANTS

PHYSIOLOGY AND BEHAVIOUR OF PLANTS

Peter Scott

University of Sussex

John Wiley & Sons, Ltd

Other Wiley Editorial Offices

John Wiley & Sons Inc., 111 River Street, Hoboken, NJ 07030, USA

Jossey-Bass, 989 Market Street, San Francisco, CA 94103-1741, USA

Wiley-VCH Verlag GmbH, Boschstr. 12, D-69469 Weinheim, Germany

John Wiley & Sons Australia Ltd, 42 McDougall Street, Milton, Queensland 4064, Australia

John Wiley & Sons (Asia) Pte Ltd, 2 Clementi Loop #02-01, Jin Xing Distripark, Singapore 129809

John Wiley & Sons Canada Ltd, 6045 Freemont Blvd, Mississauga, Ontario, L5R 4J3

Wiley also publishes its books in a variety of electronic formats. Some content that appears in print may not be available in electronic books.

Library of Congress Cataloging-in-Publication Data

Scott, Peter.
  Physiology and behaviour of plants / Peter Scott.
    p. cm.
  Includes bibliographical references and index.
  ISBN 978-0-470-85024-4 (cloth) – ISBN 978-0-470-85025-1 (pbk.)
  1. Plant physiology. 2. Plants. I. Title.
  QK711.2.S36 2008
  575.9–dc22   2007041618

British Library Cataloguing in Publication Data

A catalogue record for this book is available from the British Library

ISBN 978-0-470-85024-4 (HB)   ISBN 978-0-470-85025-1 (PB)

Typeset in 10/12 Times Roman by Thomson Digital, New Delhi, India
Printed and bound in Malaysia by Vivar Printing Sdn Bhd

3   2013

*This book is dedicated
to Helen and Edmund Scott,
whose love and support through
many plant hunts made this work possible.*

Contents

Preface

We live in an age where we are more dependent upon our understanding of plants than ever before. With the world human population exceeding 6 billion, never has it been more important to be able to feed reliably so many people. Yet, paradoxically, it is an age when interest in plant biology is at an all-time low. The plant biology content of GCSE and A level biology in the UK has become minimal as more and more human biology is added to the curriculum. Funding of plant science research has become minuscule compared to that commanded by the medical sciences. As a consequence, universities employ fewer plant biologists and as a knock-on effect plant biology is scarcely offered as part of a biology degree any longer. Where plant biology is offered, the lecture material can be badly handled and few students are prepared to study it into their final year. But plant biology has to be one of the most interesting subjects in the whole of biology. Plants are central to life and play a crucial role in dictating the diversity of life we currently enjoy on this planet.

To address this problem, what is not needed is another plant physiology textbook. There are many volumes which do the job of teaching plant form and function very well. But what often comes across in these texts is that plants are stale and irrelevant. Somewhere, the interest, appeal and relevance to the real world has been lost in the detail. If you don't think this is true, then look no further than the most amazing enzyme of plants, RUBISCO. Most books dive into the sub-unit arrangement and its carboxylase/oxygenase properties, but there is nothing to impress the reader that we are discussing the enzyme that supports virtually all of life currently on this planet! RUBISCO is the marvel enzyme of the universe, and yet most students never realize this. Then consider photosystem II, a complex enzyme which splits water to provide an electron donor for a chlorophyll molecule and allows non-cyclical photophosphorylation – but the fact that it generates all of the oxygen we need to breathe and supports the only other known biological autotrophic pathways known in bacteria is lost. So what is, on the face of it, merely a tricky biochemical pathway, suddenly becomes a vital component of our everyday life. Plant biology is amazing in so many ways that this book barely scratches the surface. Having said that, the most interesting and remarkable topics on plant biology have been hand-picked to give students the best chance possible to study plants in all of their glory. The book deliberately steers clear of in-depth discussion of molecular biology and the explosion of knowledge this is providing. This is because much of this knowledge is incomplete and as yet does not give an interesting insight into the mechanisms of plant life. Where molecular mechanisms are discussed, they are always related to the whole plant function, as this is primarily what the text is concerned with – plants.

Peter Scott
July 2007

1

Introducing plants

In this chapter, we look at how plants originated, what floral diversity there is today and the make-up of the plant and its ultrastructure.

The beginning: the evolution of plants and the major divisions

In the beginning, it is most probable that plants evolved from photosynthetic bacteria. From these bacteria the red and green algae evolved; and from freshwater-dwelling green algae the simple lower plants, such as mosses and ferns, evolved; and so on, up to the higher plants. A phylogenetic tree is shown in Figure 1.1 and Table 1.1 to demonstrate the relationship between the members of the plant kingdom and their relative abundance through the history of the planet.

Conquering the land

The origin of plants was in water, where both photosynthetic bacteria and then algae originated. Light penetration of water reduces with depth, and on average only 1% of incident light reaches to a depth of over 15 m. As a consequence, there is a body of water at the surface known as the photic zone, where all of the photosynthetic activity in oceans occurs. A great deal of this photosynthetic activity still occurs at the shores of the oceans, where more complex algae have evolved; as a group these are commonly referred to as seaweeds. Algae are restricted to the oceans and freshwater bodies since, as part of the life cycle of algae, gametes (sex cells) that swim through water are required for sexual reproduction. This is thought to have been a major hindrance to plants attempting to colonize the land surface of the Earth and to overcome this, new reproductive systems needed to evolve.

For the colonization of the land, methods of gamete transfer that were independent of water needed to evolve. Bryophytes (the mosses and liverworts), the first land-dwelling plants, still depend on moisture to complete their reproductive cycle (Figure 1.2). Sperm is released from haploid male gametophytes (a gamete-producing individual involved in the life cycle of bryophytes) which must swim to fuse with the egg cell of the haploid female gametophyte. This fusion yields a diploid zygote that divides to form a stalked cup-like structure, which releases haploid spores. These spores then form haploid male and female gametophytes. As a direct consequence of this life cycle and its requirement for water, mosses and liverworts are restricted to growing in moist habitats. In addition, these plants have no waterproof cuticle or vascular tissue, and are therefore very limited in their ability to transport water and carbohydrate made during photosynthesis to the rest of the plant. This makes it necessary for these plants to have a prostrate growth habit (rarely exceeding 2 cm in height), colonizing banks near to areas of water. Bryophytes do possess root-like structures known as rhizoids, but these are thought to have a function of anchorage rather than for transporting water to the aerial tissues. As a result of their inability to regulate water in the plant, bryophytes are poikilohydric; therefore, if the moisture declines in their habitat it also begins to decrease in their tissues. Some mosses can survive drying out but others need to be kept wet to survive. However, the ability to tolerate temporary drying of a habitat is a first step to colonizing dry land.

The mosses exhibit little ability to control water loss and if plants were ever to colonize a greater area of the land, the non-vascular plants needed to evolve solutions to this problem by developing a water transport system and a means of regulating water loss from the plant surface. The hornworts possess stomata on their leaf surfaces and therefore took the first steps to regulating water loss while maintaining gaseous exchange, which is essential for photosynthesis. These structures are absent in the mosses

Physiology and Behaviour of Plants   Peter Scott
© 2008 John Wiley & Sons, Ltd

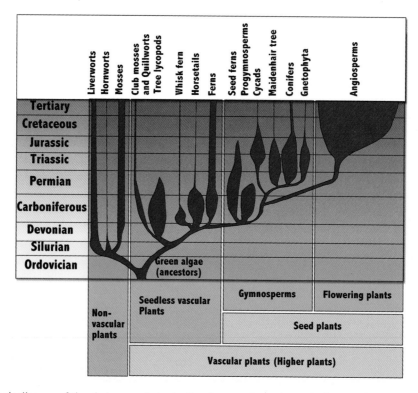

Figure 1.1   Schematic diagram of the phylogeny of plants. The diagram shows the evolutionary relationship between the different species and the relative abundance in terms of species numbers (shown by the width of the dark red lineage tree). On the *y* axis the different time periods of the evolutionary history of plants are shown. The diagram is based on that presented by Ridge 2002.

Table 1.1   Estimates of numbers of species occurring in plant divisions.

Plant division	Common name	Approximate number of species
Hepatophyta	Liverworts	6000
Anthocerophyta	Hornworts	100
Bryophyta	Mosses	10 000
Lycophyta	Club mosses and quillworts	1000
Tree lycopods		Extinct
Psilotum	Whisk fern	3
Equisetum	Horsetail	15
Pterophyta	Ferns	11 000
Peltasperms	Seed ferns	Extinct
Progymnosperms		Extinct
Cycadophyta	Cycads	140
Ginkophyta	Maidenhair tree	1
Coniferophyta	Confers	550
Gnetophyta	Vessel-bearing gymnosperms	70
Angiosperms	Flowering plants	235 000

These estimates are based on there being around 260 000 different species of plant in the world. Species of plants based on data from Ridge 2002.

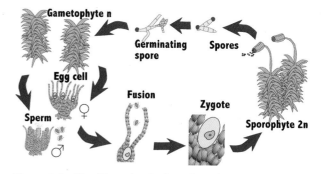

Figure 1.2   The life cycle of a bryophyte (mosses). There are around 10 000 different bryophyte species. The life cycle of bryophytes occurs in two phases, the sporophytic phase and the gametophytic phase. In the sporophytic phase the plant is diploid and develops as sporophyte. This releases spores which form gametophytes that are haploid. These form into separate sex gametophytes. The male gametophyte releases sperm, which is motile and swims towards the archegonium on the female gametophytes that bears the egg cells. These fuse to yield a zygote that then forms the sporophyte.

but present in more advanced plants. The prostrate growth of the mosses does not use light effectively and makes the damp habitat very competitive. As a consequence, there must have been a great selective pressure for structures to evolve in plants which raised the plants off the ground, which will have led to the evolution of a limited upright shoot in the bryophytes.

The evolution of the pteridophtyes (ferns) marked the development of a simple vascular tissue, allowing long-distance transport of water and carbohydrate around the plant. It also permitted the evolution of the upright shoot, thereby making taller plants possible. Ferns use a method for sexual reproduction similar to that of mosses (Figure 1.3). The adult fern plant is diploid and releases haploid spores, which divide and form male and female gametophytes. The male releases haploid mobile gametes, which swim and fuse with the female gametes to form diploid zygotes, which then divide to form the adult ferns. Pteridophytes possess distinct leaves, which enhance their ability to photosynthesize. Although the pteridophytes possess a water-resistant cuticle, they exhibit poor control of water loss from their leaves and in most instances are still restricted to moist habitats. The spores released by the adult ferns are tolerant of desiccation but movement of the male gamete still requires water. The ferns were the first plants to evolve lignin as a defence and support structure (see later). Plants that contain vascular tissue are frequently referred to as 'higher plants'.

A small number of ferns and lycophytes exhibit heterospory (separate sex spores). In bryophytes and

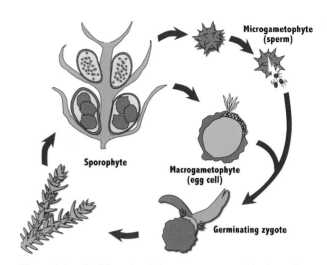

Figure 1.4   The life cycle of a heterosporous pteridophyte. In a small number of instances species of ferns are heterosporous and this is thought to be a crucial evolutionary step in the formation of the flowering plants. In Figure 1.3 the egg cell and the sperm are the same size, but in heterosporous pteridophytes, the female gamete (macrogametophyte) is much larger than the male gamete (microgametophyte).

pteridophytes discussed so far, the spores are all identical. However, with heterosporous species the male sperm are formed from a microgametophyte and the female gamete from a macrogametophyte (Figure 1.4). The formation of separate sex gametophytes is considered to be one of the major steps towards the formation of plants that bear seeds.

The habitat range of plants on land was widened considerably with the evolution of plants that produce seeds. The transition from being wholly aquatic to wholly terrestrial is considered complete in such plants. The seed-bearing plants are divided into two divisions, the gymnosperms (Pinophyta) and the angiosperms (flowering plants, Magnoliophyta). The angiosperms form the largest and most diverse plant division. Angiosperms produce reproductive structures in specialized organs called flowers, where the ovary and the ovule are enclosed in other tissues. The gymnosperms do not form a flower and the reproductive structures are in the form of cones, in which the ovule is not enclosed at pollination. On formation of a seed, the embryo is covered by a specialized scale leaf rather than the ovary and this gives rise to the name 'gymnosperm', which means 'naked seed'. Gymnosperms were widespread in the Jurassic and Cretaceous periods of the Earth's history but there are now fewer than 800 different species. They occupy a range of different habitats from temperate forests to more arid habitats. Most of the species are trees and shrubs, which are adapted very well to temperate areas of the world, where water availability is limited over extensive periods of the year (due to ground frosts).

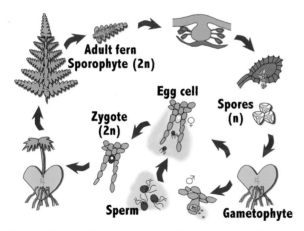

Figure 1.3   The life cycle of a pteridophyte (ferns). The adult fern, or sporophyte, is diploid and on maturity releases spores from sori on the underside of leaves. These spores fall to the soil and germinate and from a gametophyte, also known as a thallus. This matures and forms an archegonium (where egg cells form) and an antheridium (where the sperm develops). The sperm are released and swim to the egg cells, with which they fuse to form diploid zygotes. The zygote then develops into an adult fern.

Species such as pines are well adapted to cold climates through features such as small needle-shaped leaves (which hold little snow and are resistant to weather damage), sunken stomata (which reduce water loss) and a thick waxy cuticle. There are a few examples of gymnosperms which live in desert habitats, such as *Welwitschia*, and in tropical habitats, such as the cycads, but these species are limited in number.

The gymnosperms were once the dominant division of plants on Earth, but in the Cretaceous period there was a huge expansion in the number of angiosperms, which led to the steady decline of gymnosperms. The failure of gymnosperms to maintain the dominant position appears to be the result of improvements in reproductive biology in the angiosperms. Gymnosperms use wind as a means of transferring pollen from the male cones to the female cones (Figures 1.5 and 1.6). The use of air currents for pollination will be discussed in greater detail in Chapter 9, but suffice it to say here that this process is inefficient and wasteful of

resources. Moreover, gymnosperms rely mainly on the scale leaf around the seed to disperse it, with a few species such as yews using primitive fruits to attract animals. Angiosperms, however, evolved a huge range of different methods for attracting insects and animals for flower pollination and for the subsequent seed dispersal. The evolution of broader leaves may have been a disadvantage in colder climates, but everywhere else this allowed increases in efficiency of photosynthesis and hence growth rates.

Although gymnosperms are not as successful as angiosperms, they still have an important place in the ancestry of the flowering plants. Most of the structures thought of as typical in flowering plants can be found in individuals of the gymnosperms, but no species contains them all. The closest relative to the gymnosperms is thought to be the phylum Gnetophyta, which contains species such as *Welwitschia mirabilis* and *Ephedra viridis* (see Chapter 19). Angiosperms first appear in fossil records in the early

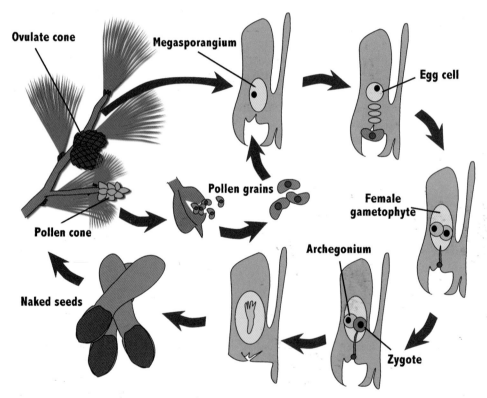

Figure 1.5   The life cycle of a gymnosperm. Mature gymnosperms such as conifers form two types of cones; the female cone is called the ovulate cone and the male cone is called the pollen cone. The female egg cell forms from a macrospore mother cell, which undergoes meiosis to form a tetrad of macrospores. Only a single one of these survives and undergoes many mitotic divisions to form ultimately the female gametophyte. An egg cell forms which can be up to 3 mm in diameter, making it the largest egg cell in the plant kingdom. The male gametes are released and are carried by air currents to the ovulate cones. The pollen germinates and fertilizes an egg cell. This goes on to develop into a zygote and ultimately a seed. This seed is not enclosed by an ovary, which gives the gymnosperms ('naked seeds') their name. There are four distinct divisions of the gymnosperms: Coniferophyta, Cycadophyta, Ginkgophyta and Gnetophyta. These are split into less than 800 different species recognized currently.

Figure 1.6 Gymnosperm cones. Here the male (B) and female cones (C) of the stone pine (*Pinus pinea*) are shown.

Cretaceous period, around 140 million years ago, but despite their late appearance they have been spectacularly successful (Willis and McElwain, 2002). The features that identify angiosperms are:

1. The presence of an enclosed ovary (covered by carpels and two layers of integument).

2. The formation of flowers.

3. The presence of specialized vascular tissues for nutrient and water movement.

4. The presence of double fertilization for the formation of the embryo and endosperm.

There are around 350 different families of flowering plant and 260 000 different species. The life cycle of a typical angiosperm is shown in Chapter 9. The angiosperm division is subdivided into two classes of plants, the Magnoliopsida (dicotyledons, around 180 000 species) and the Liliopsida (monocotyledons, around 80 000 species, about 33% of these belonging to the family Orchidaceae). The principal characteristics that allow easy identification of these classes are shown in Table 1.2. Monocotyledons produce pollen with a single pore or furrow, which is the place where the pollen tube will emerge on the stigmatic surface upon germination. This means that if the pollen grain lands with this furrow facing away from the stigma, fertilization is more difficult to achieve. In dicotyledons there are three such furrows and this is thought to result in a furrow always being in contact with the surface of the stigma, therefore making fertilization easier (Endress, 1987). However, both classes are very successful in terms of land area covered and number of different species. From the earliest fossil records there is compelling evidence that the dicotyledons were the first angiosperms to evolve, with very few monocotyledon tissues being preserved as fossils at this time. However, many of the monocotyledon species known today are

Table 1.2 The usual distinguishing characteristics between monocotyledons and dicotyledons.

Feature	Monocotyledons	Dicotyledons
Flower structure	Petals and sepals in groups of three	Petals and sepals in groups of four or five
Pollen	Monocolpate (one pore or furrow)	Tricolpate (three pores or furrows)
Cotyledons	One	Two
Vascular tissue arrangement in leaves	Parallel	Net-like
Vascular bundles through stem	No apparent order in arrangement	Encircling the stem just under the epidermis of the stem
Secondary growth	Absent	Present

Redrawn from Raven *et al.* (1992) and Magallon *et al.* (1992). There are exceptions to these characteristics, but those listed are the common features that allow easy identification.

thought to have evolved from species originating in the early Cretaceous period. It is thought that the absence of monocotyledons from the fossil record may be a result of there being fewer monocotyledon species and the fact that they are generally herbaceous and do not possess woody tissues, which would make them less likely to be preserved as fossils.

The most recent evolutionary change to have occurred, judging from the plant fossil record, is the appearance of plants possessing modifications to the usual C3 pathway for photosynthesis, i.e. C4 and Crassulacean acid metabolism (CAM) plants. These plants are discussed in greater detail in Chapter 12.

The evolution of lignin

Lignin is a complex polymer of phenylpropane units, containing very variable bonding, bound to cellulose to make the cell wall of plants (Figure 1.7). The structure is very rigid and forms a strong support or defence structure in various plant tissues. The evolution of lignin synthesis in plants was essential for the transition from prostrate plants to upright plants. In order to move water through tissues, osmosis can be relied on for short distances, but if the plant is to become upright, reinforced transport vessels are needed in the form of the xylem (see Chapter 4). It has been proposed that plants could not grow taller than a few centimetres in height without the evolution of lignin (Niklas, 1997). Lignin is thought to have evolved around 430 million years ago and the key to its appearance has been proposed to be the high oxygen levels present in the atmosphere at that time (Willis and McElwain, 2002). Lignin played a central role in the evolution of plants because it permits the support of upright plants, which optimized the use of light levels. It allowed the evolution of the upright ferns and ultimately the first trees. The evolution of upright plants also permitted the movement

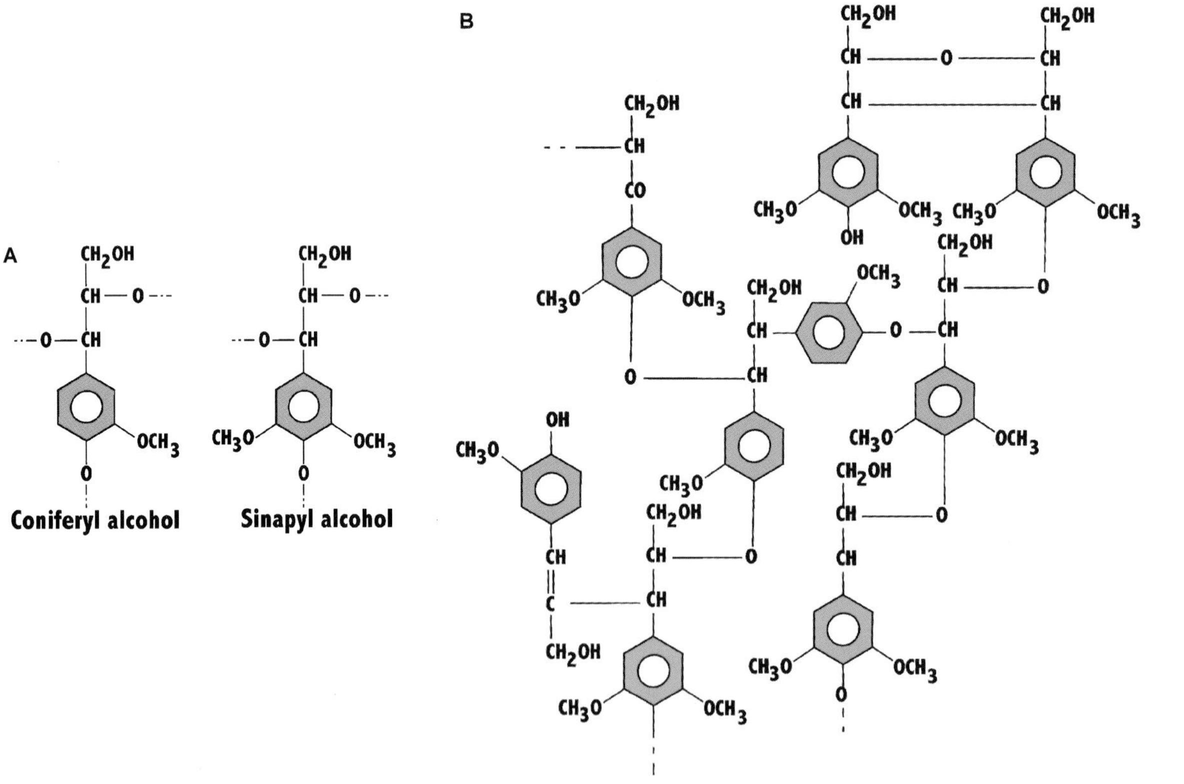

Figure 1.7 The structure of lignin. The group of compounds known as lignin are among the most abundant organic chemicals on earth. Lignin makes around one-third of the dry weight of wood and is especially used by plants to reinforce the xylem for support. It is also a very effective defensive compound, being very difficult to break down and metabolize. It is a biopolymer made up of several different monomers, two of which are shown, coniferyl and sinapyl alcohol (A). These monomers are polymerized to make the complex structure of lignin, an example of which is shown in (B).

of plants into a wider range of habitats and hence allowed plants to conquer the land. In addition, lignin is not easily broken down by other organisms; it was therefore an excellent defence compound. As a consequence of this, large amounts of plant material could not be degraded and gradually became buried in the earth, yielding what we now know as oil and coal. These reserves provided a huge sink for taking CO_2 out of the atmosphere and storing it in an organic form. This supported a fall in global CO_2 levels, which in turn influenced evolution of all the other organisms on the planet. It is only in the present day that this stored CO_2 is being re-released into the atmosphere through the actions of humankind, with the knock-on effect of potential global warming.

Plants and mass extinction

Over the course of the Earth's history, there are five recognized periods of mass extinction in the marine fauna records. These are times when 20–85% of all living marine fauna species are recorded to have disappeared in the fossil record. The most well-known of these events is at the boundary between the Cretaceous and Tertiary periods, when all the dinosaurs are thought to have become extinct. At this point, around 80% of all animal species became extinct. How did plants fare at this time? There is a marked reduction in the abundance and range of species represented in the fossil record at this point in time (Saito *et al.*, 1986). There was then a striking, but temporary, rise in the abundance of fern spores, which was followed by a major reduction in the diversity of fossil plants compared with that prior to the boundary event. It is thought that this may represent major destruction of much of the vegetation, followed by a rapid process of revegetation. However, different areas of the world seem to have been affected in different ways, such that in some regions the vegetation cover appears scarcely to have changed. The extinction rate of plants at this time has been estimated as 5–10% (Halstead, 1990). This period was therefore one in which a great deal of the Earth's biomass was lost, but there were no mass extinctions, as observed with the fauna species. Similar observations were made at the other points of mass extinction noted in the fauna fossil records.

Why did mass extinctions not occur in the plant kingdom? The answer to this question reveals something about the versatility of plants and why they are so successful. Most animal species are very mobile and if their habitat becomes less amenable to survival, they can simply move to a new region. However, if the changes are far-reaching, it may not be within the organism's capacity to move

beyond the stressed habitat and death or extinction is likely. These major extinction events are all likely to have been global catastrophes resulting in major changes in the world's climate, therefore escape would have been impossible. Plants, however, have always had to tolerate the inconvenience of not being able to move: therefore they have evolved to be adaptable to survive local catastrophes. These survival mechanisms appear to have been so effective that even after global catastrophes plants appear to have bounced back very rapidly. Put simply, shooting all the white rhinoceros in Africa will cause the species to become extinct since animals cannot regenerate or give birth to young once dead; however, cut down all the *Acacia* trees in Africa and they will resprout and grow again within a season. If all the plants were ripped out of the ground, they could rely upon the bank of seeds in and on the soil surface to regenerate the population. Seeds, as discussed in Chapter 14, can be very resistant to environmental stresses, such as long-term drought, high temperatures, darkness and other adverse conditions. In consequence, plants are much more difficult to eradicate than animals, as any gardener struggling with bindweed or ground elder can testify! It is estimated that around 1% of the angiosperm species that have ever existed are still alive today. This is a difficult figure to estimate from the fossil record but nevertheless this is a very high percentage compared with the survival of animal species, and is testimony to the resilience of plants in the face of adversity.

The majority of plants are angiosperms and therefore most of the discussion in this book will be restricted to the higher plants, and mainly the angiosperms.

Floristic kingdoms, biogeography and biomes

As plants evolved and time passed, the continents of the Earth shifted. It is therefore logical that certain later-evolving plants will be located only in specific regions of the world, whereas other species, which were particularly successful and evolved at an early stage in the Earth's history, will be found over many regions of the world. For example, the sundew *Drosera rotundifolia* is found over a wide area of the Northern Hemisphere, from the Far East to North America (Figure 1.8). This species is extraordinarily successful and is probably the most widespread of all of the *Drosera* species. The logical explanation for its distribution is that it evolved at a time when Asia, Europe and North America were contiguous. However, another carnivorous plant, *Sarracenia*, has a very wide range on the eastern side of North

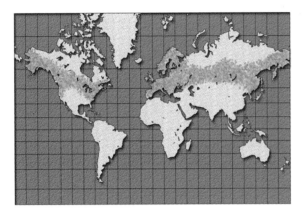

Figure 1.8 Global distribution of the round-leaved sundew, *Drosera rotundifolia*. The distribution map emphasizes that this species of sundew is found extensively in the temperate regions of the northern hemisphere, but is completely absent from the southern hemisphere. This genus is therefore highly likely to have evolved prior to the major continental drift had occurred, which isolated the Sarraceniaceae.

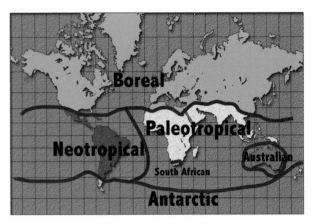

Figure 1.10 Floral kingdoms of the world. Ronald Good (1947) proposed a means of dividing the globe into six floristic kingdoms, based on the distribution areas of unrelated taxa. The location and extent of these kingdoms are shown on the map; they have since been subdivided into subkingdoms or provinces and then into regions.

America but is found nowhere else in the world (Figure 1.9). These species are also very successful and in trial planting in the UK certain species have proved to be very invasive. Thus, there are many potential habitats across the world in which *Sarracenia* could thrive, and yet it is restricted to North America. The most likely reason for this failure to spread is that this genus evolved in North America after the continents separated, and thus never had the opportunity to colonize Europe.

A traveller around the globe would encounter very different plant species in different continents. This led Ronald

Good 1947 to divide the world into six floristic kingdoms (Figure 1.10), based on the distribution areas of unrelated taxa. These kingdoms were subdivided into subkingdoms or provinces and then into regions (Table 1.3). Despite an unfortunate use of the word 'kingdom', Good had defined a very useful way of looking at the flora of the planet. The significance of this work meant that our hypothetical traveller could move westwards from China all the way to Alaska and encounter very similar species of flora (the Boreal kingdom). For example, the orchid genus *Cypripedium* can be found over just that range. Many other species could also be detailed in a similar way. However, if

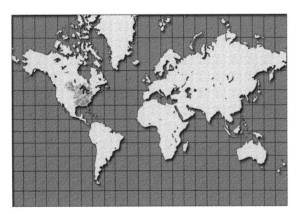

Figure 1.9 Global distribution of the North American pitcher plant, genus *Sarracenia*. The distribution map emphasizes that this genus is only found on the continent of North America. Therefore this genus this highly likely to have evolved once this continent had split from other continents.

Table 1.3 Proportions of flowering plant species found in the different floral kingdoms.

Floristic kingdom	Proportion of world species (%)	Location
Boreal (Holoarctic)	42	North America, Europe, Central Asia
Paleotropical	35	Central Africa
Neotropical	14	South America
Australian	8	Australia, Tasmania, and New Zealand
Antarctic (Holantartic)	1	Southern tip of South America
Cape (Fynbos)	0.04	Western Cape within South Africa

The densest area for flora in the world is the Fynbos (1300 species/ 10 000 km²); the next densest is the Neotropical Kingdom (400 species/ 10 000 km²).

our traveller moved southwards towards South America the vegetation would gradually change to that of the Neotropical kingdom. There have been many further attempts to redefine the floristic kingdoms into bio-geographical areas (Takhtajan, (1986)) or into biomes (major regional ecosystems, e.g. desert, characterized by dominant forms of plant life and the prevailing climate) but the underlying message is the same. Over certain areas of the world specific plant species are   endemic.

One particular feature that stands out in Good's floristic kingdoms is the presence of the Fynbos region in South Africa. This is the smallest of the floristic kingdoms, measuring only 90 000 km^2 and yet possessing over 8600 different plant species, around 5800 (68%) of them being endemic (Cowling and Hilton-Taylor, 1994). Conse-quently, in terms of flora, this is the most biologically diverse area in the world. Table Mountain above Cape Town has more different plant species growing on its hillsides than are found in the whole of the UK! The main vegetation of the Fynbos region is hard-leafed evergreen shrubland, which is prone to frequent fires. Many of the species present in the region use fire as a cue to trigger germination of the seeds (see Chapter 14). The main species are members of the families Restionaceae (Cape reeds), Ericaceae (heathers and their relatives), Fabaceae (pea family), Proteaceae (the largest shrubs in the area) and Iridaceae (bulb plants) (Figure 1.11). The other remarkable feature of the area is that more than 10 of the genera are represented by more than 100 different species, which is very diverse indeed. Many of the plants that are frequently found in gardens across the world were originally from this area, e.g. *Gladiolus, Freesia, Pelagonium, Ixia* and *Sparaxis*, to name but a few. Nowhere else on Earth can such floral diversity be seen.

What makes a plant?

Structure of the whole plant

What makes plants so remarkable and why are they any different from animals? To appreciate this, we need to look at the overall general structure of a plant and then at the ultrastructure of the plant cell.

The general structure of a plant is shown in Figure 1.12. Plants possess a root system for anchorage, water uptake and mineral ion uptake. The role of the root system in plants is discussed in more detail in Chapter 4. At the tip of the root is an apical meristem, which generates further root axis and controls the development of root growth. As mentioned earlier, the acquisition of a root system was crucial for supporting the movement of plants onto dry

land. Water is essential for the functioning of a living cell and therefore an efficient mechanism for moving water from the soil to all of the cells of a plant is necessary for the transition to drier habitats. But roots should be seen as a support tissue and, as mosses have been successful for millions of years with just a rudimentary rhizoid system for water and mineral uptake, they are obviously not essential in damp habitats. It is the presence of the root system which dictates that plants are fixed in position. For an efficient root system, there needs to be considerable branching through the soil; this makes withdrawal of the root system from the soil impossible and hence plants can never be mobile. This factor steers the rest of the evolution of plants, because they need to be able to deal with anything their environment can throw at them or they will die.

Connected to the roots is the stem, which in some plants is so short that it is barely visible without dissection. However, this is a result of internode lengths. Along the stem, leaves are attached at nodes and the stem spacing between the nodes is known as an internode. Short internodes yield a plant which is short and produces a rosette of leaves. Longer internodes yield a plant which is tall and can take advantage of light at several different levels. Which habit a plant adopts is dependent upon what the plant has evolved to compete with. A tall plant may optimize light usage but also risks damage. In addition, height does not necessarily mean that light levels are optimized. Measurements from my own laboratory show that the tongue orchid (*Serapias lingua*) is tall to compete with tall grasses in its habitat. However, it is slower-growing and less efficient at producing a tuber at the end of a growth season than the rosette-forming green winged orchid (*Anacamptis morio*), which competes with lower-growing grasses (Figure 1.13). Height may there-fore be an advantage, or it may just be a necessity of competition.

Leaves are attached at the nodes along the stem. At each leaf attachment point there is frequently an axillary bud. The leaves form the major sites on most plants for photo-synthesis. The first land plants possessed mainly photo-synthetic stems, but upright structures are inefficient at absorbing the available light and leaves evolved to form a structure perfectly adapted to capturing light while mini-mizing water loss. Leaves come in so many varying shapes and forms that there is apparently no specific leaf shape which is optimal for photosynthesis.

At the tip of the stem is the apical meristem, which generates further aerial plant structures, such as new stem, leaves, flowers and fruits. The axillary buds are usually dormant until the apical meristem has grown a sufficient distance away from the axillary bud; however, if

Figure 1.11   The smallest floral kingdom, the Fynbos region in South Africa, which covers 0.04% of the Earth's surface but possesses 3.4% of the recognized plant species, with around 68% of the species being endemic. To put this in context, the Fynbos region has 8500 different plant species with around 5700 endemics, whereas the UK only has 1500 different species with 20 endemics. This makes the Fynbos the most biologically diverse area on the planet. (A) A view from the top of Table Mountain, which is situated in the south-western corner of the Fynbos region. The other photographs show examples of some of the species which can be found in this region; (B) *Erica plukenetii*; (C) *Polygala myrtifolia*; (D) *Protea lepidocarpodendron*; (E) *Elegia capensis*; (F) *Leucospermum cordifolium*; (G) *Euryops abrotanifolius*.

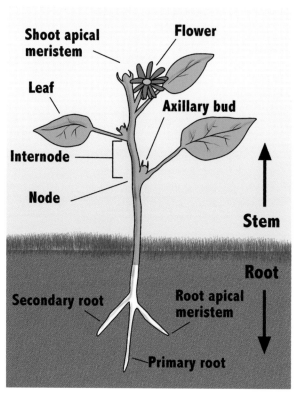

Figure 1.12 Schematic diagram of the general stucture of a dicotyledonous plant, labelled with the major structures, which are referred to later in the text.

the apical meristem is damaged the apical dominance is released and the other buds on the stem can then begin to grow.

Structure of the plant cell

The plant cell displays numerous differences from an animal cell. Typical plant and animal cells are shown in Figure 1.14. The rest of the discussion here is limited to the structures specific to plants.

The most obvious difference is the degree of compartmentation in a plant cell. Plant cells contain more organelles and microbodies than animal cells. In addition, up to 95% of a plant cell can be devoted to a large membrane-bound space, the vacuole. All of these will be discussed in turn, but the most important difference between an animal cell and a plant cell is the presence of chloroplasts.

Chloroplasts and other plastids

Chloroplasts, like mitochondria, are around the same size as a bacterial cell. Both contain DNA, which is not arranged in the form of a chromosome but as a circular piece of DNA known as a plasmid (Figure 1.15). The gene sequences possess no introns and are controlled by

Figure 1.13 Growth habits of *Anacamptis morio* and *Serapias lingua*. (A) *A. morio* in its natural habitat at the Piddinghoe Reservoir, East Sussex, UK. (B) Isolated *A. morio* plant, showing the flattened leaves and short stem which is perfect for competing with short grasses. (C) *S. lingua* plant shown in its natural grassy habitat in Portugal. (D) Isolated *S. lingua* plant, showing the raised leaves and slightly longer stem. This allows the plant to compete with longer grasses, which is its preferred habitat.

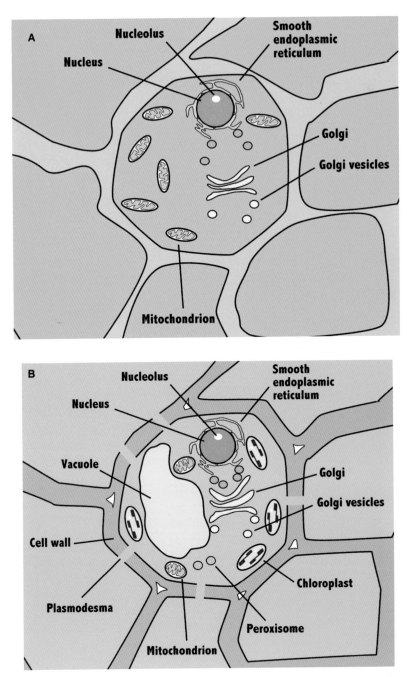

Figure 1.14 Schematic diagram of the general stucture of an animal cell (A) and a plant cell (B). There are numerous differences, which are labelled.

promoter sequences which are very similar to those observed in bacteria. In addition, the ribosomes used by both of these organelles are smaller than those used for protein synthesis in the cytosol of plant cells and are again very similar to those used by bacteria. This compelling evidence has led to the suggestion that these organelles in animals and plants were once free bacteria, which at some point in the evolution of life fused with some larger form of bacterium or eukaryote. During invagination into the larger bacterial cell, a second membrane covered the organelles to give the double-membrane-bound chloroplast (a photosynthetic bacterium) and mitochondrion (a non-photosynthetic bacterium). Animal cells never acquired chloroplasts. The photosynthetic system used by

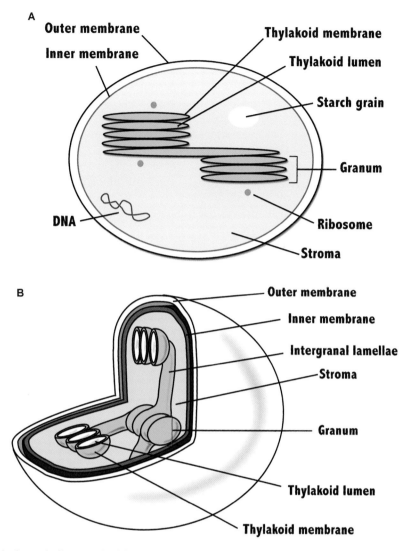

A

Outer membrane
Inner membrane
Thylakoid membrane
Thylakoid lumen
Starch grain
Granum
DNA
Ribosome
Stroma

B

Outer membrane
Inner membrane
Intergranal lamellae
Stroma
Granum
Thylakoid lumen
Thylakoid membrane

Figure 1.15 Detailed schematic diagram of a chloroplast, shown simply as a cross-section and as a three-dimensional representation.

plants is therefore essentially that of a bacterium which has evolved for millions of years within the plant cell and hence has changed with time. This accounts for why the photosynthetic mechanisms in cyanobacteria and plants are so similar. It is not known whether there was just a single union of a bacterium with a large bacterium or whether this occurred many times in the Earth's history.

The mechanism of photosynthesis and the ultrastructure of the chloroplast are discussed in greater detail in Chapter 2, but the chloroplast is just one of a family of organelles unique to plants, collectively known as plastids. In flowers there is a rearrangement of the internal thylakoid membrane to produce a wave effect, and in this membrane coloured pigments, such as carotenoids, are borne; these plastids are called chromoplasts and they give rise to colour in some flowers and other plant structures, e.g. the

colour of carrot roots is formed from carotenoids in the chromoplasts. In non-photosynthetic plant tissues the plastids are frequently used for the synthesis of starch, and these are known as leuco- or amyloplasts (see Chapter 3). The basic structure is the same in all of these plastids and in many instances they can transform from one to another, e.g. carrots can become green, and many flowers begin green but turn another colour as they mature. Many synthetic reactions occur in the plastids of plants cells, such as starch and lipid synthesis, and many amino acids are also synthesized in them. The presence of plastids in plant cells varies. In some cells, such as pollen cells, there are few plastids. In non-photosynthetic plant cells there can be 5–10 amyloplasts, whereas there can be several hundred chloroplasts in a photosynthetic cell in a leaf. They can occupy as much as 20% of the cell volume in certain cells, which highlights their importance in metabolism.

The vacuole

Another obvious feature in the plant cell is the presence of a vacuole. The volume which this 'bag' occupies varies from cell to cell and species to species. The vacuole in a Crassulacean acid metabolism plant may occupy up to 98% of the cell volume, but a typical cell will have a vacuole which occupies 70–95% of the cell. The vacuole is bound by a single membrane, known as the tonoplast. In immature cells the vacuole is frequently divided into a number of smaller vacuoles but as the cell matures these tend to fuse to form one. The vacuole is a reserve for storing certain compounds but, more importantly, it is a means of generating cell volume without diluting the contents of the cytosol (the volume of cell outside the membrane-bound organelles, vacuole and microbodies in the cell). Through regulation of the solute concentration in the vacuole, a plant can regulate its volume. The vacuole can then be used to apply pressure on the cell wall of the plant cell, and thus in certain circumstances cause the cell to expand and in others allow the cell to maintain turgidity. This permits cell growth with little investment in *de novo* synthesis of new cellular materials. It also allows plant cells to be much larger than animal cells in general.

In many tissues, pigments such as anthocyanins are dissolved in the vacuole to give the cell a colour. This is used in some plants to give rise to flower or fruit colour. Mineral ions, sucrose and secondary metabolites are also stored in the vacuole.

Microbodies

Microbodies are single-membrane bound organelles in the cytoplasm of plant cells. They are small in size, around 1 µm in diameter. They are often associated with activity of the endoplasmic reticulum but there are two very important microbodies found in plants, the peroxisome and the glyoxysome.

Peroxisomes are used mainly in the metabolism of hydrogen peroxide and most of the cellular reactions which generate this compound are carried out in this microbody. Hydrogen peroxide can be potentially very damaging to cells, as it is a powerful oxidizing agent. The peroxisome contains the enzyme catalase, which metabolizes this reagent to water. There is still some debate as to the function of the peroxisome, as certain reactions which generate hydrogen peroxide also occur in the cytosol.

However, the detoxification hypothesis is the best to date. The role of the peroxisome in photorespiration is discussed in Chapter 2.

The glyoxysome is found in tissues such as oil-bearing seeds which are metabolizing lipids and converting them to carbohydrate. They catalyse the breakdown of fatty acids to acetyl co-enzyme A (acetyl-Co A), which is exported from the microbody and used in respiration or the synthesis of sucrose. Glyoxysomes have rarely been observed to occur in animal cells and hence animals have great difficulty converting fats into carbohydrate.

The cell wall

Animal cells generally do not possess a rigid cell wall, but plant cells do. Rather than being like brick wall, the cell wall should be viewed as a molecular filter and a dynamic part of the cell. The cell wall occupies the space between neighbouring cells and is made up of complex polymers of cellulose, hemicellulose, pectin, lignin and proteins (Figure 1.16). In growing cells the primary cell wall is flexible and thin, but as the cells age a secondary cell wall develops, which is much more rigid and plays a major role in support of the whole plant. The primary cell wall is made up mainly of cellulose but the secondary cell wall contains a large proportion of lignin. As mentioned earlier, the evolution of lignin as a strengthening compound in plants played a major role in enabling plants to becoming upright (i.e. attaining the strength to hold aerial tissue off the ground) and therefore in vastly broadening the habitat range of plants. Lignin is resistant to degradation by other organisms and is very strong. The strength of lignin is best appreciated in the wood of trees. The trunk of a tree can support many tons in weight and strong winds, and lignin is probably one of the strongest natural materials. This is precisely why it has been so useful as a building material for most human cultures.

Plasmodesmata

Many plant cells possess links between cells known as plasmodesmata (singular: plasmodesma), which offer a continuum between the cytosol of one cell and that of a neighbouring cell (Figure 1.17). This allows the movement of solutes between the cells. The role of the plasmodesmata in sucrose movement within leaves and developing sink organs is discussed in detail in Chapter 3. The frequency of their occurrence between plant cells is usually an indication of the traffic between cells. There is some evidence that there is control of movement between cells through plasmodesmata, since they can be used to load the phloem

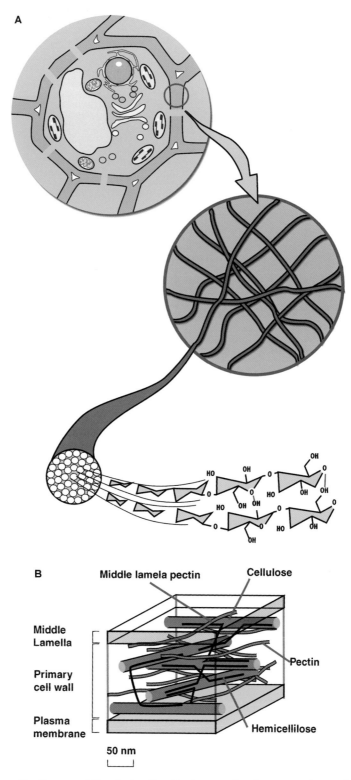

Figure 1.16 Detailed schematic diagram of the plant cell wall, shown as it relates to a plant cell in a leaf, showing a series of magnifications of the structure. The actual composition of the cell wall is also shown, and how the structure is layered around each plant cell.

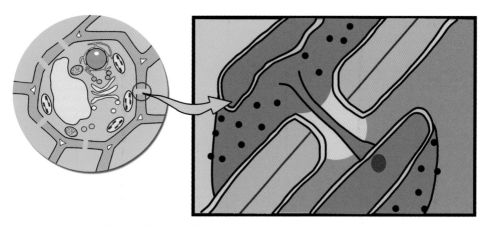

Figure 1.17 Detailed schematic diagram of an individual plasmodesma. The plasmodesma forms a symplastic link between two adjacent plant cells.

in certain plant species. Larger molecules, e.g. raffinose and stachyose, cannot get through plasmodesmata (see Chapter 3).

This chapter has outlined the evolution of plants, their types and distribution in different parts of the world and the basic structure of a plant and its cells, setting the scene for the following chapters. Plants dominate the planet – but how did they get to this position and what are the underlying physiological characteristics that allow them to be so successful?

References

Cowling RM and Hilton-Taylor C (1994) Plant diversity and endemism in southern Africa and overview, in *Botanical Diversity in Southern Africa* (ed. Huntley BJ) National Botanical Institute, Kirstenbosch, 31052.

Endress PK (1987) The early evolution of the angiosperm flower. *Tree*, **2**, 300–304.

Good R (1947) *The Geography of Flowering Plants*, Longmans/Green, New York.

Halstead LB (1990) Cretaceous–Tertiary (terrestrial), in *Palaeobiology: A Synthesis* (eds Briggs DEG and Crowther PR) Blackwell Science, Oxford, UK, pp. 203–207.

Magallon S, Crane PR and Herendeen PS (1999) Phylogenetic pattern, diversity and diversification of eudicots. *Annals of the Missouri Botanical Garden*, **86**, 297–372.

Niklas KJ (1997) *The Evolutionary Biology of Plants*, University of Chicago Press, Chicago, IL.

Raven JA, Evert RF and Eichorn SE (1992) *Biology of Plants*, 5th edn. Worth, New York.

Ridge I (2002) Plant evolution and structure, in *Plants*, Oxford University Press, Oxford, UK, Chapter 1.

Saito T, Yamanoi T and Kaiho K (1986) End-Cretaceous devastation of terrestrial floral in the boreal Far East. *Nature*, **323**, 253–255.

Takhtajan A (1986). *Floristic Regions of the World*, [transl. TJ Corvello and A Conquist University of California Press, Berkeley, CA.

Willis KJ and McElwain JC (2002) *The Evolution of Plants*, Oxford University Press, Oxford, UK.

2

Photosynthesis: the ultimate in autotrophy

Plants make up more than 99% of the biomass of the earth and this dominance is based upon one feature of plants – the ability to photosynthesize. Without any doubt the most important ability plants possess is the capacity to photosynthesize, converting light energy into chemical energy in the form of carbohydrate molecules. Photosynthesis is crucial for supporting the life of the plant and all other living organisms.

Photosynthesis is remarkable not only for the facility to convert light energy into chemical energy but also for the capacity to take CO_2 from the air and convert it into a six-carbon ring carbohydrate (glucose or fructose, etc.). The CO_2 in the atmosphere is present largely as a result of recycling from rotting organic material. However, billions of years ago CO_2 appeared in our atmosphere as a result of volcanic activity. These levels have reduced over time through carbohydrate synthesis by plants. The six-carbon carbohydrate is crucial for the survival of most living organisms today, for by the action of enzymes on glucose, compounds such as the five-carbon molecule, ribose can be made (the basis for the sugar component of DNA), together with the three- and four-carbon compounds, which are the precursors of amino acids, and repeated two-carbon units, which make up the structure of fatty acids. The six-carbon ring is central to the production of all of these molecules and the pivotal reason why plants can make these is the presence of the enzyme ribulose 1,5-bisphosphate carboxylase/oxygenase (RUBISCO). At the same time as absorbing CO_2 in the atmosphere and producing carbohydrates, plants also release oxygen as a by-product of photosynthesis. The reduction of atmospheric CO_2 levels and the release of oxygen has shaped the evolution of life on the planet. Plants are essential for life to go on as we know it and therefore it is important that we understand as thoroughly as possible how they manage to photosynthesize, which is the purpose of this chapter. Some of the biochemical reactions and pathways described are complex but it is necessary to understand them in order to appreciate the significance of the processes and the way they affect plant growth and reproduction.

Light harvesting

Leaf form

It is not unusual for compounds to interact with and absorb light. Any coloured compound reflects and absorbs light; it is therefore not surprising that plants can do so. However, in plants it is the unusual property of the pigment chlorophyll that not only allows the absorption of light energy but also converts it into a form that can be converted into chemical energy.

Leaves are the major photosynthetic organs of most plants, and there are three major considerations which dictate the shape and form of a leaf (Figure 2.1). First, leaves need to offer a wide area for the capture of light energy. Second, they need to be thin, to provide efficient gas exchange for fuelling CO_2 fixation and maximizing the interception of light per unit of material invested in the leaf structure. Third, the leaf requires a means of transporting the products of photosynthesis to heterotrophic tissues. These requirements may dictate the basic form of a leaf, but the shape of leaves from different plants varies considerably and there is obviously no optimum shape.

The epidermis of the leaf usually has no photosynthetic function and lacks chloroplasts. However, it does play a role in providing a waterproof coating and frequently the cells are arranged such that they act as lenses, focusing the light on chloroplasts in the photosynthetic cells of the mesophyll (Figures 2.2 and 2.3). In addition, the epidermis also plays a role in absorbing UV-B light, and thus protecting the photosynthetic cells from this potentially damaging radiation.

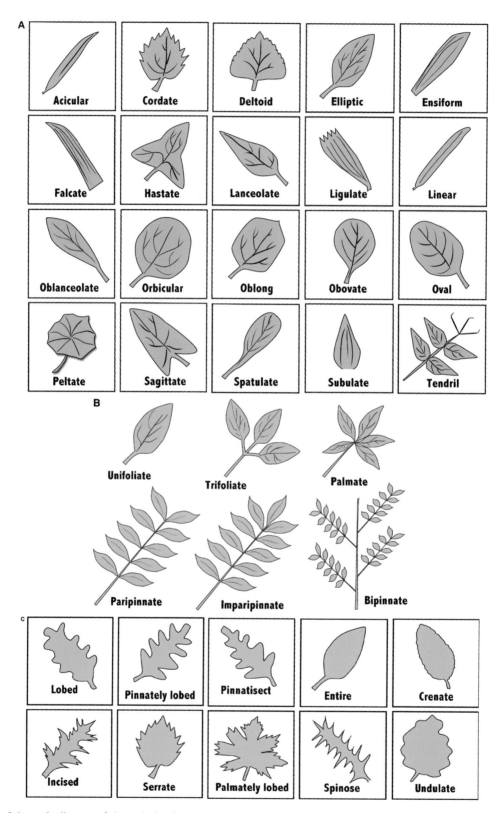

Figure 2.1   Schematic diagram of the variation in leaf shape in plants. (A) A range of overall leaf shapes and the names given to describe the leaf form. (B) Compound leaves are considered, with the varying arrangement of the leaflets. (C) The leaf margin reviewed, with the names given for the different forms the leaf margin can take in plants.

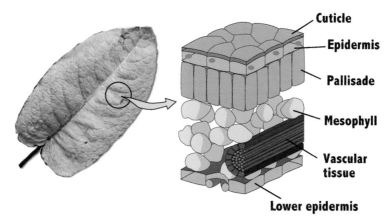

Figure 2.2 Leaf section showing detail of the cell components. A leaf of a *Rumex* species is shown, with a schematic section of the leaf. The cuticle protects the epidermis of the leaf and acts as a defensive barrier and as a barrier to limit water loss. The epidermis layer of cells produces the cuticle and further protects the photosynthetic cells within the leaf. The palisade layer of cells is part of the photosynthetic mesophyll of the leaf. The palisade cells are arranged in vertical files and contain large numbers of chloroplasts. This layer can be very variable; in some leaves there can be more than one layer, and in leaves from other species this layer can be completely absent. Beneath this is the spongy mesophyll layer. The large air spaces between the cells permit efficient gaseous exchange between the cells and the air around them. The stomatal pores (controlled by the guard cells) allow gaseous exchange between the leaf and the surrounding atmosphere.

Chlorophyll and the chloroplast

The mesophyll cells in the centre of the leaf contain chloroplasts, which can occupy up to 20% of the cell volume. Chloroplasts are made up of an inner and outer membrane which surround a space called the stroma (Figure 2.4). The outer membrane of the chloroplast appears to have little role in plant metabolism. It may be a vestige of the origin of chloroplasts, which are thought to have been absorbed into larger cells within membrane invaginations. In contrast, the inner chloroplast membrane plays a major role as a barrier between the stroma and the cytosol. Within the stroma is a complex series of membranes known as the thylakoid membrane, which form sacs. The inner space of these sacs is referred to as the thylakoid lumen. The thylakoid membrane bears a large number of proteins associated with the capture of light energy. It is to the thylakoid membrane that the chlorophyll molecules are attached. The enzymes associated with the Calvin cycle and the storage of harvested light energy (in the form of chemical energy in a carbohydrate) are present in the stroma of the chloroplast.

The thylakoid membrane is arranged in two different ways in the chloroplast (Figure 2.4). In many sections, stacks of membranes are visible and these are known as grana (singular: granum). Extending from these and linking granal stacks together there are intergranal lamellae. This arrangement is not always observed and sometimes the granal stacks are absent, for example in the bundle sheath cells of C4 plants.

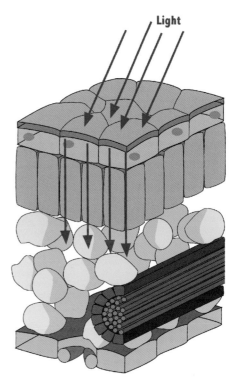

Figure 2.3 Light striking the leaf surface; the light first needs to penetrate the epidermis cells, which are usually transparent and have no photosynthetic function. These cells help to refract light into the mesophyll cells, so that light capture is optimized.

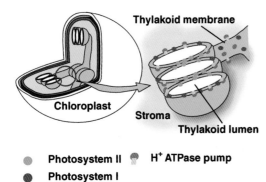

● **Photosystem II**   ● **H⁺ ATPase pump**
● **Photosystem I**

Figure 2.4   Detail of the chloroplast. The photosynthetic cells in the mesophyll contain large numbers of chloroplasts. Light strikes the antennae complexes linked to photosystems I and II, which are present in the thylakoid membrane within the chloroplast. The arrangement of these protein complexes and the H⁺ ATPase are shown.

Specific pigments in plants contained within the chloroplast capture light energy. The major pigments are chlorophyll (which is present in higher plants in two forms, chlorophylls *a* and *b*), and carotenoids (β-carotene being the major pigment of this class of compounds). The structure of these molecules is shown in Figure 2.5 and their light absorption characteristics in Figure 2.7. These pigments are bound to proteins on the thylakoid membrane to form antennae complexes. The chlorophylls form the major pigments for light absorption and their interaction with light depends on the presence of a porphyrin ring with a magnesium ion bound to the centre of the molecule. Chlorophyll absorbs light mainly at the blue and red ends of the visible spectrum of light and reflects mainly green wavelengths of light (Figure 2.6). Carotenoids absorb light of wavelengths of 400–500 nm and reflect yellow to orange wavelengths of light. It is these characteristics of the pigments that cause plants to be green. The yellow light reflected by carotenoids is

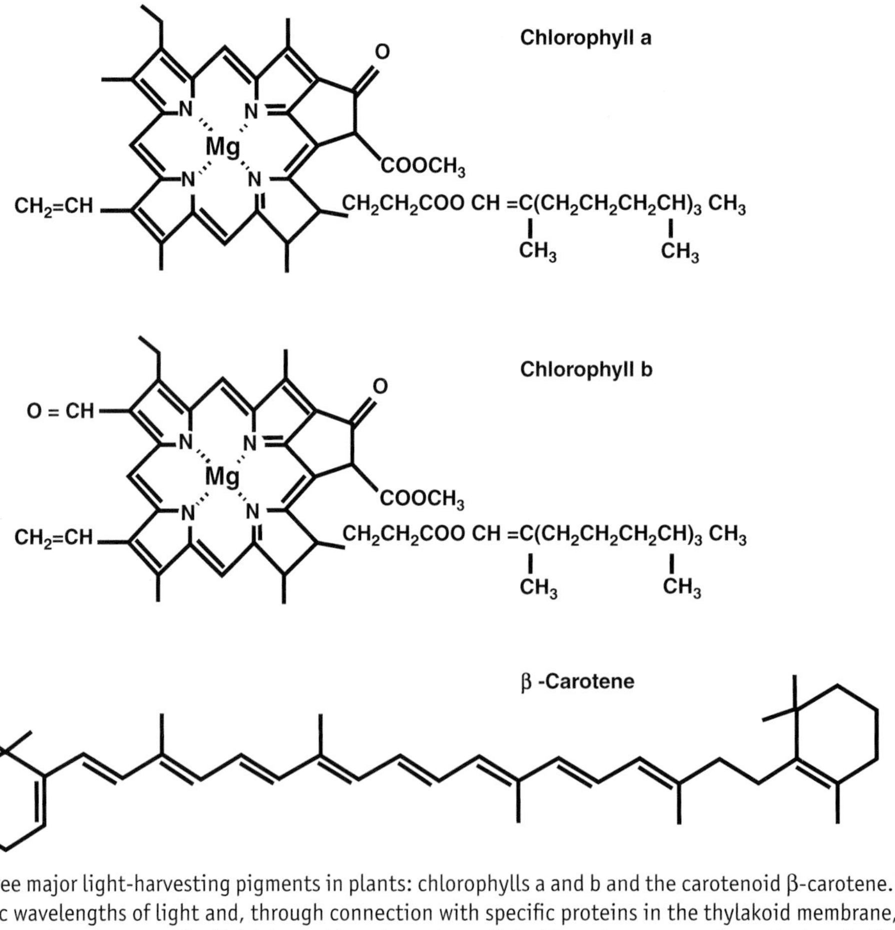

Chlorophyll a

Chlorophyll b

β -Carotene

Figure 2.5   The three major light-harvesting pigments in plants: chlorophylls a and b and the carotenoid β-carotene. The molecules interact with specific wavelengths of light and, through connection with specific proteins in the thylakoid membrane, they can pass this absorbed energy to other pigments. The light-harvesting pigments are linked to antennae complexes that are further linked to the photosystems in the thylakoid membrane.

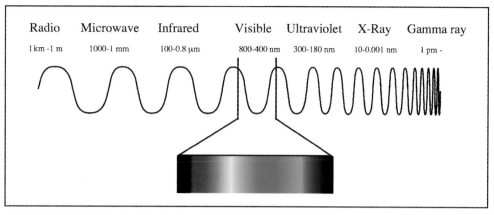

Figure 2.6 Electromagnetic radiation wavelengths. There are many different wavelengths of electromagnetic radiation that strike a leaf surface, but only a very limited range of wavelengths that make up visible light, as shown here, actually play a role in exciting the light-harvesting pigments.

usually masked by the reflected green light of the chlorophyll. Plants cannot absorb all of the light which is incident upon the leaves and hence they are not black. The carotenoids help to broaden the range of wavelengths of light that the plant can absorb and are known as accessory pigments. The ability of the photosynthetic pigments to absorb light of different wavelengths can be assessed through measuring an action spectrum. This is a measure of the response of a plant to light and can be assessed through measuring oxygen evolution. A typical action spectrum is shown in Figure 2.7. Such measurements give an assessment of the ability of a plant to photosynthesize with any given wavelength of light. As can be seen from Figure 2.7, a plant can absorb a small percentage of the light in the green region of the spectrum and this is a direct result of the accessory pigments.

The chlorophyll molecules and carotenoids are bound to proteins which make up the antennae complexes; 200–300 chlorophyll molecules per complex are bound to these complexes, in addition to the accessory pigments, which are arranged around a central chlorophyll molecule known as the reaction centre chlorophyll. When struck by photons of light, electrons within the pigment molecules enter an excited state. Much of the light energy absorbed is lost as heat as the electron returns to the ground state, but some can be passed on to neighbouring pigments through a process of resonance transfer. This in turn excites an electron in the next pigment, and so on, within the antenna complex. The pigments are arranged such that the energy required for an electron to attain an excited state decreases, the closer the pigment is to the reaction centre of the antenna. In practice, this means that energy is usually channelled from the carotenoids to chlorophyll *b*

molecules, then from chlorophyll *b* to chlorophyll *a* molecules, and ultimately to the reaction centre. This arrangement funnels the energy from the captured light to the reaction centre and therefore maximizes the conversion of captured light energy into chemical energy. There is some energy loss during this transfer; however, this loss is essential to maintain the direction of flow of captured energy to the reaction centre chlorophyll molecule. In addition, the presence of the antenna complex around each reaction centre chlorophyll increases the efficiency of photosynthesis. Without an antenna complex, each reaction centre chlorophyll molecule would need to be struck by a photon of light energy. This would lead to a huge increase in the number of reaction centres required and the associated protein machinery, but the majority of these reaction centres would be redundant most of the time, waiting to interact with photons of light.

Converting light energy into chemical energy

When the reaction centre chlorophyll *a* molecule enters an excited state it loses its electron completely, and it is this loss that is at the heart of the conversion of light energy into chemical energy within the plant (Figure 2.8). The antennae complexes are linked to two photosystems on the thylakoid membrane, photosystems I and II, which have different roles during photosynthesis and interact with one another. Photosystem I is mainly located in the intergranal membrane region of the thylakoids, whereas photosystem II is mainly located in the granal stacks, where the membranes are sandwiched together (appressed membranes).

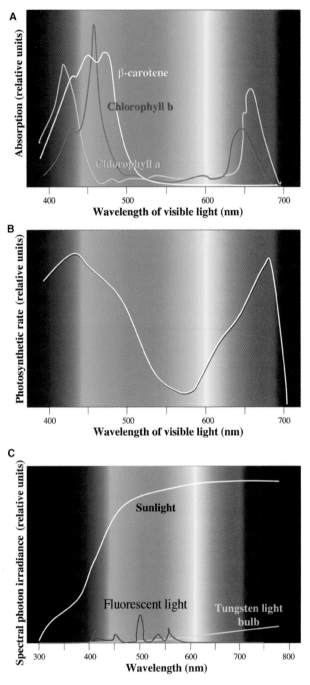

Figure 2.7   Three graphs showing the effects of different wavelengths of light on different aspects of photosynthesis. (A) The absorption properties of the three major photosynthetic pigments. All three pigments, on their own, show an absorption peak in both the blue and the red ends of the spectrum. Green light is mainly reflected by a leaf, hence its colour. (B) The photosynthetic rate for a plant as a function of the wavelength of light shone on the leaf. These data show that, although the absorption properties of the isolated pigments show two clear peaks, inside the leaf these peaks are still evident, but there is considerable broadening on the peak. This means that the leaf is more able to harvest light in the green section of the spectrum than would be anticipated from the absorption characteristics of the isolated pigments. This broadening is a function of the pigments being associated with membranes and proteins in the leaf, and other minor pigments will be present also. (C) The relative frequencies of different wavelengths of light in varying light sources. These data clearly show that sunlight contains the broadest array of different wavelengths of light, with fluorescent lighting being a poor second. Tungsten lights are very poor for supporting photosynthesis.

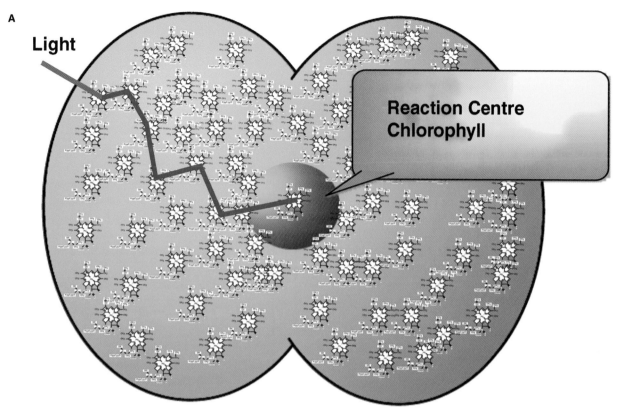

Figure 2.8 The capture of light energy. (A) Schematic representation of the antenna complex. Around 200 chlorophyll molecules are arranged on a protein complex, so that they can interact with light. The array of chlorophyll molecules is such that, if a suitable wavelength of light strikes any molecule, then an electron will move into an excited state in a higher energy level. When the electron returns to the unexcited state, the energy is passed on from one chlorophyll molecule to another [B, (a)–(d)] by transduction. At each step energy is lost, but this allows the whole antenna complex to act as a funnel, channelling the captured light energy to the centre of the array, where the reaction centre chlorophyll lies, linked to a photosystem complex.

Photosystem I

The reaction centre of this photosystem is called P700, as this is the maximum wavelength of light it can absorb. When the reaction centre chlorophyll associated with this photosystem loses an electron, it is passed on to a series of electron acceptors, as shown in Figures 2.9 and 2.10A. In cyclical electron transport, the electron is passed to ferredoxin and then to cytochrome b_6 in the cytochrome b/f complex in the thylakoid membrane. This then ferries the electron back to the P700 via a plastocyanin electron carrier. Therefore, the electron has travelled through a cycle back to the reaction centre chlorophyll. However, during this process H^+ ions are moved by the Q-cycle (part of the electron transport complex) from the stroma to the thylakoid lumen. This raises the pH of the stroma and also generates an electrochemical proton gradient. It has not yet been established how many protons are pumped from one side of the thylakoid membrane to the other during the passage of a single electron through the cyclical electron transport chain, but the Q-cycle would suggest that two protons are moved for every two electrons.

The electrochemical proton gradient is then used to make ATP (Figure 2.10C). The protons diffuse through an ATPase/H^+ pump driven by the electrochemical gradient. During this movement ATP is synthesized in the stroma of the chloroplast. It has recently been suggested that four H^+ ions need to move through the ATPase/H^+ pump to synthesize a single molecule of ATP (Haraux and De Kouchkovsky, 1998).

If electrons can be donated from another source to P700, then non-cyclical electron transport can occur. The source of donated electrons is photosystem II. In this scheme, instead of ferredoxin reducing the cytochrome b_6, it reduces NADP. This then generates NADPH in the stroma of the chloroplast. However, without the activity of

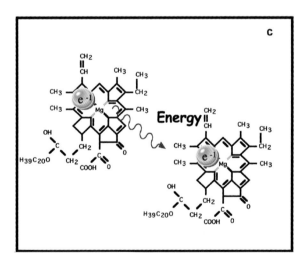

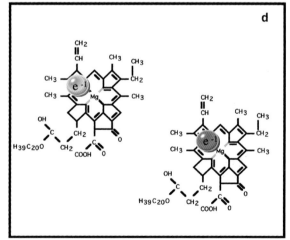

Figure 2.8   (*Continued*)

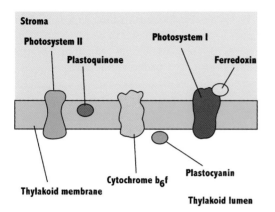

Figure 2.9 The arrangement of proteins in the thylakoid membrane, showing the arrangement of the various components of the energy-transducing system in the chloroplast. These proteins must convert the energy captured by the chlorophyll molecules into a form of chemical energy. This process is depicted in Figure 2.10.

photosystem II to replace electrons in the P700, photosystem I can only generate ATP.

Photosystem II

The reaction centre of this photosystem is called P680, as this is the maximum wavelength of light it can absorb. The loss of an electron by P680 does not generate as powerful a reducing agent as P700 (Figure 2.10B), but photosystem II can oxidize water via the following reaction:

$$2H_2O \rightarrow O_2 + 4H^+ + 4e^-$$

This reaction is the only known biological reaction that can split water into its component parts. It is the activity of this reaction which has generated the 21% oxygen in the earth's atmosphere and supports virtually all of the living organisms dependent upon aerobic respiration. In addition, the O_2 serves as a substrate for the only other biological autotrophic pathways (chemosynthesis).

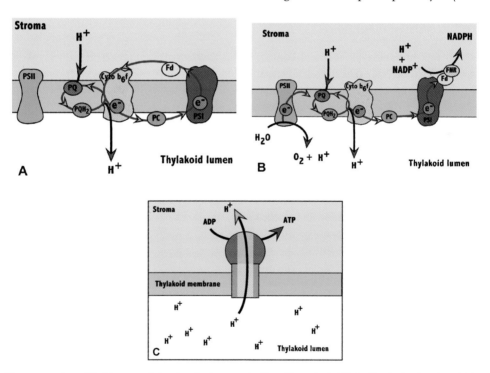

Figure 2.10 The conversion of light energy into chemical energy by the chloroplast. (A) Cyclic photophosphorylation. The reaction centre chlorophyll molecule in photosystem I loses its electron as a result of excitation but, instead of the electron falling back to the ground state, it is passed on through a chain of reactions that lead to the net movement of H^+ ions into the thylakoid lumen. Ultimately the electron is cycled back to the chlorophyll molecule for further excitation (hence the name 'cyclic photophosphorylation'). If photosystem II is active (B), then instead of the electron being cycled back it is replaced by an electron generated from photosystem II. The lost electron is replaced in the reaction centre chlorophyll molecule in photosystem II by the splitting of a water molecule. This non-cyclical photophosphorylation causes the movement of H^+ ions into the thylakoid lumen and the synthesis of NADPH in the chloroplast stroma. (C) The high concentration of H^+ ions in the thylakoid lumen causes a build-up of positive charge in this space, and this causes the movement of H^+ ions back into the stroma through the H^+/ATPase complex, generating ATP in the stroma.

The splitting of water generates electrons to replace those lost by P680 during the absorption of light energy. As a consequence of this, the electrons lost by P680 can be passed on in a non-cyclical manner and used to replace electrons lost by P700 (Figure 2.10B). Therefore, the interaction between photosystems I and II permits non-cyclical electron transport in the chloroplast.

The water-splitting reaction generates more H^+ ions in the thylakoid lumen, which further drives ATP synthesis. The light reactions of photosynthesis produce around three ATP molecules for every two NADPH molecules and this is precisely the ratio required by the Calvin cycle to make carbohydrate. However, under certain circumstances this ratio needs to be varied to allow reactions which require a different ratio of ATP and NADPH, and this can be achieved through modification of the relative activity of the cyclical and non-cyclical electron transport chains. These activities are adjusted through alterations in the degree of association of the antennae complexes with the two photosytems. A severe example of this is observed in the bundle sheath cells of C4 plants, which have completely lost the ability to carry out non-cyclical electron transport and can only synthesize ATP. The chloroplasts in these cells lack granal stacks as a result of loss of photosystem II activity.

The Calvin cycle

The cycle and its functions

The enzymes of the Calvin cycle (or photosynthetic reduction cycle) are contained within the chloroplast. However, the majority of the enzymes are also present elsewhere in the cell, but the crucial enzyme for carbohydrate synthesis, RUBISCO, is only located in the chloroplast. Light striking the chlorophyll molecules in the chloroplasts gives rise to two major products, ATP and reduced NADP. The amount of these two compounds which can accumulate is limited to a low concentration, since they are costly to make in terms of energy, are reactive and not stable for long-term storage, and use up phosphate for their synthesis (which is in limited supply). Therefore, ATP and NADPH cannot continue to accumulate indefinitely and, if all of the available ADP and NADP in a cell were consumed, there could be no further product of photosynthesis. It is the primary role of the Calvin cycle to recycle these compounds and at the same time to use the energy within the bonds of the molecules to make compounds that can accumulate and do not strain a plant's resources.

The actual biochemical pathway of the Calvin cycle is shown in Figure 2.11A, but it can be simplified to three sections (Figure 2.11B): regeneration, CO_2 fixation or carboxylation, and synthetic flux.

Regeneration

By its very name, the Calvin cycle is a cycle, and therefore as much or more ribulose 1,5-bisphosphate as is consumed by the cycle needs to be regenerated again or the cycle will grind to a halt. Therefore, one of the major roles for the cycle is to regenerate the capacity to continue functioning, and this is usually thought of as regenerating the substrate for RUBISCO activity. RUBISCO forms the three-carbon phosphorylated intermediate phosphoglyceric acid (3PGA), which is not easy to convert back into the five-carbon compound ribulose 1,5-bisphosphate. This feat is the most complex part of the cycle but this complexity is solely to achieve this regeneration. When the cycle is inactive at night, the concentration of ribulose 1,5-bisphosphate is barely detectable, so, as light hits the photosynthetic cells, one of the first responses of the cycle is rapidly to increase the intermediates in the cycle. As a consequence, the first products of photosynthesis are intermediates of the Calvin cycle and not carbohydrates. Once the flux of fixed carbon through the cycle exceeds a certain level, synthetic fluxes can begin.

The cytosol of plant cells contains a similar biochemical pathway known as the oxidative pentose phosphate pathway, which is just as complex and contains many of the same enzymes. The role of this pathway is to generate pentose sugars and generate NADPH for synthetic pathways in the cell.

CO_2 fixation or carboxylation

As has been emphasized earlier, the crucial feature of photosynthesis is the ability to fix CO_2 and incorporate it into a form which is useful for further metabolism. RUBISCO is the key enzyme in this process and is the only enzyme known that can achieve this in plants. The role of the enzyme PEP carboxylase in C4 and CAM plants is discussed in Chapter 12. This enzyme can use dissolved CO_2 in the form of carbonic acid and bind it onto a phospho-enol pyruvic acid (PEP) molecule generating a four-carbon compound, oxaloacetic acid, which is rapidly reduced to malic acid. However, malic acid cannot be used to synthesize a six-carbon carbohydrate, as there is no known metabolic pathway that allows the retention of all the carbons in the molecule. Malic acid cannot be converted into a three or four-carbon carbohydrate (which then could be used for glucose synthesis) without the

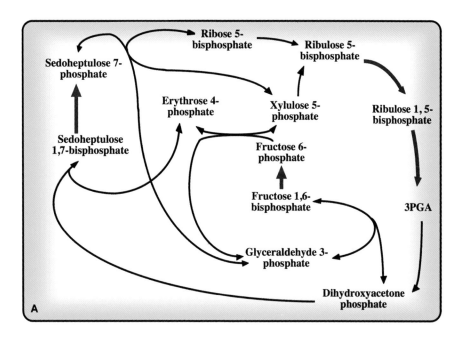

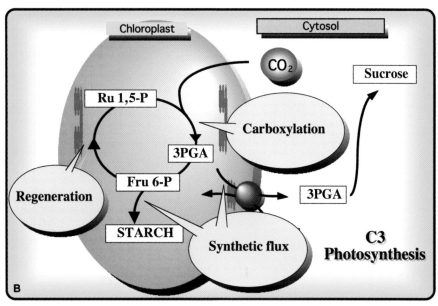

Figure 2.11   (A) The entire Calvin cycle. This pathway is remarkable in that it is autocatalytic and it generates the substrate needed to drive it. The only requirements it has are the input of ATP and NADPH, generated in the light reactions of photosynthesis. The complexity of the pathway is required in order to convert three-carbon compounds into five-carbon compounds without losing any carbon atoms in the process. (B) A simplified version of the Calvin cycle, summarizing its three major roles: (a) to fix carbon dioxide; (b) to generate a synthetic flux (carbohydrate synthesis); and (c) to regenerate substrate in the form of ribulose 1,5-bisphosphate to continue driving the cycle.

release of CO_2, which defeats the original objective. PEP carboxylase can only fix CO_2 temporarily, which is precisely what is required for the C4 and the CAM photosynthetic pathways. RUBISCO reacts one molecule of ribulose 1,5-bisphosphate with a CO_2 molecule and synthesizes two molecules of 3PGA. This three-carbon phosphorylated intermediate is readily converted into a six-carbon carbohydrate. The Calvin cycle is regulated such that once this intermediate is made, it is metabolized by the cycle and used for either the regeneration of ribulose 1,5-bisphosphate or the synthesis of carbohydrate.

Synthetic flux

The purpose of the Calvin cycle cannot be solely to regenerate its intermediates and regenerate ADP and NADP. All the intermediates of the cycle have phosphate bound to the molecules, e.g. 3PGA, triose phosphate, etc. Therefore ADP and NADP can be recycled for supporting the light reactions of photosynthesis, but the phosphate now becomes bound to a sugar and produces a phosphorylated intermediate (Figure 2.11). As has been pointed out earlier, phosphate availability frequently limits the growth of a plant and phosphorylated intermediates of the Calvin cycle cannot build up indefinitely. Storage compounds are required that do not contain phosphate groups or other limiting ions and the solution to this problem is the synthetic fluxes. In plants the synthetic fluxes lead to the formation of carbohydrates, compounds which contain only carbon, hydrogen and oxygen. These elements are readily available from gaseous CO_2 in the air and water in the soil. The majority of plants produce two major carbohydrates, sucrose and starch (whose structure and function are discussed later), and to a lesser extent glucose and fructose also accumulate. In a limited number of species the raffinose series oligosaccharides, such as raffinose, stachyose and verbascose (see Chapter 3 and the transportation of such sugars) accumulate, or complex polysaccharides called fructans. As these carbohydrates are made, the phosphate molecules joined to the substrates are released and can be used for further ATP synthesis by the light reactions of photosynthesis.

Regulatory enzymes

The Calvin cycle is frequently described as part of the dark reaction of photosynthesis because its activity does not directly require light. However, several of the enzymes are switched on indirectly by the activity of the light reactions of photosynthesis. The key regulatory steps of the Calvin cycle are inactive in the dark (Figure 2.12). Since there is no light at night, there is no requirement for photosynthetic fluxes and the cycle shuts down. When a chloroplast is illuminated, four major changes occur. First, due to the activity of the light reactions, H^+ ions are moved into the thylakoid lumen. This causes a rise in the pH of the stroma of the chloroplast, and this activates the enzymes RUBISCO, ribulose 1,5-bisphosphate kinase, sedoheptulose 1,7-bisphosphatase and fructose 1,6-bisphosphatase.

Second, as the H^+ ions enter the thylakoid lumen, other ions move to counter the charge imbalance across the membrane. One of these ions is Mg^{2+}. This ion acts to activate the enzymes listed earlier. Third, reduced ferredoxin generated on the thylakoid membrane is used to produce reduced thioredoxin (see Figure 2.13), which then acts to reduce disulphide bonds on both sedoheptulose 1,7-bisphosphatase and fructose 1,6-bisphosphatase, which activates the enzymes. Finally, the rise of ATP concentrations in the stroma of the chloroplast stimulates the activity of RUBISCO.

The remarkable role of RUBISCO

The role of RUBISCO in photosynthesis is so critical that it merits special mention. It is one of the few enzymes known that can take inorganic CO_2 and convert it into organic matter. This means that virtually all of the carbon in every living organism was once fixed by RUBISCO. Most of the CO_2 currently in the atmosphere is derived from the decay of organic materials. However, billions of years ago the main generator of CO_2 was volcanic activity of the planet. It has taken those billions of years for plants to reduce CO_2 levels to those currently observed, and RUBISCO has played the dominant role in this transformation of our atmosphere. RUBISCO accounts for 50% of the soluble protein in most leaves, which makes it the most abundant protein on earth (Ellis, 1979). That there is such a large amount of RUBISCO also leads us to the conclusion that there is far more of it than is necessary in a leaf. This has led some researchers to suggest that RUBISCO could additionally be acting as a storage protein in leaves. However, this huge amount of the protein in leaves could be a result of the very low specific activity of RUBISCO, which reacts with a few molecules of CO_2 per second, 1000 times less than most other enzymes, such as PEP carboxylase. The enzyme has a high affinity for ribulose 1,5-bisphosphate but a low affinity for CO_2, and this affects the rate of reaction between the two substrates (Spreitzer, 2002). It is known to be activated by an enzyme in *Arabidopsis* called RUBISCO activase and is inhibited by nanomolar amounts of a compound that mimics an intermediate state of the carboxylation reaction, carboxyarabinitol 1-phosphate.

RUBISCO can make two products. It can react with CO_2 and ribulose 1,5-bisphosphate and form two molecules of 3PGA. Alternatively, it can bind onto ribulose 1,5-bisphosphate and react with oxygen, generating one molecule of 3PGA and a molecule of phosphoglycollate; this second reaction is known as photorespiration. 3PGA formed by RUBISCO is rapidly reduced by the Calvin cycle and used

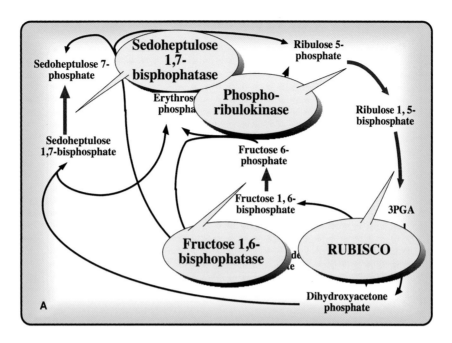

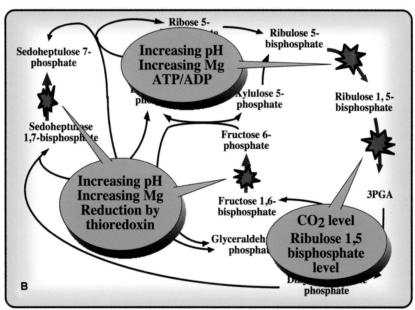

Figure 2.12 (A) Regulation of the Calvin cycle. There are four major enzymes which appear to control the activity of the cycle. (B) These enzymes are activated by metabolic changes initiated by the light reactions of photosynthesis, summarized here.

for regeneration of the cycle intermediates or for carbohydrate synthesis. However, the two-carbon molecule is much more difficult to deal with, since there are no two-carbon intermediates of the Calvin cycle. Plants possess a whole pathway, known as the photorespiratory pathway, to deal with this intermediate.

RUBISCO may seem inefficient, since it suffers from photorespiration and low activity, but it is one of a handful of enzymes, coupled with the Calvin cycle activity, that possesses autocatalytic ability. That is to say, it regenerates one of its own substrates while having a synthetic capacity at the same time. Attempts have been made to modify RUBISCO genetically to remove the oxygenase capacity of the enzyme, but these have all failed because reducing the affinity of the enzyme for oxygen has a knock-on effect of reducing the affinity of the enzyme for CO_2. It seems

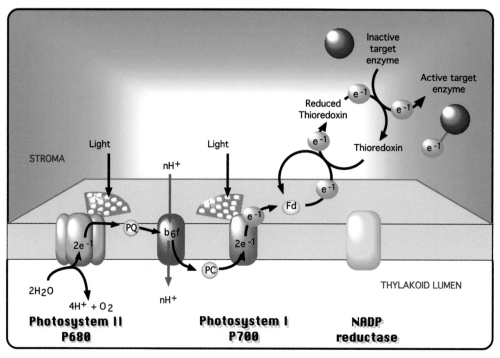

Figure 2.13   Reduction of thioredoxin. One of the factors which activates the fructose 1,6-bisphosphatase and sedoheptulose 1,7-bisphosphatase enzymes in the Calvin cycle is reduction of the enzyme. This is achieved by the reduction of thioredoxin, which is used to reduce these two enzymes. The process for the synthesis of reduced thioredoxin is shown.

that to disable the interaction of the oxygen molecule with the reactive site of RUBISCO also interferes with the interaction with CO_2. These two substrates are so similar in size and electron configuration that it is highly unlikely that this approach will ever change the reactive nature of RUBISCO. If it were possible, then plants would have already evolved a RUBISCO enzyme without an oxygenase activity. As has been mentioned, there are three other pathways known that permit CO_2 fixation and carbohydrate synthesis; these are fuelled by oxidation of chemicals generated in geothermal features. Such habitats are now rare and, while the chemosynthetic bacteria may dominate them, they make little impact on the rest of the world. For plants there is a huge quantity of light energy available to fuel photosynthesis. Hence, inefficient as RUBISCO is, it is pivotal to the success and dominance of plants on this planet.

There is a small group of autotrophs that use chemosynthesis for carbohydrate synthesis and they are apparently self-perpetuating, relying on the availability of compounds such as hydrogen sulphide in thermal vents of geothermal features in bodies of water. The net reaction catalysed by these bacteria is:

$$CO_2(aq.) + H_2S(aq.) + H_2O(l) + O_2(aq.) \rightarrow (H_2CO) + H_2SO_4(aq.)$$

However, this mode of nutrition is dependent upon the availability of O_2 dissolved in the water. As this O_2 is a by-product of the activity of photosynthesis, and as plants are the dominant producers of O_2, the chemosynthetic pathways are dependent on plants for their survival. These life forms also possess the three other known pathways that utilize geothermal energy to produce chemical energy in a living cell for the fixation of CO_2 and the synthesis of carbohydrate. However, in comparison to the total CO_2 fixation by plants, the contribution of these bacteria is very minor.

Photorespiration

The troubles with photosynthesis

The problem with using RUBISCO as the enzyme to synthesize carbohydrate is that it has dual activity and can act as a carboxylase or an oxygenase. The oxygenase activity gives rise to two products, 3PGA and 2-phosphoglycollate. This activity uses up ribulose 1,5-bisphosphate but does not result in CO_2 fixation. In addition, the 2-phosphoglycollate is not an intermediate of the Calvin cycle, and thus is lost to the cycle, but this will lead to a failure of regeneration of ribulose 1,5-bisphosphate and the cycle will collapse. It is vital that the 2-phosphoglycollate is reconverted back into 3PGA, and this is the role of the photorespiratory pathway.

The solution to the problem

The photorespiratory pathway is shown in Figure 2.14. In photosynthetic cells, where there is high activity of the photorespiratory pathway, the chloroplast, peroxisome and mitochondria are frequently observed to be very closely sandwiched together. All three of these cellular compartments are involved in the pathway. It begins with the dephosphorylation of the 2-phosphoglycollate to give glycollate. Phosphate is always conserved in the chloroplast and hence the phosphate residue bound to glycollate needs to be removed before export. Then the glycollate is converted to glyoxylate, followed by glycine in the peroxisome. The peroxisome is required at this point, as the reaction generates hydrogen peroxide, which is very reactive and needs to be metabolized rapidly before it damages any cellular processes. Glycine is then exported to a mitochondrion, where it is converted to serine with the release of ammonia and CO_2. Serine is a four-carbon amino acid formed from two glycine molecules joining together. The ammonia needs to be rapidly reassimilated into amino acids in the chloroplast or it will begin to affect the pH of the cell. The serine is transported back into the peroxisome, where it is converted into pyruvate and then reduced to glycerate. Glycerate is finally transported back into the chloroplast, where it is phosphorylated to produce 3PGA and therefore is reintroduced into the Calvin cycle. However, during this pathway ammonia has been released, ATP and NADH have been consumed, and 25% of all fixed carbon entering the pathway has been released as CO_2. This pathway is very wasteful indeed; it is costly in energy and resources and, moreover, a whole pathway with all associated enzymes is required to correct this problem. Under severe conditions it is estimated that, at temperatures greater than 30 °C, over 40% of the CO_2 fixed is lost again through photorespiration (Lea and Leegood, 1993). This is precisely why there is so much research interest in limiting the interaction of RUBISCO with oxygen.

There are certain conditions which increase the activity of photorespiration by RUBISCO. Under conditions where the concentration of CO_2 in the air is low and the

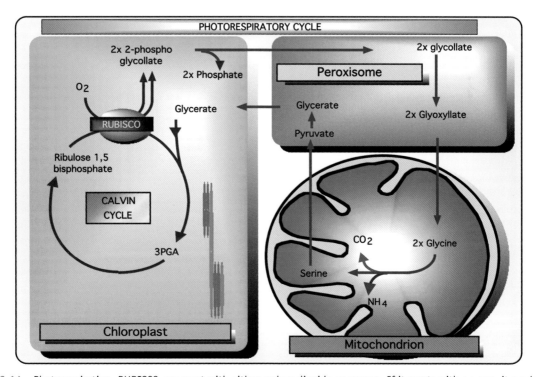

Figure 2.14 Photorespiration. RUBISCO can react with either carbon dioxide or oxygen. If it reacts with oxygen, it produces one molecule of 3PGA and one molecule of phosphoglycollate. 3PGA can be metabolized by the Calvin cycle but phosphoglycollate cannot, and poses a problem to the plant. In order to metabolize the phosphoglycollate and recycle as much as possible of the carbon within the molecule, plants possess the photorespiratory pathway, shown here. Even with this pathway, 25% of the carbon in the phosphoglycollate molecule will be lost as carbon dioxide. This is an appreciable loss to plants and photorespiration can be a severe problem in certain environments.

Table 2.1 The differing equilibrium solubilities of oxygen and carbon dioxide in water at varying temperatures.

Temperature (°C)	Solubility (mM)		
	$21\%\ O_2$	$0.035\%\ CO_2$	$O_2{:}CO_2$ ratio
10	350	17	20.5
20	284	12	23.0
25	258	10	24.7
30	236	9	25.5
40	201	7	28.0

Leave based on Lea and Leegood (1993).

concentration of oxygen is high, the oxygenase activity of RUBISCO is favoured (see Table 2.1). This occurs at times when plants are very close together and the rates of photosynthesis are high. In addition, as the temperature to which a plant is exposed is raised, gases become less soluble in the fluids in the leaf. The solubility of CO_2 reduces more rapidly than that of oxygen and therefore, at high temperatures, the ratio of CO_2 to oxygen dissolved in the cellular fluids falls and the interaction between RUBISCO and oxygen becomes more frequent. As a consequence, plants grown in close proximity to one another and in warm climates are much more likely to experience photorespiration and suffer large losses of fixed CO_2. As will be discussed later, one group of plants have evolved a mechanism to avoid photorespiration, and these are the plants which use the C4 photosynthetic pathway which is discussed in Chapter 12. Only a small percentage of plants use the C4 pathway but most of those are found in tropical areas of the world, where photorespiration is a major problem.

Carbohydrate synthesis and storage

The major carbohydrates of photosynthesis

In the majority of plants the main carbohydrates synthesized are sucrose and starch (Figure 2.15). Sucrose is a soluble disaccharide which is made in the cytosol of the photosynthetic cells in the leaf. For most plants, sucrose is the form in which carbohydrate is transported around the plant. Starch is an insoluble carbohydrate which is made in the chloroplast. Glucose and fructose also accumulate as minor products of photosynthesis but these occur as breakdown products of sucrose, mainly in the vacuole of

photosynthetic cells. These sugars cannot be translocated by the phloem and are solely for storage purposes or exist as intermediates during sucrose metabolism.

In some plants substantial quantities of fructans accumulate. Fructans are polymers based on the sucrose molecule. By adding extra fructose molecules to sucrose, long-chain fructans are produced, reaching a maximum of around 250 separate hexose molecules (Pollock, 1986). Examples of short-chain fructans are shown in Figure 2.18. Fructans are water-soluble and are stored in the vacuole of both photosynthetic and non-photosynthetic cells. In the leaf, fructans are short-term carbohydrate reserves; however, in roots and tubers fructans are frequently long-term storage carbohydrates that act as a replacement for starch. Fructans accumulate as two major groups, which are defined by the bonding pattern between the fructose molecules. The levans are formed from β(2–6)-links and the inulins (such as in Jerusalem artichokes, *Helianthus tuberosum*) are formed from β(1–6)-links.

For some plants, the aldehyde or ketone groups within the sugar molecule can be modified to form alcohols or carboxylic acids. The reduction of these groups yields sugar alcohols, such as sorbitol and mannitol. The sugar alcohols are only observed in a few of the higher plants, such as celery and apple trees, but are more common in lower plants. Oxidation of the carbohydrate hexoses yields acids such as glucuronic and galacturonic acid, which can be used for the synthesis of polymers such as pectic acid.

Sucrose

Sucrose possesses numerous remarkable physical and chemical properties. It is highly soluble in water, with over 200 g of sucrose dissolving in 100 ml of water. It is not a reducing sugar, unlike the two hexose sugars of which it is comprised. Both glucose and fructose can act as reducing agents, as a result of the reactivity of the aldehyde and ketose groups formed when the hexose ring disassociates (Figure 2.18). Sucrose also has little apparent effect on plant metabolic processes, even at high concentrations. It is stable and its metabolism is heavily dependent on the presence of specific enzymes. As a consequence, sucrose is the perfect carbohydrate for transport around the plant and this will be explored further in Chapter 3.

Sucrose is synthesized in the cytosol of the photosynthetic plant cells during photosynthesis. In order for sucrose to be made, the phosphorylated intermediates of the Calvin cycle in the stroma must be transported to the cytosol (Servaites and Geiger, 2002). This is performed by the membrane transporter triose phosphate/phosphate translocator, located on the inner chloroplast membrane. This flow of photosynthate is so important that this protein

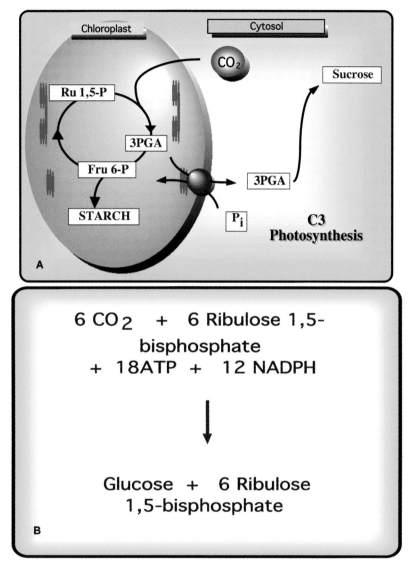

Figure 2.15 The products of photosynthesis. In order to maintain the phosphate balance in a photosynthetic cell, the Calvin cycle cannot continually accumulate more and more sugar phosphates. Ultimately, the phosphate groups need to be recycled. To do this, the plant makes carbohydrates from readily obtainable hydrogen, oxygen and carbon. The major soluble carbohydrate is sucrose, which is synthesized in the cytosol of the plant cell (A). Starch is the major insoluble carbohydrate and is stored in the chloroplast. A net equation for the synthesis of products is shown (B). A glucose molecule is shown, to represent carbohydrate synthesis; in fact, glucose is rarely a major product of photosynthesis but it does serve as a representative of the six-carbon unit that goes to make up sucrose and starch.

makes up over 30% of the protein content of the chloroplast membrane. The activity of the triose phosphate/ phosphate translocator is summarized in Figure 2.16. Put simply, the translocator can move three-carbon (3C)-phosphorylated intermediates from the stroma of the chloroplast to the cytosol in exchange for inorganic phosphate molecules. The 3C-phosphorylated intermediates are then converted into six-carbon (6C)-phosphorylated intermediates via a process which is the opposite of glycolysis,

known as gluconeogenesis. This pathway is summarized in Figure 2.17. Ultimately, UDPglucose and fructose 6-phosphate are fused to give sucrose phosphate, which is in turn dephosphorylated to yield sucrose. There are two major enzymes which regulate the synthesis of sucrose, fructose 1,6-bisphosphatase and sucrose phosphate synthase.

Fructose 1,6-bisphosphatase is inhibited by fructose 2,6-bisphosphate, but the regulation of sucrose phosphate

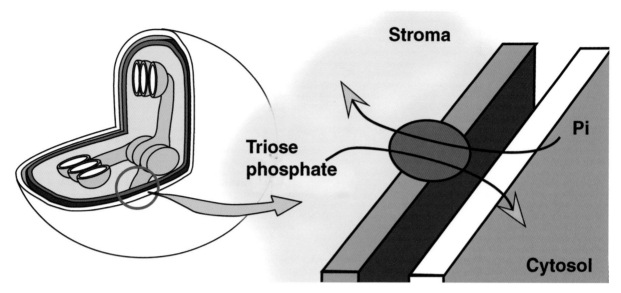

Figure 2.16   Transport of three carbon units to the cytosol. As sucrose is synthesized in the cytosol of plant cells, the substrate for its synthesis needs to be transported out of the chloroplast. The substrate is a three-carbon phosphorylated sugar and the enzyme which carries out this movement is known as the triose phosphate/phosphate translocator protein. It moves three-carbon phosphorylated sugars out of the chloroplast in a strict exchange for phosphate ions moving in the opposite direction. Thus, only if there is availability of phosphate ions in the cytosol does this movement occur.

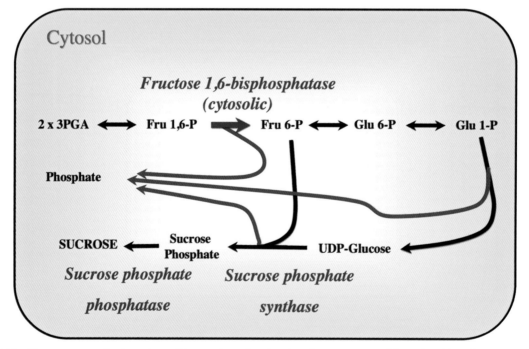

Figure 2.17   The sucrose synthesis pathway. Three-carbon phosphorylated sugars exported from the chloroplast are converted to sucrose through a reversal of the glycolytic pathway. As this pathway progresses, phosphate ions are released and these can then be used to export further three carbon-phosphorylated sugars from the chloroplast.

synthase is less clear. Recent evidence suggests that sucrose phosphate synthase and sucrose phosphate phosphatase form a complex which has regulatory properties that are different from those of the individual enzymes. Evidence suggests that sucrose phosphate synthase is stimulated by high glucose 6-phosphate (Glc-6-P) concentrations and inhibited by high inorganic phosphate. The enzyme is also subject to phosphorylation. In the phosphorylated state, the enzyme is less sensitive to Glc-6-P and more sensitive to phosphate, and therefore its activity is reduced. The cytosolic concentration of sucrose influences the degree of phosphorylation, and this is discussed later in relation to the regulation of sucrose and starch synthesis.

Sucrose accumulates in the cytosol of the photosynthetic cells during photosynthesis. It also can be stored in the vacuole, where it moves by free diffusion mediated by the presence of a tonoplast sucrose transporter. In addition, sucrose can be transported out of the photosynthetic cells to the apoplastic space via a passive sucrose transporter. Only a finite concentration of sucrose can build up in the cytosol and vacuole of the cells, beyond which the synthesis of sucrose becomes inhibited. At this point the only

sucrose made diffuses out of the photosynthetic cells for translocation to other tissues.

Starch

Starch is an insoluble carbohydrate and therefore can accumulate in large amounts without affecting the osmotic potential of a plant cell. Starch is made up of glucose monomers to make a long-chain polysaccharide. There are two types of starch present in plants: amylose, which is made up of long chains of glucose molecules bound by $\alpha(1-4)$ linkages; and amylopectin, which is made up of glucose molecules bound together by $\alpha(1-4)$ linkages and also some bound by $\alpha(1-6)$ linkages (Figures 2.18 and 2.19). In amylopectin these $\alpha(1-6)$ linkages form branches in the starch molecule and they occur every 20–30 glucose residues. The bonds between the glucose molecules in starch prevent the glucose molecule acting as a reducing sugar in all but the last residue on the glucose chain.

Starch accumulates as a product of photosynthesis in the chloroplast (Figure 2.20). Through a number of conversions, fructose 6-phosphate from the Calvin cycle is

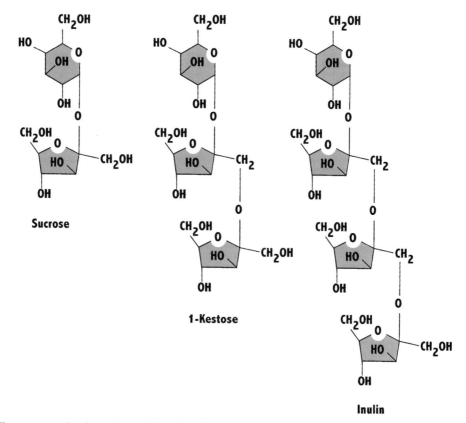

Figure 2.18 The sucrose molecule and some derivatives of sucrose. The major soluble carbohydrate in most plants is sucrose, whose structure is shown. Some plants also produce fructans, which are derived from sucrose; kestose and inulin are shown.

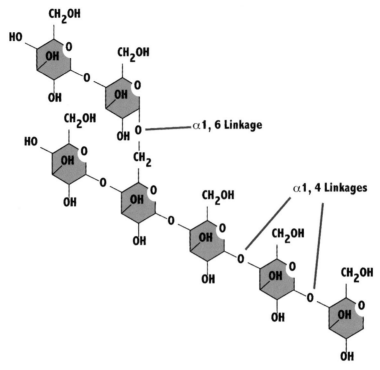

Figure 2.19   The starch molecule. Starch is a polymer of glucose and is made up of many thousands of glucose molecules. The two major bonds between the monomers, 1–4 and 1–6 linkages, are shown.

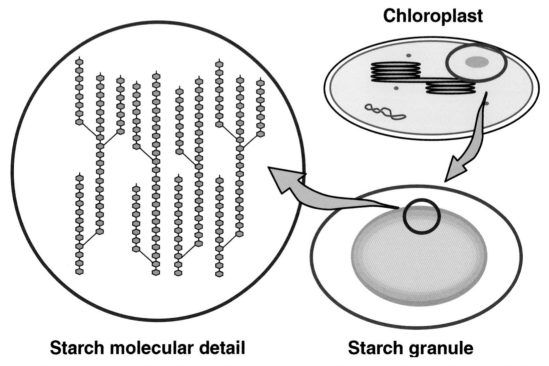

Figure 2.20   The starch molecule and the chloroplast. Starch granules are visible under the electron microscope. This figure shows how these granules are made up of radial chains of glucose chains. Periodically in the granules there are radial patterns, which seem to result from major synchronized changes in the granule structure.

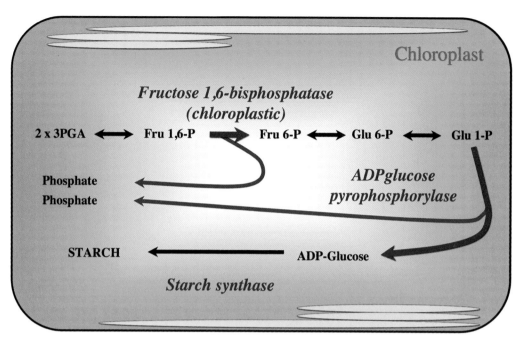

Figure 2.21 The starch synthesis pathway. If three-carbon phosphorylated sugar is not exported from the chloroplast, one possible option for the plant is to convert it into glucose 1-phosphate and then use it for starch synthesis. This pathway releases phosphate ions progressively and starch is synthesized. These can then be used for further ATP synthesis. A build-up of three-carbon phosphorylated sugars and a fall in stromal phosphate ions activate the enzymes involved in starch synthesis.

converted into ADPglucose by the enzyme ADPglucose pyrophosphorylase (Figure 2.21). The enzyme starch synthase uses ADPglucose as a substrate to add glucose residues onto the starch molecule. Periodically, branches in the starch molecule are introduced by starch branching enzyme, which forms the $\alpha(1-6)$ linkages. This branching of the starch molecule increases the number of points where the enzyme starch synthase can function, to add further glucose residues to the growing starch molecule. Starch synthesis is regulated by the activity of ADPglucose pyrophosphorylase. This enzyme is stimulated by high concentrations of 3PGA and low stromal concentrations of phosphate. The enzyme is freely reversible, but the presence of alkaline pyrophosphatase in the stroma causes any pyrophosphate generated by the reaction to be broken down rapidly and thus the reaction is essentially irreversible.

The fate of carbon fixed during photosynthesis

The metabolic fluxes to sucrose and starch are competing for the same reserves from the Calvin cycle. The metabolic pathways are very similar, since they both really have the same objective, i.e. to remove the bound phosphate and produce compounds that store the chemical energy generated by the light reactions of photosynthesis. The extent to which sucrose and starch are made by plants varies between different species and the time of day. Generally, sucrose is the dominant carbohydrate made in the first few hours of the day. This is because sucrose is soluble and can be translocated to other plant tissues, where it is needed for growth and metabolism.

Once the phosphorylated intermediates of the Calvin cycle have built up to sufficient concentrations for efficient photosynthetic activity, the rising stromal concentrations of 3PGA and high cytosolic phosphate concentrations cause movement of 3PGA out of the chloroplast in exchange for phosphate. As a consequence, carbon fixed by photosynthesis is first moved to the cytosol. High night-time levels of fructose 2,6-bisphosphate inhibit gluconeogenesis, but the fall of phosphate and rise in 3PGA concentrations affects the activity of the enzymes that regulate fructose 2,6-bisphosphate levels in the cytosol, causing it to fall (Figure 2.22). This reduces the inhibition of fructose 1,6-bisphosphatase and therefore the gluconeogenic pathway is switched on and sucrose synthesis occurs in the cytosol. At this point the majority of the CO_2 fixed in the Calvin cycle is converted into sucrose. As sucrose is made, the phosphate groups linked to the phosphorylated

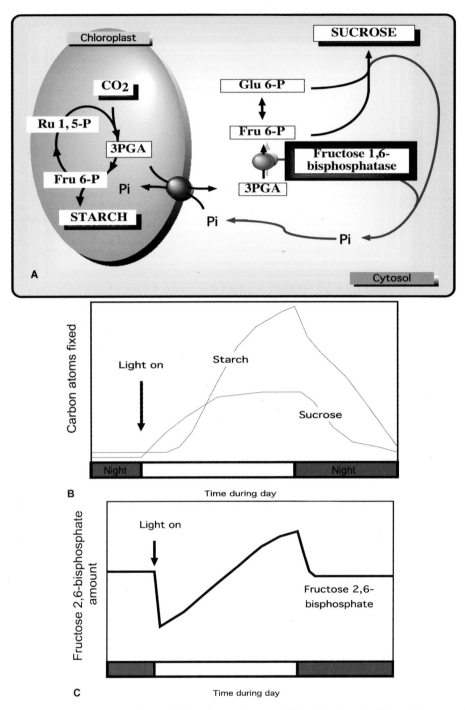

Figure 2.22   The control of sucrose and starch synthesis pathway. The metabolism of carbon dioxide and its conversion into sucrose or starch is known as partitioning. The carbon fixed can be partitioned into either sucrose or starch. This partitioning is controlled by a cytosolic sugar phosphate, called fructose 2,6-bisphosphate (fru 2,6-P), through the inhibition of the enzyme fructose 1,6-bisphosphatase (A). This enzyme is crucial for the synthesis of sucrose. Fru 2,6-P levels vary throughout the day (B). When a leaf is first illuminated, the levels of fru 2,6-P fall and thus sucrose can be synthesized (C). But gradually, throughout the day, fru 2,6-P levels rise and thus fructose 1,6-bisphosphatase is inhibited and sucrose synthesis slows down or halts. Carbon dioxide fixed after this point is converted into starch.

intermediates are progressively released as inorganic phosphate in the cytosol. This supports the continued depression of fructose 2,6-bisphosphate concentrations and the further export of 3PGA from the chloroplast (Scott and Kruger, (1995)). As the day progresses, sucrose concentrations rise in the cytosol and this leads to export of sucrose from the photosynthetic cells to the apoplastic space and to storage of sucrose in the vacuole. The limited storage capacity for sucrose in the cytosol and the vacuole is soon filled, and at this point any sucrose that can be made needs to be exported from the cells. It has been mentioned already that sucrose can be accumulated to relatively high concentrations before metabolic pathways are affected; however, sucrose is not permitted to accumulate to a great extent in the photosynthetic cells. Sucrose accumulating in the cytosol gradually begins to affect the phosphorylation state of sucrose phosphate synthase and sucrose synthesis is inhibited. As a consequence, sucrose accumulation reduces, which leads in turn to Glc-6-P and Fru-6-P concentrations rising in the cytosol. The rise in these compounds competes with the earlier signal of 3PGA and phosphate, which drove down the concentration of fructose 2,6-bisphosphate. Therefore fructose 2,6-bisphosphate begins to rise as the maximum capacity of the cytosol for sucrose accumulation is reached. Fructose 2,6-bisphosphate then, in turn, inhibits fructose 1,6-bisphosphatase and prevents further accumulation of Glc-6-P and Fru-6-P. Inhibition of fructose 1,6-bisphosphatase then causes the accumulation of 3PGA in the cytosol and, since the gluconeogenic pathway activity is reduced, organic phosphate is no longer released from the compounds coming out of the chloroplast (Stitt, 1990). This leads to low cytosolic phosphate and high 3PGA concentrations. These changes affect the equilibrium across the inner chloroplast membrane mediated by the triose phosphate/phosphate translocator, and 3PGA ceases to be exported from the chloroplast.

Once 3PGA is no longer exported from the chloroplast, another means of storing the chemical energy is required, and this is where starch is used. 3PGA concentrations rise in the stroma as a result of a fall in export, but continued photosynthesis leads to a fall in inorganic phosphate in the stroma. These two factors activate the enzyme ADPglucose pyrophosphorylase, which catalyses the conversion of Glc-1-P to ADPglucose. The ADPglucose generated is used for starch synthesis, and starch then accumulates.

Therefore, at the start of the day most plants make sucrose as the dominant carbohydrate, as a product of CO_2 fixation in photosynthesis. Sucrose accumulates in the leaves and is transported around the plant throughout the day to provide energy for heterotrophic tissues. However, the capacity for sucrose accumulation in the cytosol and vacuole is limited and, as this limit is reached, the rate of sucrose synthesis declines. The plant then gradually switches from sucrose to starch being the dominant carbohydrate accumulating as a product of photosynthesis, and starch will continue to accumulate throughout the rest of the day. At night time, starch is metabolized to provide energy for leaf metabolism and for export to the rest of the plant. Thus, by the end of the night starch levels in the chloroplast have fallen to very low levels, so that the cycle can begin again. It is a combination of fructose 2,6-bisphosphate and the feedback inhibition loop of sucrose synthesis that controls whether sucrose or starch will be made as a product of CO_2 fixation.

It is estimated that around 70% of the starch made during photosynthesis is retained in and used by the leaf to support its metabolism at night, with the remainder being exported. Therefore, flows of carbohydrate around the plant are much reduced at night. The pathway of exit of starch from the chloroplast is not via a 3C phosphorylated intermediate, since the high concentrations of fructose 2,6-bisphosphate in the cytosol at night mean that the gluconeogenic pathway is inactive (Figure 2.23). Starch is broken down by amylase enzymes to form glucose and maltose. In the inner chloroplast membrane there is a glucose and maltose transporter, which permits glucose and maltose to leave the chloroplast and enter the cytosol with an H^+ ion. This sugar is then phosphorylated by hexokinase and can be converted to sucrose for translocation, or Glc-6-P enters glycolysis for respiration or other metabolic pathways. In the daytime there is a rise in the pH of the stroma of the chloroplast and therefore there will be little glucose or maltose leaving the chloroplast. However, during the night the stroma pH falls and as starch is degraded, sugars are released to the cytosol.

Fructose 2,6-bisphosphate may not play a role in the actual pathway for starch degradation at night but it may still play a role in regulating the rate of starch metabolism in plants (Scott and Kruger, (1995)). A rise in the sucrose concentration in a leaf at night can result from a fall in sink demand for carbohydrate. This in turn raises the hexose phosphates and triose phosphates in the cytosol. A rise in the triose phosphates in the cytosol can lead to a rise in the triose phosphates in the stroma of the chloroplast through equilibration by the triose phosphate/phosphate translocator. This rise can then stimulate the synthesis of starch and hence reduce the net rate at which starch is mobilized at night.

The fate of sucrose once it leaves the photosynthetic cells is the subject of the next chapter.

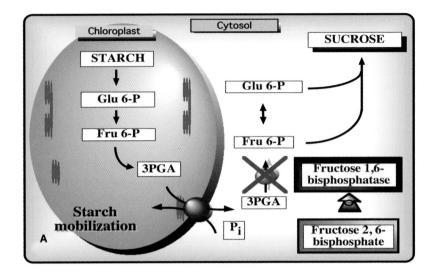

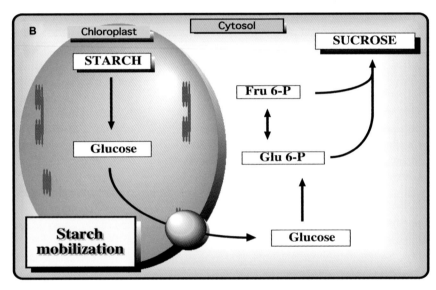

Figure 2.23   The metabolism of starch at night. Starch accumulates in the chloroplast during the day but is metabolized at night and converted into sucrose for transport to the rest of the plant or for metabolism in the leaf. The movement of three-carbon phosphorylated sugars out of the chloroplasts is restricted at night, due to the low activity of the cytosolic fructose 1,6-bisphosphatase (this is inhibited by fru 2,6-P) (A). Thus, starch is thought to be metabolized to glucose and then this is exported to the cytosol for sucrose synthesis (B).

The efficiency of photosynthesis

The actual capture of light energy by chlorophyll is very efficient, with over 99% of the energy being channelled to the reaction centre chlorophyll molecules. However, only around 60% of the available light is actually useful for photosynthesis, i.e. the wavelengths of light are sub-optimal for photosynthesis. There are substantial losses in energy in the biochemical reactions in the stroma and the pathways for sucrose and starch synthesis. Under optimal conditions, the theoretical maximum efficiency of photo-

synthesis is around 35%, i.e. 35% of the light energy is converted into chemical energy in carbohydrate. However, conditions in nature are never optimal, and for most plants the efficiency is less than 1%. Maize appears to be an exception in possessing an efficiency of around 8%. These efficiency values may seem poor, but two facts need to be considered. First, in order to maintain fluxes, metabolic pathways often need steps that release large quantities of energy – this is certainly so during photosynthesis. Energy is released during the metabolic pathways for sucrose and starch synthesis and in the translocation of carbohydrate around the plant. These losses are necessary in order to maintain the direction of metabolic fluxes and are all part of the physiology of plants. Second, plants produce enough carbohydrate for growth and they are the dominant autotrophs on Earth. If greater efficiency was needed, then plants would have evolved to be more efficient, and there are probably other more important restrictions on plant growth than efficient energy interconversion.

References

Ellis RJ (1979) The most abundant protein in the world. *Trends in Biochemical Sciences*, **4**, 241–244.

Haraux F and de Kouchkovsky Y (1980) Measurement of chlorophyll internal protons with 9AA. Proton binding, dark proton gradient, and salt effects. *Biochimica et Biophysica Acta*, **592**, 153–168.

Lea JL and Leegood RC (1993) The Calvin cycle and photorespiration, in *Plant Biochemistry and Molecular Biology*, Wiley, Chichester, UK.

Pollock CJ (1986) Tansley Review No. 5. Fructans and the metabolism of sucrose in vascular plants. *New Phytologist*, **104**, 1–24.

Scott P and Kruger NJ (1995) Regulation of starch mobilisation in leaves in the dark by fructose 2,6-bisphosphate. *Plant Physiology*, **108**: 1569–1577.

Servaites JC and Geiger DR (2002) Kinetic characteristics of chloroplast transport. *Journal of Experimental Botany*, **53**, 1581–1591.

Spreitzer RJ (2002) RUBISCO: structure, regulatory interactions, and possibilities for a better enzyme. *Annual Review of Plant Biology*, **53**, 449–475.

3

Non-photosynthetic metabolism

Leaves of a plant are autotrophic and can make their own carbohydrates from carbon dioxide and water. However, tissues that cannot photosynthesize, known as the non-photosynthetic tissues, are heterotrophic and must rely on the leaves to supply carbohydrate. These heterotrophic tissues are not merely a drain on the plant's resources, because in return they fulfil some other important function; for example, roots provide support and harvest water and minerals from the soil, but to do this they need energy from leaves in the form of carbohydrate. The majority of plants studied produce two major carbohydrates during photosynthesis: starch and sucrose. Starch accumulates in leaves during photosynthesis and is then used as a carbohydrate supply at night, where it is mobilized for respiration in the leaves or is converted to sucrose. Sucrose, being soluble, is the main form in which carbohydrate is moved around the plant during the daytime and at night.

How carbohydrates are moved around plants is a very important area of research, because humans consume very few leaf crops, e.g. lettuce, cabbage and spinach. Leaves may be rich in certain vitamins and mineral ions but they are low in carbohydrate, fats and protein. Furthermore, most of the carbohydrate in leaves is present as cellulose, which humans cannot digest. However, many non-photosynthetic plant organs contain large concentrations of these essential dietary components. Examples of such organs are tubers, seeds, roots, bulbs and stems. These plant products form a major component of our diet. As a consequence, it is essential that we have a thorough understanding of what dictates the rates of growth of such organs, so that plant productivity can be maximized. In this chapter we look at how plants move carbohydrate around the plant and how they control how much carbohydrate is delivered to a particular organ.

Phloem transport

One of the basic requirements of the structure of a leaf is that the products of photosynthesis generated in the photosynthetic cells are made readily available to the rest of the plant. This is achieved through the vascular tissue, which branches through the leaf structure forming finer and finer networks, such that for many leaves each photosynthetic cell is within four cell lengths of the vascular tissue. The vascular tissue is made up of xylem tissue (this will be considered in detail in Chapter 4) and phloem tissue. It is the latter which acts as the transport mechanism for carbohydrate and amino acids to travel around the plant.

During photosynthesis, sucrose is made in the cytosol of the photosynthetic cells. Sucrose can build up to a certain concentration and then there is a gradual switching off of the gluconeogenic pathway by the feedback inhibition of sucrose phosphate synthase by sucrose itself. Photosynthetic cells have two means of allowing sucrose to leave them. First, there are plasmodesmatal links between the photosynthetic cells, which allow sucrose to diffuse freely between cells. Therefore, if there is a difference in concentration of sucrose between two photosynthetic cells, sucrose will move from the higher concentration to the lower concentration by diffusion. This is known as symplastic movement of sucrose (Figure 3.1). This occurs to some extent in all plants, but in most leaves there is little reason why there should be a gradient of sucrose concentration between the photosynthetic cells; therefore, in most leaves it is unlikely to be the major driver for moving sucrose to the phloem tissue. Second, in the plasma membrane of photosynthetic cells there are passive protein transporters, which permit movement of sucrose out of the cell into the apoplastic space (Figure 3.1). This movement is protein-mediated, but all it will do is establish equilibrium in the concentration of sucrose within and outside the

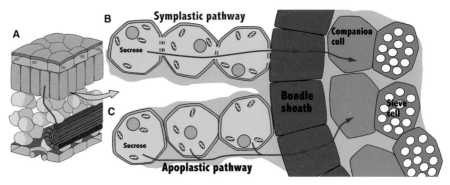

Figure 3.1   Apoplastic and symplastic loading of the phloem. Sucrose made in the photosynthetic cells needs to provide carbohydrate for all of the non-photosynthetic tissues. To do this, sucrose needs to move to the phloem in the vascular tissue. (A) The location of the vascular tissue in a leaf section. (B) Schematic showing the symplastic pathway, where sucrose diffuses from cell to cell in the direction of the phloem tissue. However, sucrose does need to leave the symplast prior to entering the phloem tissue. (C) The apoplastic pathway, where sucrose leaves the cell and freely diffuses towards the phloem tissue.

cell. However, the apoplastic space is in contact with the phloem tissue and can therefore allow diffusion of sucrose in its direction. In order to appreciate how the phloem obtains sucrose from the photosynthetic cells, we need to look in greater detail at the phloem tissue itself.

Structure of the phloem

The phloem is made up from a number of different cell types. There are the sieve tube cells, which transport sucrose and other dissolved solids around the plant. These cells are living but contain few cellular organelles (Figure 3.2). In addition, they also contain long strands of a protein running from cell to cell, called phloem protein (P-protein). At the end of each cell is a sieve plate which contains large holes allowing access to the next cell in the sieve tube. The P-protein runs through these holes. Adjacent to these there are companion cells which contain numerous organelles, particularly mitochondria, and are very active metabolically. It is the metabolic function of these cells that allows the sieve tube cells to survive. Joining the companion cells with the sieve tube cells there are numerous plasmodesmata, allowing the exchange of dissolved solutes and metabolites. Around these cells there are fibres which reinforce and protect the phloem. As the vascular tissue carries vital carbohydrates, amino acids and dissolved mineral ions around the plant, it is important to protect the tissue from potential parasitic predators, such as aphids and scale insects. The phloem is on the upper and lower side of the vascular tissue in leaves of many plant species and on the outer parts of the stem, and as a consequence it is very exposed to such predators and needs protection. When a plant is grown under poor conditions,

stressed by lack of water and nutrients, this protection is less well developed and so such weak plants are very prone to attack by insect parasites.

The vascular tissue within the leaf fuses together along the midrib and then travels down into the petiole until the stem is

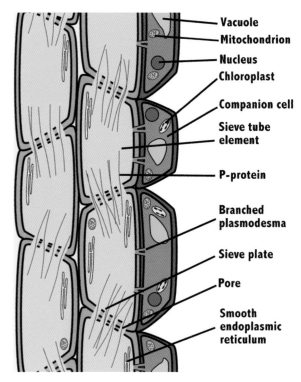

Figure 3.2   Structure of the phloem. A simplified cross-section of the phloem tissue shows the individual cell components. The companion cells are marked in red and the sieve element cells are marked in orange. The long files of sieve elements make up sieve tubes.

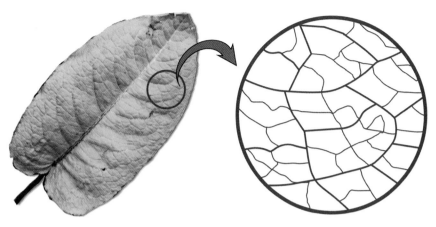

Figure 3.3 The vein network in a leaf. A *Rumex* leaf is shown, with a schematic diagram of the vein network in one section of the leaf. This network links to the major veins and then to the midrib of the leaf. From there, carbohydrate in the phloem can be transported anywhere within the plant.

reached (Figure 3.3). Here the vascular tissue is arranged in bundles which run both up and down the stem. An individual sieve tube should never be seen as running all the way from the leaf down into a non-photosynthetic organ. The sieve tubes periodically cease and the contents need to leak out in order to progress further through the vascular tissue. In the roots the pattern of organization changes again, with the phloem lying in the stele of the root, just outside the xylem tissue. In this position the phloem is well protected and there are few predators that can suck sap from the root system. For every organ coming from the roots or stem there is a system of vascular bundles which runs through the organ to permit carbohydrate, water and nutrient supply to the organ. If it were not so, the organ would never obtain enough nutrition to develop in the first place. It is this network of phloem tissue through the vascular bundles that connects metabolism in every part of the plant together.

Phloem loading

In the majority of plants the phloem in the leaves obtains carbohydrate from the apoplastic space. The companion cells contain a large number of mitochondria and, through respiration of carbohydrate, these cells produce an electrochemical gradient of H^+ ions between the cytosol of the companion cells and the apoplastic space. This movement of H^+ ions gives rise to a low pH in the apoplastic space. This gradient of H^+ concentration is then utilized by a H^+/sucrose symporter protein pump (Figure 3.4). The high concentration of H^+ ions outside the cell fuels the pumping of sucrose from the apoplast into the cytosol. This is a secondary active transport system, so called because it does not rely on ATP hydrolysis directly for the pumping

mechanism but on the H^+ gradient established across the plasma membrane. This mechanism allows the companion cell to accumulate a higher concentration of sucrose in the cell than is present in the apoplast (Figure 3.5). The high metabolic activity of the companion cells allows them to achieve this. Moreover, the pumping of sucrose out of the apoplastic space causes the concentration of sucrose locally around the phloem to fall. Sucrose therefore diffuses to this low concentration from the apoplastic space around the photosynthetic cells. As a consequence, the concentration of sucrose around the photosynthetic cells is also lowered and so sucrose can freely diffuse from the cells into the apoplastic space. Thus, there is a constant flow of sucrose from the photosynthetic cells to the phloem which is driven by the metabolic activity of the companion cells. This is known as apoplastic movement of sucrose into the phloem.

As has been mentioned earlier, sucrose can diffuse freely through plasmodesmata between the photosynthetic cells. There is nothing to prevent a gradient in concentration of sucrose being set up between neighbouring photosynthetic cells because of the metabolic activity of the companion cells. However, there are no plasmodesmatal links between the photosynthetic cells and the companion cells in the majority of plants; therefore, at one point sucrose would have to leave the cell to enter the apoplast before entering the phloem (Figure 3.5).

As sucrose is imported into the companion cells, its concentration rises and free diffusion through plasmodesmata allows the sucrose to enter the sieve tube cells. This build-up in sucrose concentration also causes the water potential of the phloem tissue to fall. Water therefore enters the tissue by osmosis from the apoplast and cells neighbouring the phloem tissue, and this leads to a rise in

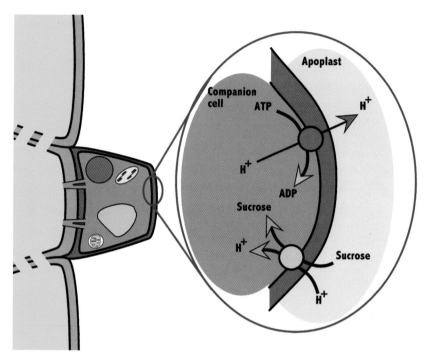

Figure 3.4   Uptake of sucrose into the companion cell. In a majority of plants there is no direct symplastic link between the photosynthetic cells and the companion cells in the phloem. Sucrose diffuses down a concentration gradient into the apoplastic space. The companion cells have a high activity of H^+/ATPase pumps in their outer membranes, which pump H^+ ions from the cytosol of the cells into the apoplastic space. This creates a high concentration of H^+ ions in this region of the leaf. The H^+ ions diffuse back into the companion cells with a sucrose molecule, a process known as secondary active transport.

pressure in the phloem sieve element cells. Since the sieve element cells are linked to neighbouring cells, the pressure results in a motive force which pushes the solution through the tubes to an area where there is a lower pressure in the system. Inevitably, this lower pressure region is going to be somewhere in the plant where sucrose is not being synthesized for export to other tissues.

A small percentage of plants have evolved a symplastic mechanism for loading the phloem tissue with carbohydrate. The difficulty encountered with symplastic loading is that the concentration of sucrose in the phloem needs to build up to a higher concentration than the photosynthetic cells for the pressure-flow mechanism of sucrose transport to function. However, if sucrose is loaded into the phloem through plasmodesmatal links between the photosynthetic cells and the companion cells, then the build-up of sucrose in the phloem would favour diffusion in the reverse direction. Certain plants possess cells near to the phloem known as intermediary cells, which possess plasmodesmatal links with the photosynthetic cells (Haritatos and Turgeon, 1995). Once sucrose is taken up into these cells, it is converted into a raffinose series oligosaccharide, e.g. raffinose, stachyose or verbascose (Figure 3.6). These are made by sequentially adding galactose residues onto the

sucrose molecule. This increases the size of the carbohydrate molecule and, as a consequence of the size of the aperture of the plasmodesmata, the larger molecule cannot get back through to the photosynthetic cells. This then acts as a carbohydrate-concentrating mechanism, which allows the build-up of carbohydrate and the generation of a low water potential per unit weight in the phloem. The raffinose series oligosaccharides tend to be mainly the shorter-chain molecules, since the longer-chain molecules generate a smaller fall in water potential and therefore a smaller pressure can be generated for carbohydrate movement around the plant. The raffinose series oligosaccharides are generally not easily digested by the human digestive system and as a result large quantities of carbohydrate reach the lower intestine and are broken down by bacteria there. This results in large quantities of gases being liberated by the bacteria, consequently causing flatulence. Plants which are high in raffinose content are certain bean species, melons and cucumbers.

Phloem unloading

There is not a continuous sieve tube leading from the photosynthetic cells all the way to a sink tissue such as a

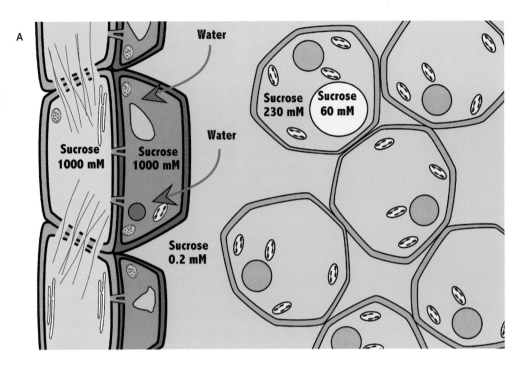

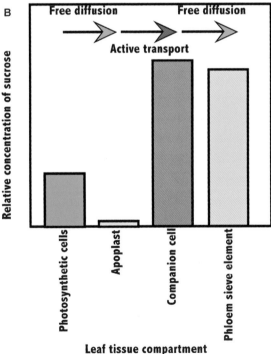

Figure 3.5 The concentrations of sucrose in different leaf compartments. For sucrose to diffuse from the photosynthetic cells to the phloem, there needs to be a concentration gradient of sucrose, with sucrose being at a high concentration in the photosynthetic cells and a low concentration in the apoplastic space (A). The concentrations of sucrose in these compartments of a leaf are shown. The high concentration of sucrose in the phloem causes water to enter the cells. This increases the pressure in the cells, and it is this pressure that pushes the phloem contents around the plant. The contents move from an area of high pressure to an area of low pressure. (B) Bar chart representing these concentrations, emphasizing the large increase in concentration of sucrose in the companion cells compared with the photosynthetic cells and the apoplast. This can only be achieved through the expenditure of energy.

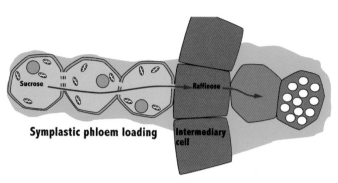

Figure 3.6 The transport of raffinose series oligosaccharides. There is a small group of plants which transport raffinose series oligosaccharides instead of sucrose, e.g. melons and cucumbers. These carbohydrates are basically sucrose molecules with galactose molecules added on. In such plants sucrose diffuses through the symplastic space into the companion cells. In these cells the sucrose is converted into raffinose or stachyose. These new carbohydrate molecules are too large to diffuse through the plasmodesmatal links back into the mesophyll cells; thus, they are trapped, which then generates a low concentration of sucrose in the companion cells and therefore drives the net movement of sucrose into the phloem by diffusion.

tuber. The sieve tubes are not continuous; there are breaks at intervals, at which sucrose leaves one tube and is reloaded into the phloem sieve element cells in the next sieve tube. This constant reloading of sucrose is a process requiring energy and may seem wasteful, but without this there would need to be an individual sieve element leading from specific photosynthetic cells to every non-photosynthetic organ. This would require a large number of phloem sieve tubes to fuel growth and would be more costly in energy than the system described above. The constant leakage of sucrose

from the phloem allows the translocated carbohydrate to be transported efficiently to where it is required most. If there is no demand for sucrose in the tissue region where the sucrose leaks from the phloem, it is taken up again by the companion cells and they are reloaded (Figure 3.7), so there is little fall in pressure in the phloem. However, if the tissue has a high demand for carbohydrate (e.g. a growing meristem of a expanding leaf), then the sucrose will not build up around the phloem tissue and be reloaded into the phloem. Such tissues are known as sink tissues

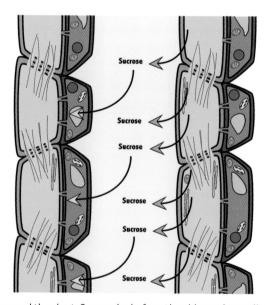

Figure 3.7 The transport of sucrose around the plant. Sucrose leaks from the phloem tissue all around the plant. This sucrose can then be picked up by neighbouring tissues to fuel their metabolism or it can be taken up again by the phloem. This reloading of sucrose avoids the necessity for there to be a single tube leading from a leaf to a particular organ in the plant. Sucrose can be reloaded into the phloem at any point and transported to where it is needed most.

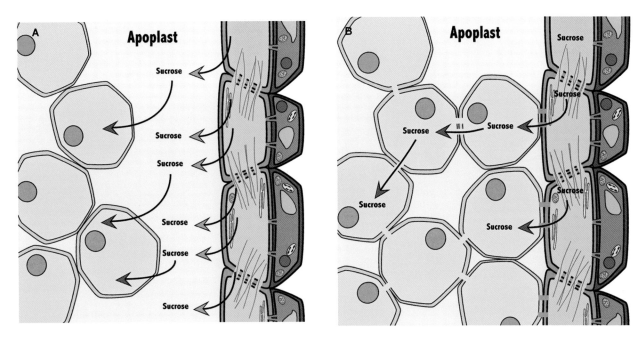

Figure 3.8 Unloading of sucrose. There are two ways for sucrose to leave the phloem in the region of a sink tissue. It can be lost through diffusion into the apoplastic space (A); the sucrose here must be metabolized rapidly by the sink cells in order to prevent sucrose reloading and to facilitate further sucrose diffusion from the phloem. Alternatively (B), sucrose can diffuse directly from the phloem cells into sink cells through the symplast. Again, sucrose needs to be metabolized or stored in order to permit its continued diffusion from the phloem tissue into the sink cells.

(Figure 3.8). The sucrose must be metabolized by the sink tissue in order to permit continued diffusion from the phloem tissue, and this is the subject of a later section in this chapter.

As sucrose diffuses out of the phloem, the concentration of sucrose in the phloem falls. This leads to a rise in the water potential of the phloem and water is drawn from it into the neighbouring tissues. With water leaving the phloem, the pressure drops in the sieve elements and this creates a low-pressure region in the phloem sieve tube. The contents of the phloem are pushed towards this low-pressure region. The extent of the pressure difference in the phloem between the leaf and the sink tissue dictates the scale of the movement of sucrose to that particular sink. Thus, the more metabolically active a tissue is, the more carbohydrate it receives.

Coping with damage to the phloem

This whole transport system is very susceptible to attack and damage by other organisms. One example is the damage caused by grazing, where mammals such as cows tear at grasses, ripping their leaves apart. Given that pressures within the phloem are estimated to be around 100 times that found in the aorta of the human heart, such damage

would be expected to have spectacular results. In fact very little happens at all. The phloem system can shut down very quickly, and this confounded studies of the system for a long time (Figure 3.9). If the phloem is damaged and a large pressure drop occurs, such as would happen if the leaf or stem were broken, then the P-protein in the phloem sieve tubes rapidly coils up and blocks the sieve plates. This prevents wholesale loss of the phloem contents when the plant is under attack. The sieve plates are subsequently blocked completely by callose deposition across the endplates. A damaged sieve tube can only be repaired through cell division and the production of new phloem cells.

As mentioned earlier, the phloem is protected from attack by sap-sucking organisms, but some organisms can get past these defences. One example is the aphid. These insects can stab the leaf surface, hunting for the phloem (Figure 3.10). The sharp stylet can puncture the reinforced tissues which protect the phloem and, as a consequence, they can parasitize the carbohydrate supply in the plant. The stylet is very fine and, when inserted into the phloem, only causes a small drop in pressure in the sieve tubes. This does not trigger the coiling of P-protein and hence the flow in the phloem continues. However, the pressure drop is sufficient to create a false sink tissue in the plant and the translocation system will transport sucrose to the insects.

A   B

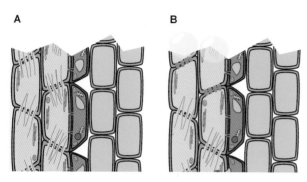

Figure 3.9 Herbivore attack of the phloem tissues. The contents of the phloem are under extremely high pressure (over 100 times that in the arteries of the human body). Thus, if the phloem is damaged there is great potential for the loss of the contents of the phloem. This damage occurs regularly when herbivores damage the leaves of a plant. This is shown in (A), where a leaf has been ripped open, exposing the phloem sieve tubes. To prevent sap squirting out of the phloem, the plants use the P-protein. This protein, on the release of pressure in the phloem that occurs during the rupturing process, instantly coils up and forms plugs, which block the sieve tubes (B). This effectively prevents wholesale loss of the phloem contents.

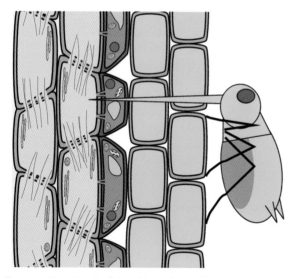

Figure 3.10 Attack of the phloem by sucking insects. The are numerous insect species which attack the phloem and can successfully get access to the phloem contents. One such group are the aphids. In order to prevent the collapse of the pressure in the phloem, the insect uses a very fine proboscis, which causes only a small change in pressure in the phloem (similar to what would occur at a sink tissue). Therefore, there is no collapse of the P-protein and the insect can feed on the phloem contents. This has proved useful to scientists studying the contents of the phloem, as the proboscis can be cut off from the insect and it continues to function as a tap on the phloem.

The pressure in the phloem forces the sap contents through the insect and out of its body again. The aphid gains carbohydrate and amino acids from the sap and can live off this mixture.

As an interesting aside, the ability of aphids to tap the phloem contents has permitted analysis of the different components transported in the phloem. The body of the aphid can be parted from the stylet. In this form the stylet continues to exude the phloem components without going through the body of the insect. The components are summarized in Figure 3.11. The major components in the phloem are sucrose and different amino acids. As a result of the high concentration, these compounds could dictate the movement of the phloem contents around the plant. However, the other components are present in low concentrations and will follow the movement of sucrose and amino acids. Hexose carbohydrates are rarely detected in the phloem. The plant hormones present in the phloem will follow the translocation of sucrose around the plant and are needed most by the tissues which have a high metabolic activity, which is precisely where the sucrose is also required.

The sink tissues

Sink tissues, by definition, cannot synthesize sufficient carbohydrate to supply their needs and are dependent on other tissues for their energy supply. Some sink tissues may be photosynthetic but their ability to photosynthesize is limited. Sink tissues include roots, the stem, growing shoots, flowers, tubers, fruits, developing seeds and expanding leaves. All of these tissues receive their carbohydrate through the phloem. The extent to which they import carbohydrate depends on their capacity to convert incoming sucrose into other products in the cells and therefore maintain a gradient in concentration of sucrose between the phloem (high concentration) and the sink tissue itself (low concentration) (Figure 3.12). There is a net diffusion of sucrose from the phloem into the apoplastic space or directly into the cells of the sink tissue. The phloem does not play an active role in this process and it is the activity of the sink tissue which dictates the extent of unloading of carbohydrate from the phloem. The structure of the phloem does tend to be slightly different in a sink tissue from that observed in a source tissue such as a leaf. Generally, the companion cells are more prominent in the source tissues than in the sink tissue. However, despite this, a sink tissue can also act as a source tissue. In temperate orchids, which develop tubers, in some species, as a new tuber develops for next year's growth, if many of the flowers are fertilized then the new tuber can be used as a carbohydrate reserve to

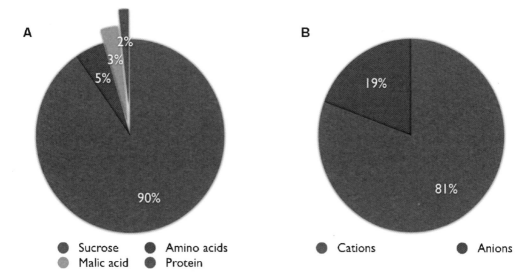

Figure 3.11 The contents of phloem exudates, based on data from *Ricinus communis* given in Hall and Baker (1972). (A) The percentage components present in phloem exudates based on weight of the component (average total weight = 110 mg solids/litre of exudate). (B) Total ionic components present in exudates (average total meq = 130/litre of exudate). Based on Hall and Baker (1972).

fuel flower and seed development. This has led to the observation that species such as the UK bee orchid (*Ophrys apifera*) are monocarpic, i.e. they flower only once and then die. In tissue such as developing potatoes, there is evidence that some sucrose is transported out of the developing tuber at night. In addition, a mature tuber or seed also becomes a source tissue as the new plant shoots or germinates. Therefore, whether a tissue is a sink or source tissue frequently depends on its developmental stage (Figure 3.13).

How is sink strength dictated?

Maintaining the gradient of sucrose concentration between the phloem and the sink tissue is the crucial factor in dictating the capacity of a sink tissue to draw carbohydrate to itself. This will dictate the yield of a particular plant organ but may not dictate the yield of a particular plant, since the number of sink organs will also influence overall yield. It is these two factors that govern important parameters such as total crop yield, and it is crucial that humankind understands these factors if the world's growing population is to be fed.

We need to consider how carbohydrate enters the cells of the non-photosynthetic cells. Sucrose leaks out of the phloem tissue into the apoplastic space or, if plasmodesmatal links are present, sucrose diffuses directly into the non-photosynthetic tissue. If sucrose enters the apoplastic space, there are two major methods for its uptake. Sucrose may be taken up directly into some non-photosynthetic cells via a sucrose/H$^+$ symporter membrane pump present in the plasma membrane (Figure 3.14). To energize this pump, H$^+$ ions are pumped out of the non-photosynthetic

cells into the apoplastic space at the expense of ATP hydrolysis. The electrochemical gradient established is then used by the sucrose/H$^+$ symporter pump to import sucrose. This maintains a low concentration of sucrose in the apoplast and therefore promotes the continued diffusion of sucrose from the phloem into the apoplastic space. Therefore, provided that the sucrose imported into the cell is metabolized and does not interfere with the sucrose import pump, the sucrose/H$^+$ symporter mechanism can account for much of the sink tissue's ability to draw carbohydrate to it. In some tissues there is an inability of the cells to take up sucrose without hydrolysis. In such tissues there is an invertase enzyme which breaks down the sucrose molecule into glucose and fructose. Then there is a range of transporters which can transport glucose, fructose or both of these molecules into the non-photosynthetic cell. These transporters are all linked to the electrochemical gradient of H$^+$ ions established by the H$^+$/ATPase pump, as is the sucrose symporter mechanism. For tissues which break down sucrose prior to entry into the sink cells, the ability of the tissue to hydrolyse sucrose and import glucose and fructose has a major impact on the sink strength of the tissue. Most seeds import their sucrose via the apoplastic pathway described here.

These protein pumps bring sucrose or hexoses (glucose or fructose) into the cytosol of the non-photosynthetic cells. If sucrose leaves the phloem via a symplastic pathway, then sucrose can diffuse from cell to cell in the sink tissue through plasmodesmata. This movement would establish a small gradient in concentration of sucrose between the phloem and the non-photosynthetic cell, but this gradient is vastly increased through metabolism of the

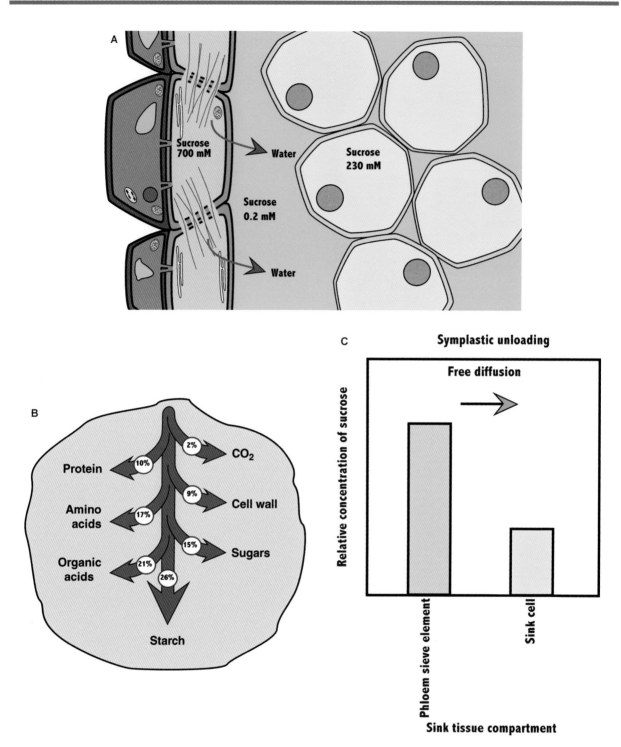

Figure 3.12   The movement of sucrose at sink tissues. As sucrose leaves the phloem tissue in the region of a sink organ or group of cells, the concentration of sucrose in the phloem falls (A). This means that water moves out of the phloem by osmosis, which decreases the pressure in the phloem. This creates a pressure difference in the phloem tissue between the source tissues (high pressure in the photosynthetic tissues) and the sink tissues (low pressure, due to the exit of sucrose and water). Using this pressure difference, the plant can supply sucrose to the tissues in direct proportion to their demand for the carbohydrate (B, C).

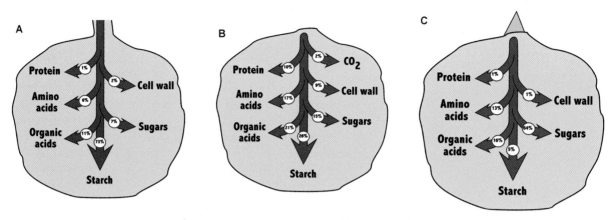

Figure 3.13 The metabolic cycle in a potato tuber, showing the fate of ^{14}C-labelled sucrose supplied to developing (A), mature (B) and a sprouting (C) potato and the percentages of sucrose metabolized to different cellular components. When a potato tuber (a strong sink tissue) is developing, the sucrose is rapidly metabolized, mainly to starch. However, when the tuber is mature the metabolism changes and a futile cycle set up, whereby starch is made and broken down continually. However, in a mature potato metabolism is much reduced compared with the developing potato. Once the potato begins to sprout, the starch content of the tuber is metabolized and converted to sucrose for translocation to the growing shoot, hence the major changes in the metabolism of any added sucrose. In all of these measurements sucrose cannot be identified as a product of metabolism, as it is added to the tubers in the first place and *de novo* synthesized sucrose would be indistinguishable from that originally supplied to the tubers.

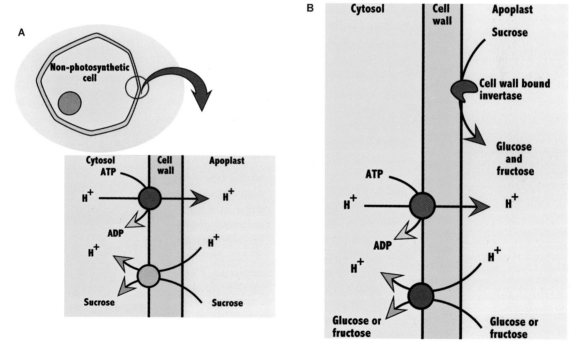

Figure 3.14 Uptake and metabolism of sucrose in the apoplastic space. Sucrose in the apoplastic space around a sink tissue can be taken up by a protein transporter very similar to that in the plasma membrane of the companion cells. H^+/ATPase pumps in the outer membranes of the sink cells pump H^+ ions from the cytosol of the cells into the apoplastic space. This creates a high concentration of H^+ ions in this region of the apoplast around these cells. The H^+ ions diffuse back into the companion cells with a sucrose molecule, the process known as secondary active transport (A). For some sink tissues, sucrose needs to be metabolized to glucose and fructose prior to uptake (B). However, the movement of the hexose sugars into the cells is very similar to that of sucrose.

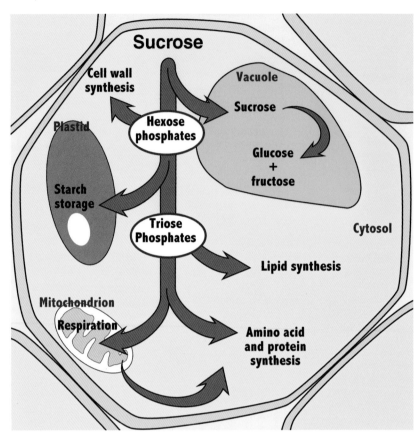

Figure 3.15   Fates of sucrose in sink tissues, summarizing the fates of sucrose after a sink cell has taken it up. In order to create a strong sink demand for sucrose, sucrose needs to metabolized or stored. It can be stored in the vacuole or be metabolized by glycolysis to create numerous products in the plant cell.

sucrose. Thus, the ability of these tissues to import carbohydrate is governed by their capacity to metabolize sucrose (Figure 3.15). Potato tubers are an example of a tissue which imports sucrose directly into the non-photosynthetic cells through the symplastic pathway.

Metabolism of carbohydrate

There are three main destinations for carbohydrate arriving at a sink tissue (Figure 3.15):

1. It can be respired.

2. It can be stored as sucrose, starch or another carbohydrate, or converted to a lipid.

3. It can be used for growth and the synthesis of cell components.

Respiration

Photosynthesis could provide all of the energy needs, in terms of ATP and NADPH, of the leaf during the day, but at night the concentrations of these compounds fall in the cells. In a leaf, the soluble sugars and insoluble starch reserves are mobilized to provide substrates for respiration, and this provides the ATP and NADPH requirements of the leaf. In other tissues the degree of autotrophy will vary, depending on the role of the tissue in the plant, but most tissues will need to respire some of the carbohydrate to continue to survive. The pathway of respiration is identical to that in animals; the only major difference is the presence of an enzyme called the alternative oxidase in the mitochondrial membrane of plants (Figure 3.16). This enzyme allows the oxidation of NADH in the mitochondria without the synthesis of ATP. At first sight this seems very wasteful, as a plant is unable to synthesize as much ATP per turn of the Krebs cycle as an animal is able to do. There have been a number of hypotheses as to why this

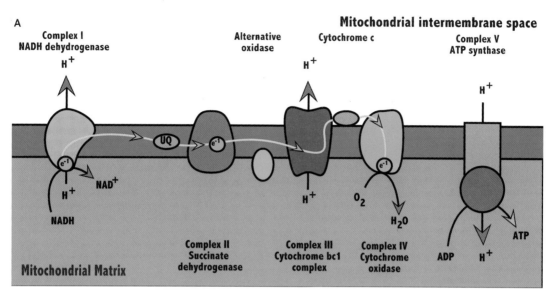

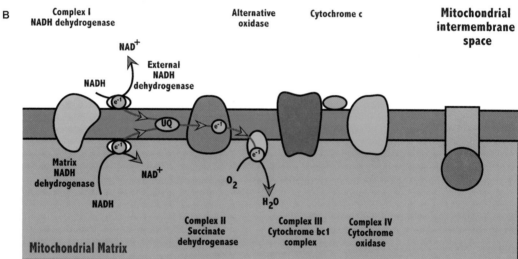

Figure 3.16 Respiration in the mitochondria. In photosynthetic tissues, much of the ATP and NAHPH demands of a cell can be met by photosynthesis, but in a non-photosynthetic tissue that is not an option. The cell respires carbohydrate that arrives in the cell to produce energy for growth and other processes. Much of this energy is released in the electron transport chains in the mitochondria (A). However, plants possess a second pathway in the mitochondria which allows the Krebs cycle to function but much less energy is formed as chemical energy and heat is lost instead. This pathway uses a membrane channel known as the alternative oxidase (B). This pathway is thought to be important when a plant needs to generate heat (e.g. in the flowering of a *Arum* plant) or when intermediates from the Krebs cycle are required at a time when there is already a surplus of ATP and NADPH.

alternative oxidase is present, but two stand out as being the most likely. In specialized tissue such as the spadix of *Arum maculatum*, during flowering a large quantity of starch is metabolized very rapidly on opening of the inflorescence chamber (Chivasa *et al.*, 1999). The temperature of the spadix rises above that of the rest of the plant during respiration of the starch and this permits the plant to volatilize aromatic compounds released by the spadix,

which act as attractants to potential pollinators. The alternative oxidase fuels this rapid release of heat energy in these plants. In other plants it is thought that the presence of the alternative oxidase permits the Krebs cycle to turn under conditions where ADP or NADP availability would limit the cycle's activity. This would be important in leaves when photosynthesis is occurring. The intermediates of the Krebs cycle are not just respired and oxidized to CO_2 in

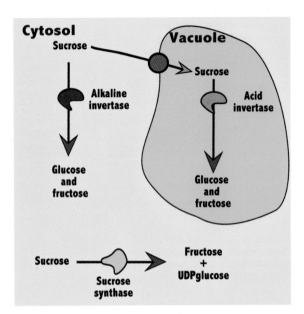

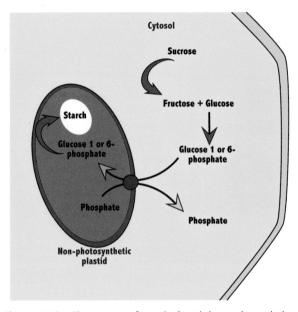

Figure 3.17 Storage of sucrose in the vacuole. Sucrose can accumulate to a limited extent in the cytosol of a sink cell, but the largest area for sucrose storage is in the vacuole. Provided that the concentration of sucrose in the vacuole never exceeds that of the cytosol, no active storage mechanism is required. However, in plants which store large amounts of sucrose, such as sugar cane and sugar beet, active storage mechanisms are required.

Figure 3.18 The storage of starch. Starch is a major carbohydrate storage product in the sink tissues of plants, and is made in the non-photosynthetic equivalent of the chloroplast in the sink tissue; for starch storage, these are normally called amyloplasts. Glucose 1- or 6-phosphate are imported into the amyloplast in exchange for phosphate and then the glucose monomers are used to make starch, just as in a leaf. However, in sink tissues the starch is usually stored for a long time and is used to fuel another period of the plant's life cycle.

plants, but are also used as substrates for the synthesis of a range of different products, such as amino acids. The alternative oxidase permits the plant to do this and it is probably its presence which reflects the remarkable synthetic ability of plants. The intermediates of glycolysis, the oxidative pentose phosphate pathway and the Krebs cycle, provide a hub for the synthesis of many different compounds required by plants.

Storage of carbohydrate or lipid

One of the major fates of carbohydrate imported into non-photosynthetic plant organs is the synthesis of storage products such as starch or fats. This capacity to metabolize and synthesize a storage product affects the ability of a sink tissue to draw sucrose from the phloem tissue. Different plants and tissues adopt different means of achieving this and it is best handled with a series of case studies.

Potatoes and starch storage

The tubers of potatoes can be regarded as one of the best understood major starch storage organs. Sucrose enters the tubers via the symplastic pathway. The sucrose is then

metabolized to hexose phosphates by the enzyme sucrose synthase or a cytosolic invertase enzyme (Figure 3.17). Sugar phosphates are then imported into amyloplasts (the non-photosynthetic equivalent of a chloroplast). The pathway of import into the chloroplast and subsequent conversion to starch is shown in Figure 3.18. As the tuber develops, over 75% of the incoming sucrose is converted into starch. The starch molecule grows and becomes a starch granule, which can just be seem with the naked eye. As the granules swell, the tuber expands outwards and grows, forming further cells to accommodate more starch. It therefore does not matter when a potato is harvested; it is always ripe and ready to eat. The different categories of potato available, such as 'first early' or 'maincrop', just describe when the tubers are produced and reach their maximum size during the growth of the potato plant. A similar pattern is followed by most root and tuber crops.

Bananas and starch storage and ripening

Bananas (*Musca* species) were first domesticated over a wide area in southern Asia. They are now grown in a wide range of tropical countries across the world. Most of the bananas grown commercially are triploid and therefore sterile, and as a consequence only small vestiges of the aborted seeds are seen in banana fruit. Fruits develop in a slightly different way from the root and tuber crops. After fertilization the flower dies and the ovaries begin to expand, along with the tissues surrounding the developing seeds. The banana fruits are initially green but will change to yellow as the fruit ripens. To deter animals from preying on the developing fruits, the cells of the banana contain high concentrations of tannins. The role of these is discussed in Chapter 16; suffice it to say here that they are a potent mechanism for deterring herbivores. The fruit initially converts the incoming sucrose to starch in amyloplasts which are distributed throughout all of the fruit cells. The mechanism for this conversion is identical to that for potato tubers. This conversion establishes the gradient of sucrose concentration between the developing fruit cells and the phloem tissue. Starch-laden bananas are not very attractive to herbivores and although some bananas are eaten as a starch crop, the majority are grown for their sweet taste. However, there is a problem, since a sweet banana would encounter many difficulties in establishing a gradient of sucrose concentration between the fruit cells and the phloem if the sucrose was not metabolized. The banana fruit acts as a strong sink organ by accumulating starch as the banana fruit develops and then the fruit cuts its links to the phloem tissue and hence with the rest of the plant (Figure 3.19). Once this tissue has senesced, the fruit begins to ripen. Bananas go through a process known as the climacteric, whereby starch is broken down and sugars begin to accumulate in the tissue. However, there is also a burst of rapid respiration which fuels the ripening process. This whole ripening process is controlled by ethylene gas. The fruit naturally generates the gas as it approaches maturity.

Sugarcane and the storage of sucrose

Sugarcane is the major sucrose-yielding crop and although sugar beet plays a similar crop role in Europe, sugarcane gives a far higher yield than sugar beet. Sugarcane is a C4 plant and is capable of high rates of photosynthesis. Nevertheless, sugarcane takes around 5 years to mature and flower. At this point the plant is ready for harvest. The stem of the plant is harvested and the lower internodes are used for the extraction of sucrose. When harvested and smashed, the internodes respond by generating brown phenolic compounds as a defence. It is these compounds that give the brown coloration to brown sugar. If the sugar is further purified, this coloration can be removed to give white sugar. The cells in the lower internodes accumulate the highest concentrations of sucrose and these are the most important part of the crop. The cells in the internode face a challenge

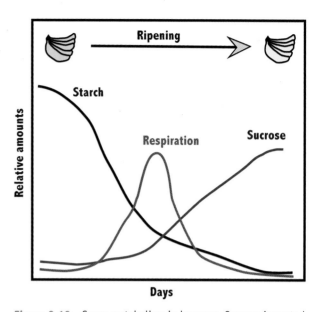

Figure 3.19 Sugar metabolism in bananas. Sucrose imported into a developing banana is mainly used for tissue growth and starch storage. Thus, the developing fruit has a high demand for sucrose from the phloem. However, as the fruit matures, import of sugars ceases and the tissue undergoes what is called a climacteric. The starch is metabolized and sugars accumulate but at the same time there is a burst in respiration, as shown. The accumulation of sugars signals that the fruit is ripening.

to accumulate sucrose in the cells and yet generate a gradient of sucrose concentrate between the phloem and the internode cells. This is achieved through the sucrose being stored in the vacuole. To allow a higher concentration of sucrose to build up in the vacuole than in the cytosol of the cell, a sucrose/H^+ antiport pump is used. H^+ ions are pumped into the vacuole by an ATPase/H^+ pump. This electrochemical gradient is then used to move sucrose molecules into the vacuole at the same time H^+ ions are exported. This active transport of sucrose into the vacuole maintains a low concentration of sucrose in the cytosol of the internode cells and therefore sucrose can easily be transported from the apoplastic space around the cells. This then facilitates the diffusion of sucrose from the phloem.

Grapes and developing fruit

Grapes are mainly grown for the wine industry and there are three major species cultivated, *Vitis vinifera* (European), *Vitis rotundifolia* (Muscadine) and *Vitis labrusca* (American). These crops are potentially some of the most valuable on earth. Certain wines can command prices of over £10 000 pounds a bottle! Grapes use a similar metabolic process to that which occurs in sugarcane. Carbohydrate is metabolized to malic acid and tartaric acid in the early stages of fruit

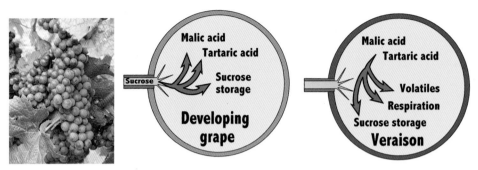

Figure 3.20   Grape ripening. As grapes develop, they use imported sucrose to make new tissues and accumulate high concentrations of organic acids. As the fruit matures, import of sucrose ceases and then the acids are metabolized for respiration and sucrose accumulation. This change in metabolism is part of the fruit-ripening process.

maturation (Figure 3.20). Then, once the fruit is near maturity, in order to draw carbohydrate to the fruit, the sucrose is transported to the vacuole and hydrolysed to glucose and fructose. This establishes the steep gradient in concentration between the phloem and the cells in the fruit tissue. The fruit also accumulates high concentrations of tannins, so the unripe fruit is very bitter and acidic. In warm weather the acidity in the fruit declines as it ripens, through respiration of the malic acid. Once the fruit is ready to ripen, the vascular tissue links between the parent plant and the fruit senesce.

Peanuts and lipid storage

Peanuts or groundnuts (*Arachis hypogaea*) are a member of the family Fabaceae and form in a similar way to peas or beans. Once a flower is fertilized, the inflorescence stem becomes negatively gravitropic and pushes the developing seed pod or peg to the ground. The pod then extends into the soil and the nut actually forms underground, hence its name. In the mature seed around 30–50% of the final weight will be made up of fats, with the

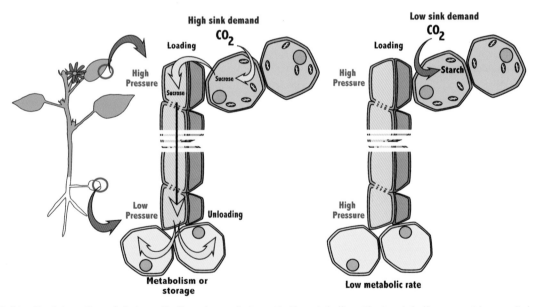

Figure 3.21   The integration of photosynthetic and non-photosynthetic metabolism. Plant metabolism cannot be seen in isolation, as the whole process within a plant is an integrated unit. If sink tissues have a high demand for sucrose, there is a fall in pressure in the phloem in the locality of the sink tissue, which causes more sugars to be translocated to that region of the plant. In addition, the leaves make sucrose to supply the demand for sugar. However, if demand should fall, the pressure difference in the phloem will fall and less sugar is transported to that plant tissue. In the leaf this will cause a reduction in sugar synthesis (since there is a fall in demand for it) and more starch will be made instead.

remainder being made up of carbohydrates, protein and water. The major fatty acids that accumulate in peanuts are stearic acid (77%), palmitic acid (14%) and oleic acid (4%) (Sreenivas and Sastry, 1995). For peanuts, or any seed which accumulates fats as one of as major storage components, the sucrose arriving at the seed is metabolized to 3C-phosphorylated intermediates and then converted to fatty acids and glycerol in the plastids to make fat. This metabolism of sucrose again allows a gradient of sucrose concentration to be established between the developing nut and the phloem.

Growth and the synthesis of cell components

Another important use of carbohydrate arriving at a sink tissue is the use of the carbon skeletons derived from metabolism for the synthesis of new cells. The glycolytic and oxidative pentose phosphate pathway and the Krebs cycle produce many intermediates, which can be used for the synthesis of other cellular components. Glc-6-P can be metabolized to produce cellulose for cell wall synthesis, or ribose sugars for DNA synthesis. 3C-phosphorylated intermediates can be used for lipid synthesis, which can be used for membranes. The intermediates of the Krebs cycle can be used for the synthesis of amino acids for protein synthesis. Therefore, there are many different ways in which sucrose can be metabolized and it is not necessary for the carbohydrate to be stored, as in the examples quoted earlier.

Sink regulation of photosynthesis

Metabolism within the sink tissues has a major impact on the photosynthetic metabolism in leaves (Figure 3.21). If the sink tissues are very active and are metabolizing large quantities of sucrose and converting it to storage compounds, then there is a demand for sucrose and there will be a low concentration of sucrose in the phloem, and hence a low pressure in the phloem sieve tubes in the vicinity of the sink tissue cells. Sucrose will be pumped to this region of the plant as a result of the pressure difference in the phloem. As a consequence of this, sucrose will be mobilized in the leaf and there will be a requirement for sucrose synthesis in the photosynthetic cells. Therefore, sucrose will be a major product of photosynthesis under such circumstances. If the demand for sucrose falls in the sink tissues, for example as a result of the tissue reaching maturity and being unable to store more sucrose, then the pressure in the phloem will rise. This will cause a fall in the pressure difference between the phloem in the leaf and sink tissues, and a reduction in the amount of sucrose translocated to the sink tissue. This will lead to a rise in sucrose in the leaf and this in turn will feedback-inhibit the synthesis of sucrose in the photosynthetic cells, so more of the CO_2 fixed during photosynthesis will be converted to starch in the leaf instead of sucrose. Thus, photosynthetic metabolism cannot be looked at in isolation from the metabolic processes occurring in sink cells because the processes are all interlinked.

References

Sreenivas A and Sastry PS (1995) A soluble preparation from developing groundnut seeds (*Arachis hypogaea*) catalyses *de novo* synthesis of long chain fatty acids. *Indian Journal of Biochemistry and Biophysics*, **32**, 213–217.

Haritatos E and Turgeon R (1995) Symplastic phloem loading by polymer trapping, in *Sucrose Metabolism, Biochemistry, Physiology and Molecular Biology*, (eds Pontis HG, GL Salerno and EJ Echeverria), American Society of Plant Physiologists, Rockville, MD, 216–224.

Chivasa S, Berry JO, ap Rees T and Carr JP (1999) Changes in gene expression during development and thermogenesis in *Arum*. *Australian Journal of Plant Physiology*, **20**, 391–399.

Hall SM and Baker DA (1972) The chemical composition of Ricinus phloem exudates. *Planta*, **106**, 131–140.

4

Roots and the uptake of water

Roots are easily neglected when studying plant organs. They are generally non-photosynthetic and are a drain on the reserves of the aerial tissues, yet they undertake vital functions for the plant. To colonize dry land effectively, higher plants needed to evolve methods to raise leaves above the ground. To achieve this, leaves needed to be held in an upright position in the air. For this to occur, plants needed to be anchored to the ground and held in position. Roots provide this anchorage and they also provide water, which is essential for the turgor pressure necessary for holding the leaf structure in position, and for ion transport and replenishing water lost by transpiration from the leaf as a result of photosynthetic activity. Gaseous exchange is vital; to photosynthesize, plants need to receive carbon dioxide and to lose oxygen. To facilitate this process, they have stomatal pores in the leaves to allow exchange of gases. However, these pores also allow the loss of water vapour. On average, for every gram of carbon fixed by a plant, approximately 1 litre of water is lost. A plant needs to be able to replace this loss and this is one of the functions of the root system. Only around 1% of the water taken up by the root system is actually used by the plant for the purposes of growth, metabolism and support; the remainder is lost as water vapour through transpiration. However, this water flow needs to be maintained in order to allow photosynthesis to proceed efficiently. Thus, plants need large quantities of water to survive. This has shaped the range of habitats in which plants can survive. Damp fresh water habitats are ideal, whereas drier areas require the plant to become much more specialized. As water is transpired, vital mineral ions are also transported to the leaves from the soil (which will be covered in the next chapter). This chapter concentrates on the role of roots in water uptake.

Types of root

The main forms of roots are shown in Figure 4.1. They can be split into two groups: the *true roots*, which develop below ground; and the *adventitious roots*, which develop above ground from the stem. Adventitious roots usually play a major role in support of the plant. Of the true roots, there are two common types of structure: taproots and fibrous roots; other root forms are specializations of these. Monocotyledons possess only fibrous roots, whereas dicotyledons can possess either taproots or fibrous roots.

Taproots form with a single primary root, which is thickened at the top and tapers to the growing base. The primary root can reach to large depths in the soil, with many lateral roots branching off. The thickened primary root often has a key function as a food storage area of the plant (e.g. Umbelliferae species, such as carrots). In addition, the taproot provides excellent anchorage for the plant.

In many monocotyledons, the primary root is short-lived and is replaced by adventitious roots originating from the stem. These roots form the fibrous root system. They are made up of many roots equal in diameter, which are shallow in the soil but are spread out widely. These root systems bind the soil particles together and prevent erosion of the soil.

Most root systems are estimated to extend not much deeper than a metre into the soil, which is where the majority of the soil nutrients are found. However, for some desert plants root systems have been known to extend over 50 m into the soil. For such plants it is vital to exploit the soil for as much water as possible and deep root systems are required to achieve this.

Physiology and Behaviour of Plants   Peter Scott
© 2008 John Wiley & Sons, Ltd

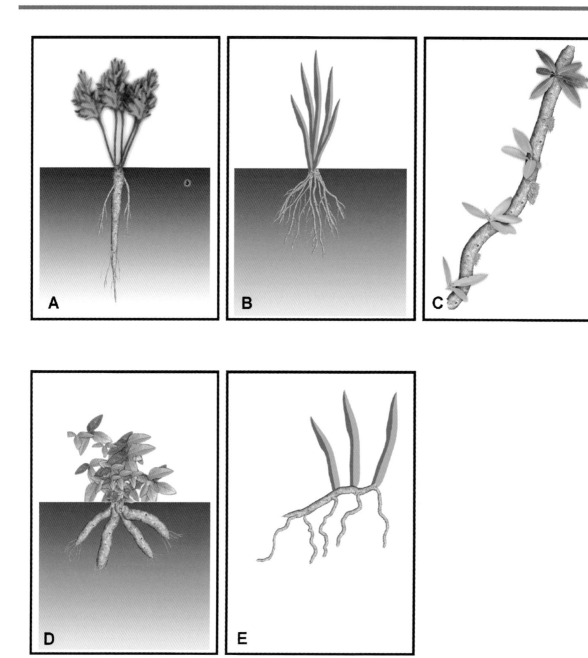

Figure 4.1   Different types of roots, summarizing some of their many functions in plants. (A) A tap root system, where one primary root forms the major centre for all of the other roots to branch. Such a root system is usually deep-rooted and offers great support to the plant and is also very useful for accessing water and mineral ions deep in the soil. (B) A fibrous root system, which has many primary roots entering the soil. This root system tends to achieve less of a depth than the tap root system but has the same functions. (C) An aerial root system, common in climbing plants. Here the root system acts a means of support for the plant on the vertical substrate on which it is growing. (D) A tuberous root system. Generally, the fat roots are used as food storage areas in the plant and are less well adapted to function as retrievers of water. (E) An aerial root system for an epiphytic plant. This root system often acts as a water store to tide the plant over during dry conditions.

Functions of roots

There are numerous functions for roots and these can be summarized as follows:

1. *Uptake and conduction of water from the soil to the rest of the plant.* This is one of the primary purposes of a root system and will form the main topic of the rest of this chapter.

2. *Absorption of minerals and water from soil.* Many mineral ions vital for the healthy growth of plants are present in the soil. Another important function of roots is to forage for these reserves and conduct them to the rest of the plant. This is the topic of the next chapter.

3. *Anchorage of the plant in the soil.* Roots systems are obviously used to anchor the plant to a particular point. This allows the stem to be maintained in an upright position, thereby allowing leaves to attain a height and position to compete with other plants for light. In some senses, anchorage is a disadvantage to a plant, because a root system prevents a plant from moving from one location to another. However, the need for water, nutrients and a stable leaf system outweigh the questionable benefits of plant mobility, and movement is only achieved through reproduction. The roots anchor the plant so firmly to the ground that it can take considerable force to uproot an established plant. Generally, the deeper the root system, the firmer the anchoring. Tree species such as pines, which adopt a shallow fibrous root system, have strong anchorage when the plant is in a stand of other such species but becomes very vulnerable to uprooting (through weathering) when isolated. Dicotyledonous trees, such as oak species (*Quercus*), form much deeper root systems and are consequently much more stable than pines when isolated. For some plants, adventitious roots form additional support structures. For climbing plants such as ivy and Virginia creeper, aerial roots form that bind onto nearby surfaces to support the plant as it grows. In other species, long arching adventitious roots form which act as props for the stem of the plant. This is particularly important in semi-aquatic plants, such as mangroves. With these species the arching roots act to give further support to the stem in changing tidal conditions.

4. *Storage of foods and mobilization of these reserves for shoot metabolism.* Roots frequently form the main storage organs of the plant and the major storage compound is carbohydrate. The carbohydrate is usually present to enable the plant to complete some part of its life cycle at a later period when rapid growth is required. For example, many biennial plants, such as carrots, use the first year of growth to establish a large root storage organ underground. The plant mainly stores carbohydrate in the form of a combination of starch and sucrose. The following year, the plant mobilizes these reserves to fuel early flowering and seed production. On completion of these tasks, the plant dies. In the poikilohydric plant *Craterostigma plantagineum*, the roots store large quantities of the raffinose series oligosaccharide stachyose (over 10% of the dry weight of the tissue is made up of this carbohydrate). During drought stress the plant mobilizes this carbohydrate reserve to produce sucrose, which is used to protect the root and shoot system during desiccation of the tissue.

5. *Production of plant growth regulators.* The root system is the site for production of several plant growth regulators. Cytokinins, gibberellins and abscisic acid are made in roots and then transported to other tissues, where they have a role in growth and development.

6. *Propagation.* For several plant species the root system is used as a means of asexual reproduction. As the root system spreads, roots near the soil surface form buds and develop clonal plants. Examples of such species are members of the mint family, poplar trees and bindweed. Some species of plant display spectacular properties of clonal propagation through their root system, such as stands of creosote and huckleberry which are thought to be over 11 000 years old (Wherry 1972, Madson, 1996). The quaking aspen (*Populus tremuloides*) can also form clonal stands and in one case a single individual is believed to cover around 43 hectares and includes over 47 000 trees (Mitton and Grant, 1996), which would make this the largest clonal tree known in the world (see Chapter 19 for a discussion on the world's largest plants).

7. *Contraction.* Some roots have a contractile function and at periods during the life of a plant, the root contracts to move the stem. This can improve anchorage of the stem through being pulled deeper into the ground. Other plants, such as *Dactylorhiza* orchids, have contractile roots, which pull the new shoots deeper underground during winter to protect them from cold, and then in the springtime relax to push the shoots upwards again to place them in an advantageous position for growth. For bulbs this occurs through two separate mechanisms. Vertical roots below the bulb grow vertically downwards. These

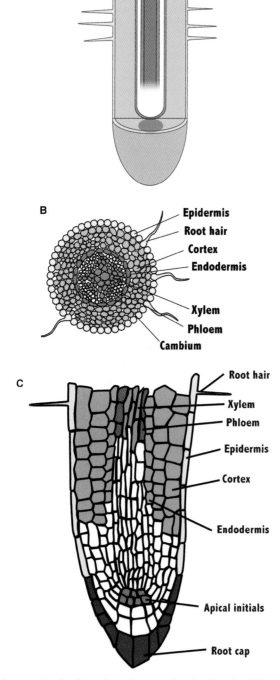

Figure 4.2 Sections through a root tip, showing the different tissues: (A) stereogram, showing the structure in three dimensions; (B) transverse section; (C) longitudinal section.

longer than the first set, and these roots contract as the soil dries in the summer. This acts to pull the bulb deeper into the soil. How the bulb senses its depth in the soil is not known.

8. *Parasitism.* Roots can also provide the structures for the interface between a host and parasite (see Chapter 04 for further details). For example, in parasitic infections of common broomrape on vetches, the haustorial link between the plants is formed from a modified root system.

Structure of roots

To a limited extent, mineral ions and water may diffuse or move by capillarity towards the roots of a plant, but generally, for efficient uptake, the roots must move towards them. To increase the surface area of the roots, both taproots and adventitious roots form lateral roots, which can then form further lateral roots, and so on. These are known as first-order, second-order and third-order lateral roots (Figure 4.3). A cross-section of a root is shown in Figure 4.2.

The root structure is formed from an apical meristem (Figure 4.4). This divides to produce more root axes and a root cap. The root cap is a reinforced tissue that acts to protect the dividing meristem. As the root cap pushes through the soil, cells are sloughed off, hence the structure needs constantly to be replaced. The root cap is covered with a thick layer of mucilage, which is secreted by the root tip (Rougier and Chaboud, 1985) and which helps to lubricate the root as it pushes through the soil. It is the

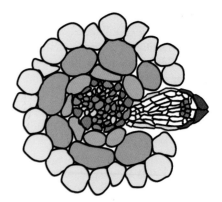

Figure 4.3 Formation of a lateral root. The huge area of a plant root system is enhanced by the formation of lateral roots, which originate from meristematic cells in the stele of the root (that region surrounded by the endodermis). The new root literally grows out from the centre of the primary root, usually breaking cells apart as is develops.

roots are broad and act to push soil below the bulb to the side. This produces a very porous region of soil below the bulb when these roots die at the end of the growing season. A second set of roots persist a little

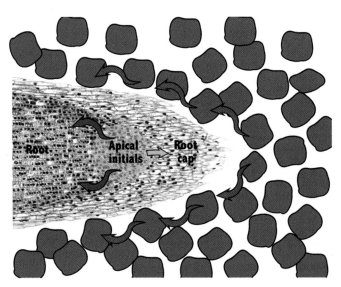

Figure 4.4 Protection of the root tip. The dividing cells for root growth are situated at the very tip of the structure. These cells are very easily damaged and thus, to protect them, a root cap forms, which allows the root tip to push through the soil, forcing soil particles aside. The root cap is sloughed away during this process but the apical initials continually form a new root cap to replace lost cells. The force for pushing the root through the soil is created by the expansion of the cells just behind the apical initials.

cells of the root cap that allow the root to sense the direction of gravity and thus dictate the direction of root growth. Cells that form the root go through a phase of elongation, and it is this process that drives the root through the soil. This growth allows the root system to extend and explore new volumes of soil, and to replace old roots which have died or no longer function well. It is in the zone of elongation that the xylem and phloem begin to develop. The phloem allows the dividing meristem and the developing root cells to obtain carbohydrate and amino acids for growth. The xylem development allows the root to begin to carry out its water-gathering function.

Just behind the zone of elongation, root hairs develop. Our current understanding of the development of root hairs is summarized in Figure 4.5. Root hairs act to increase vastly the surface area of the root system and the volume of soil that can be exploited. A hair can be very long, reaching up to 5 mm, and thus can be over 50 times the length of the cell from which it originates. An estimate of the size of the root system for a 4 month-old rye plant indicated that it

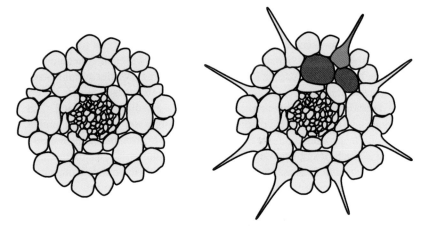

Figure 4.5 Root hair development. The large length of the root network of a plant is largely made up from root hairs in most plants. Root hairs form in epidermal cells, which form above a junction between two other cells. As shown, the two red cells in the cortex of the root form a junction just below the orange epidermal cell; hence, this cell forms a hair, whereas the neighbouring cells do not.

extends to around 11000 km! (Dittmer, 1937). Of this length, over 10 000 km was made up of root hairs. In his study, Dittmer identified around 13,815,672 branches in the root system. Thus, the contribution of the root hairs to the root structure is enormous. It is in the region of root hairs that the vast majority of the absorption of water and mineral ions from the soil takes place (Marschner, 1986). The root hair cells possess no cuticle, so there is no barrier to the diffusion of water and solutes in the soil. The root hairs are short-lived structures and estimates suggest they frequently last less than a week. This makes sense, because as the root grows, volumes of soil will be exhausted of nutrients and to some extent water, and new areas of soil need to be explored. Therefore, the root continually adds new root axes and root hairs during a plant's lifetime. For the 4 month-old rye plant, it was estimated that the plant extended its root system by around 5 km/day! The rate of growth of root hairs in *Arabidopsis* has been shown to be at around 10 μm/minute; hence a mature root hair can form and be fully mature in less than 2 hours. Such cellular growth rates are among the highest recorded for eukaryotic cells. The rate is similar to that achieved by the pollen tube as it grows through the style during fertilization, and that of fungal hyphae. This rate can be achieved because the growing structure is very fine and requires very little material to be deposited.

Once the root hairs begin to wither, the epidermis of the root can become suberized. This involves the deposition of a hydrophobic lipid structure in the cell wall of the epidermis. A suberized root is no longer functional as a structure for the gathering of water and minerals, but is protected from fungal infection and predation. From then on the root's major function is to act as a conduit between the root tip and the rest of the plant. The formation of mycorrhizal associations with fungi only occurs at the tips of the roots, where the cells are newly formed or actively growing.

As the root system ages, some roots cease to grow. These are easily identified as they have mature xylem vessels right the way up to their apex and the root cap is frequently suberized. Annuals (species which carry out their entire life cycle in a single year) do not produce such inactive roots, since their growth period is so short. However, perennial plants such as trees frequently produce such roots (Brundrett and Kendrick 1988). Much of our knowledge about root structure and growth is based on measurements from crop plants, which may not necessarily represent what happens in natural habitats. For example, many crop plants have been shown to extend an individual root by 1 cm/day or more (Russell, 1977), but measurements in natural habitats suggest that many roots grow by less than 1 mm/day (Brundrett and Kendrick, 1990). Crop plants

tend to be fast-growing annuals which produce root systems formed to last several months, whereas plants in their native habitats are mixtures of annuals and perennials, which are much more likely to grow perennial root systems made up of coarse (heavily suberized) roots. Thus, caution needs to be exercised in making generalizations concerning root growth based on data from crop plants.

The cross-section of the root in Figure 4.2B, C shows the internal structure. The epidermis of the root surrounds a loosely linked tissue known as the cortex. This surrounds a circle of cells known as the Casparian strip, which surrounds the endodermis (a layer of cells surrounding the stele). The stele contains the vascular tissue (xylem, phloem and the cambium tissue).

Shoot and root growth

During the growth of a plant there needs to be a careful balance between root and shoot growth. The shoot needs to acquire enough water and nutrients for optimal photosynthetic function, so the root system needs to be of sufficient capacity to meet these needs. If the shoot extends beyond the capacity of the root system to support it, then the plant could easily suffer great damage. However, the root system is heterotrophic and does not generate energy for itself. If the root system over-extends, there will be insufficient carbohydrate to support its function. Hence, there is careful regulation of root and shoot growth through the relative concentrations of auxins and cytokinins. The shoot tip and expanding leaves produce auxins and the root tip makes cytokinins. Careful regulation of these growth regulators controls the shoot-to-root ratio of the plant.

Interaction with the soil

The type of soil a plant grows through has a significant impact on its ability to take up water. Sandy soils containing large particles of non-porous silica show rapid drainage. The large particle size possesses a low surface area-to-volume ratio and consequently little ability to retain water hitting the soil. Conversely, the finer the particle size, the greater the water retention capacity of the soil. Soils containing clay have a very small particle size and thus excellent water retention, but such soils are likely to flood easily as a result of poor drainage. The presence of organic material gives very good water-holding capacity combined with reasonable drainage. Thus, a plant on a sandy soil will be quickly drought-stressed unless it is frequently irrigated with water. Conversely, a plant on clay will have a ready supply of water but will have a dense material to grow through and, if frequently watered, will become waterlogged and possibly die. For most plants the ideal soil for growth is a mixture of sand, clay and humus. When water hits soil, it rapidly drains downwards and it is the surface area of the particle size that dictates how much water will be retained and for how long. Plants must act quickly to obtain the water from the soil before it begins to dry.

To obtain water from the soil, the plant must be in close contact with the soil particles. This is where fine root hairs are very useful for accessing the volume of soil around the plant.

Uptake of water

The uptake of water by a plant takes place through the process of osmosis. Before we go into the mechanics of water movement in plants, we need to understand osmosis.

Osmosis

Put simply, osmosis is the diffusion of water molecules through a semi-permeable membrane from a place of higher concentration of water molecules to one of lower concentration. A semi-permeable membrane is a barrier that allows water to penetrate and pass through it, but restricts the movement of solutes. For a cell, this barrier is the cell membrane. Thus, if two volumes of water are separated by a semi-permeable membrane and one contains pure water and the other a dissolved solute, then water will move from the pure water to the solution by osmosis. This process will continue until the pressure begins to build up in the solution to such an extent that further osmosis ceases. Hence, pressure and gravity, which will become relevant later on, affect osmosis.

Water can get into roots by one of two routes. First, the water can move freely around the cells of the root epidermis and the cells of the root cortex (Figure 4.6). However, water cannot penetrate the cell barrier known as the Casparian strip. These cells have suberized cell walls as a band around the cells, so the only way for water to gain entrance to the stele is to be transported through the cells that make up the strip. This pathway is known as the apoplastic pathway up to the point of uptake into the cells of the Casparian strip. Alternatively, the water can be taken up directly into the cells of the epidermis and the root hairs. These cells lack a cuticle and can therefore readily take up water. Water itself cannot cross the plasma membrane of the cells without the aid of channels in the roots known as aquaporins. Thus, the plasma membrane acts as a

semi-permeable membrane. The high concentration of ions and other dissolved solids in the root cells means that water passes into the cells by osmosis and the root cells are said to have a lower water potential than the surrounding water in the soil. There are also plasmodesmatal links between the cells in the root cortex and so water can pass from cell to cell via osmosis until it crosses the Casparian strip. This pathway is known as the symplastic pathway. There have been many debates as to which of these routes is the main means of water movement in roots. Currently there is no persuasive evidence categorically to pinpoint one route as being more dominant than the other, and it is likely that both routes are used.

Loading and movement in the xylem

Once water has passed across the Casparian strip and enters the endodermis, it passes symplastically until the xylem tissue is reached. Movement to the xylem is via osmosis. The water potential in the xylem is maintained lower than that of the surrounding root tissues in the stele, so water moves into the xylem by osmosis.

There are two feasible ways of achieving water movement in the xylem: through roots pushing water upwards or leaves creating a negative pressure and literally sucking the water upwards. Many plants do exhibit pressure generated by the root system, and when a plant is decapitated this results in exudation of a solution from the xylem. However, this pressure is insufficient to generate the forces required to move water up the stem at the rates observed (5 mm/hour). In addition, under humid conditions root pressure would fail completely. There is good evidence that the leaves generate the negative pressure required to move water up the xylem.

Structure of the xylem

The xylem is the main water-conducting tissue in plants. It is made up of two different types of trachery elements, tracheids and vessels (so called because of their similarity to the appearance of the mammalian trachea), shown in Figure 4.7. These cells are elongated and reinforced with a perforated secondary cell wall mainly composed of lignin. The cells lack protoplasts. The arrangement of the xylem through a plant stem is shown in Figure 4.8.

Tracheids are the main water-conducting tissue in gymnosperms but they are also found in some angiosperms. They are around 1 mm long and have a cross-sectional diameter of around 40–60 μm. The cells are pointed, and frequently square in cross-section. They are surrounded by

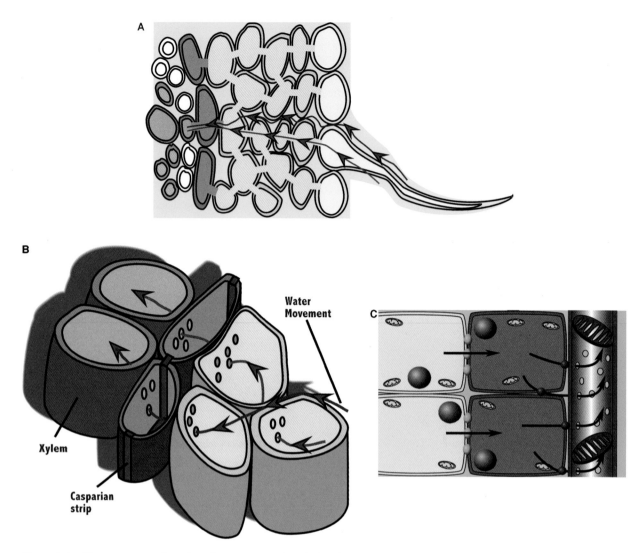

Figure 4.6   Water movement into the xylem. Water can be taken up by the root cells through osmosis or it can enter the apoplastic space between the root cells (A). However, water cannot cross the endodermis of the root without first entering the endodermis cells (shown as purple cells). This is because a water-resistant layer known as the Casparian strip surrounds these cells (shown in red) (B). Water must enter the symplast and pass to the stele of the root. Once in the stele, the water can flow symplastically or apoplastically into the xylem vessels to be transported to the leaves (C).

pits, which link individual tracheids together. In some gymnosperm species, the pits are bordered and contain a small disc of cell wall material known as a torus (see Figure 4.11). This has a role in preventing the formation of air bubbles in the xylem (see later).

Xylem *vessels* are usually rounded in cross-section and have a diameter of up to 120 μm. The cells of vessels are shorter than tracheids (0.5 mm or less) but they are fused together to form tubes ending in a perforated plate. These tubes can be several metres long. As a plant gets older, the newly formed xylem vessels are wider than the older vessels. Xylem vessels are easily distinguished from other

tissues by the appearance of the secondary cell walls (Figure 4.9), which are characteristically laid down in spirals or rings.

Cohesion–tension theory of water movement in the xylem

Dixon and Joly 1894 first proposed the cohesion–tension theory of water movement in the xylem. As mentioned earlier, it is the leaves that generate most of the negative pressure to allow water to ascend the xylem. The driving

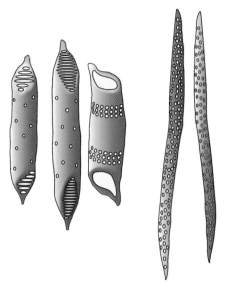

Figure 4.7 The xylem tissue is made up of xylem vessels and tracheids. On the left are shown three xylem vessels from an angiosperm. They are wide, short, and bear large perforations in their end walls to permit fast efficient water flow. Tracheids are found mainly in gymnosperms (conifers). These possess long overlapping ends and have numerous pits in their sides.

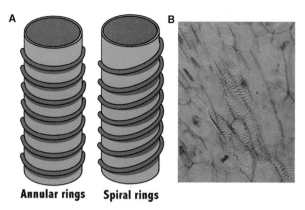

Annular rings   Spiral rings

Figure 4.9 Reinforcement of the xylem tissue. The xylem vessels are reinforced with lignin to keep the vessels rigid and open for water flow. Some are thickened with annular rings and others with spiral rings (A). In some instances the whole vessel can be covered with secondary thickening, with only holes left for pores between vessels. (B) Longitudinal section of a plant stem. Annular rings are seen in the centre of the photograph.

force for water movement is created by the surface tension of water in the apoplastic space of the leaf exposed to the air (Tyree, 1997). As water evaporates from this surface, the water withdraws into the cell wall structure. This has the effect of creating an uneven curved surface, due to menisci

forming with the cell wall fibrils. This dramatically reduces the water-potential of the fluid layer in this region of the apoplast, so water is drawn to it from the surrounding region, mainly the xylem, and this creates a negative pressure in the xylem itself. The whole process is fuelled by the sun, which causes most of the evaporation of water.

The negative pressure in the upper sections of the xylem draws water up the system, rather like water in a

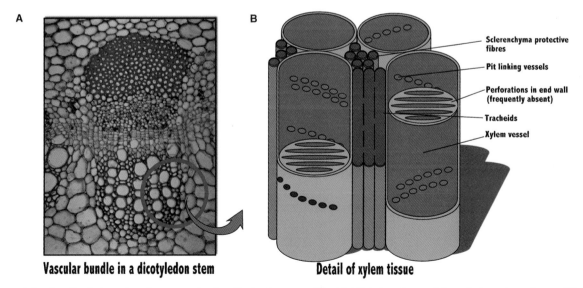

Vascular bundle in a dicotyledon stem   Detail of xylem tissue

Figure 4.8 Detail of the xylem in a vascular bundle in the stem (A), highlighting the position of the xylem tissue. (B) 3D representation of the xylem structure, revealing how individual vessels and protective fibres are linked together. The end wall perforations and pits linking the separate vessels are shown to illustrate the route by which water can travel between vessels.

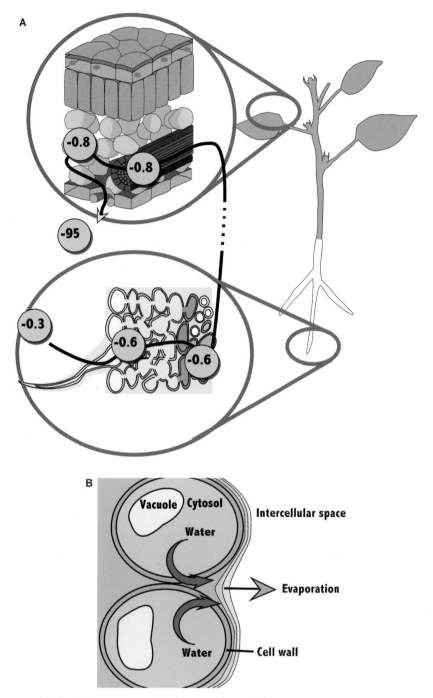

Figure 4.10   Water potentials in different areas of the plant associated with water transport. Plants cannot physically pump water from one place to another; this movement occurs by osmosis. For water to move by osmosis, there must be a progressive drop in water potential from the water in the soil all the way up to the water in the leaves, and then ultimately to the air (A). This drop is shown in the figure, with expanded areas in the roots and the leaves. (B) Gas exchange in the leaves is essential for photosynthesis; however, at the same time, water is lost through the stomatal pores. The water evaporates from the cell wall region of the mesophyll cells. As water evaporates, the water withdraws into the cell wall as shown; this causes the water potential in this region to fall and hence produces the driving force necessary to drive transpiration in the plant.

drinking straw. The water can be pulled upwards provided the column of water does not break and develop air bubbles. The column of water is held together by the cohesive properties of water molecules for each other and the adhesive properties of water molecules for the tube they are in. In this way, the negative pressure generated in the leaves also influences the roots and maintains a gradient of falling water potentials throughout the plant, from the soil to the root to the xylem and ultimately to the leaf (Figure 4.10). Thus, water evaporated at the leaf surface is replaced by water taken up by the root system.

There is a trade-off between the dimensions of the xylem tissue and water conduction. Narrow vessels develop larger cohesive forces between the water molecules and the tubes, but these also generate more friction and thus make water movement harder to achieve. It is generally believed that xylem vessels are more efficient that tracheids as a result of their greater diameter and fewer pores to impede water flow. However, it is curious to note that many of the largest plants are gymnosperms (e.g. *Sequoia sempervirens*) and therefore use tracheids for water conduction.

The whole process of water movement in plants is a very delicate balance, dictated by the creation of a falling water potential from the soil to the leaf. However, this balance is easily upset. If the plant is exposed to drought, there will be a gradual lowering of the water potential of the soil. This will reduce the availability of water to move into the plant and so, in order to continue to move water at the same rate, the leaves need to generate even lower water potentials. A similar situation occurs when a plant is supplied water containing a high concentration of dissolved solutes. Such a solution, if concentrated enough, can possess a lower water potential than the root itself and then draw water out of the plant. This has probably been the main reason why flowering plants have never recolonized the sea, as the high salinity levels of the seawater make effective transport of water very difficult. Thus, the only plants found in the sea are the brown, red and green algae. However, some higher plants, such as mangroves, manage to survive on the margins of the sea.

With drought stress there comes a point where the plant can no longer draw water from the soil and it then needs to stop transpiration, and thus photosynthesis, by closure of the stomata. It is through control of

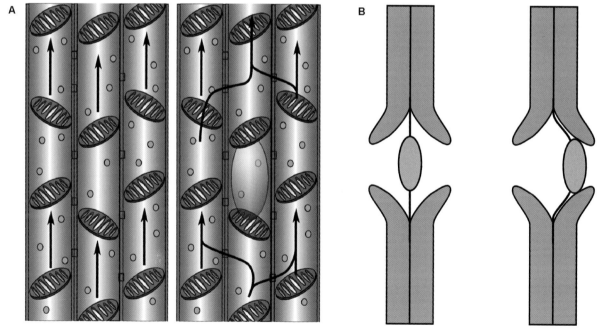

Figure 4.11 Embolisms within the xylem. Movement of water through the xylem vessels requires clear vessels. Occasionally air can block the vessels and this is known as an embolism (A). The structure of the vessels restricts the movement of the air between cells and, because of the numerous pits between the vessels, water simply flows around the obstructed vessel until it is cleared through dissolving of the air blockage. The tracheids of some species of gymnosperm possess bordered pits between the tracheids (B). The torus is a central thickened piece that can move like a valve. When a tracheid becomes blocked with air, the torus moves to seal that area of the xylem to prevent further tracheids being affected by the air.

the stomatal aperture that the whole process of transpiration is regulated.

Transpiration rates

Plants have a large requirement for water to photosynthesize efficiently, and the larger the plant the greater the loss of water. For some species of tree, such as a mature birch, the rate of water loss has been estimated to be up to 400 litres/day. This amounts to well over 16 litres/hour, assuming that most of the water is lost during the day-time. For smaller plants, however, water uptake will be considerably less. This water must be constantly taken up from the soil. As the soil is depleted of water around the roots, the water potential of the rhizosphere falls and it is replaced by a process called bulk flow or simple diffusion. The nature of the soil has a great influence on this bulk flow process. Small-particle soils, such as clay, while retaining water successfully, have a very poor conductivity for water and thus restrict movement of water to the roots. Sandy soils, on the other hand, retain less water but have much better water conductivity.

Cavitation

One the major problems with the mechanism for water uptake adopted by plants is that the water column is placed under a great deal of tension, and if it is broken then the whole process could easily collapse. The main cause of this collapse is the formation of embolisms or cavitation, where air bubbles form in the xylem and interrupt flow. Once a bubble has formed, the low pressure in the xylem expands the gas and this makes it very difficult for the leaves to generate sufficient pressure to maintain water movement. Bubbles arise from leaks through the pores of the xylem, damage to the tubes or through gases coming out of solution during transport. The structure of the xylem minimizes the damage from bubbles forming; the end walls of the vessels and the pores prevent the bubble from spreading from tube to tube and thus it is contained. The bubbles are then reabsorbed slowly by the plant to permit flow again (Figure 4.11).

Stomata

The epidermis of leaf and stem bears a cuticle which reduces water losses to a minimum. However, as has been emphasized earlier, the plant needs to have an efficient means of gas exchange with the atmosphere to fuel

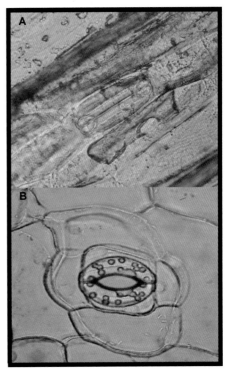

Figure 4.12   Stomata. Two forms of stomatal pore are shown. (A) A barley (*Hordeum* sp.) leaf stoma, which has bulbous ends to the cells. By swelling and contracting, the stomatal pore is respectively opened and closed. (B) This second form of stoma is from *Commelina communis*. The cells are slightly bent and have unequal thickening on their sides. This allows the stomatal pores to open when the cells swell and close when they shrink.

photosynthesis, and it is the stomatal pores which provide a channel for this process. These pores are strictly regulated by the plant through the action of guard cells, the structure of which is shown in Figure 4.12.

Guard cells regulate the gaseous exchange of the leaf. At night, little gaseous exchange is required and the stomatal pores are almost closed. The night-time temperatures are also lower than those in the day-time and therefore water loss from the leaf is minimal. However, during the day the cells open the pores and permit CO_2 to enter the leaf and O_2

Figure 4.13   Opening and closing of stomata. During the day the guard cells swell and the stomatal pores open to allow gas exchange. Carbon dioxide enters the leaf and oxygen and water leave. At night in most plants the stomata are closed and gaseous exchange is minimal.

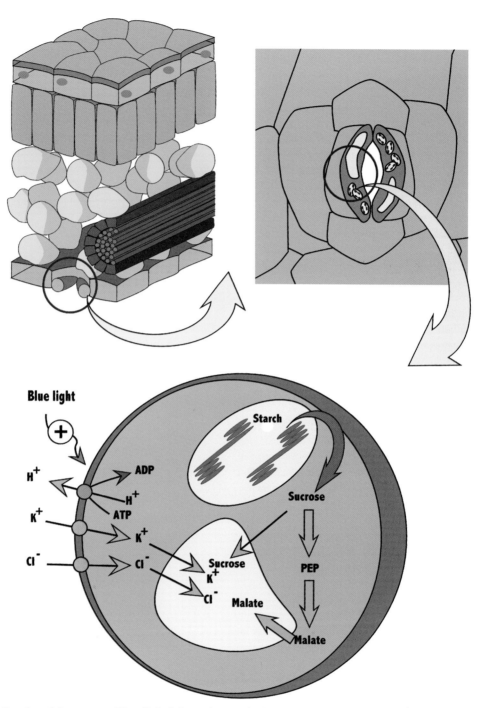

Figure 4.14 Opening of the stomata. When light falls on the guard cells (particularly blue light), a H^+/ATPase pump is activated in the plasma membrane of the cells. This causes a movement of H^+ ions out of the cells into the apoplast. K^+ ions enter the cells, and starch is metabolized to sucrose and malate; these changes bring about a fall in the water potential of the cells. Water enters the cells by osmosis and causes them to swell. The unequal thickening of the cell wall of the guard cells causes the cells to expand outwards opening the stomatal pore.

to leave. The cells also need to regulate water loss from the leaves during drought conditions. Thus, they are sensitive to the levels of abscisic acid, which signal the onset of drought stress (see later).

Stomatal function

The structural feature that makes guard cells special is the way the cell wall is laid down. In most cells, the cell wall is laid down evenly around the cell. However, in guard cells the cell wall is thicker near the opening to the pore. Thus, when the cells become turgid, the cell bends to form a curve and, as a result of the orientation of the guard cells in a pair, they bend away from each other, creating a pore (Figure 4.13). Thus, by regulating the turgidity of the cells, the pore size can be controlled and hence the gaseous exchange from the leaf.

Unlike the other cells in the epidermis, the guard cells are photosynthetic but they also bear blue-light receptors. As light intensities rise at the onset of day-time, the blue-light receptor is activated and triggers H^+/ATPase pumps at the plasma membrane (Figure 4.14). The build-up of positive charge at the plasma membrane as a result of the H^+ movement causes K^+ ions to enter the cell and the vacuole. The concentration of K^+ ions rises from around 100 mM in the dark to above 400 mM in the light, and this movement causes the water potential of the cells to fall. Thus, water enters the cells by osmosis and this causes the cells to become turgid and the stoma is opened. However, the K^+ ions only accumulate for the first few hours of the day, then the concentration falls. At this point other solutes take over the role of maintaining the low water potential of the guard cells. Starch is maintained at high levels in the chloroplasts of guard cells at night. During the day this starch is mobilized and converted into sucrose or, to a lesser extent, malic acid. This is then stored in the vacuole. Hence, movements in K^+ ions cause stomata opening and the generation of high concentrations of sucrose during the day in the guard cells that maintain the open stomata. For stomatal closure, the K^+ ions are already low in concentration and are not considered to be important at this stage, but sucrose concentrations are still high. Sucrose is reconverted to starch at the end of the day to cause the water potential of the guard cells to rise and thus cease to be turgid.

The ability to photosynthesize makes in necessary for plants to be able to exchange large quantities of specific gases with the atmosphere. This ability also necessitates the consequential loss of great volumes so water. For a plant to survive, it needs to be able to replace this water, and this is one of the major roles of the roots. The xylem tissue then distributes this water around the plant to where it is needed. Another major function of the root system is to obtain nutrients from the soil, and this is the topic of the next chapter.

References

Brundrett MC and Kendrick WB (1988) The mycorrhizal status, root anatomy, and phenology of plants in a sugar maple forest. *Canadian Journal of Botany*, **66**, 1153–1173.

Brundrett MC and Kendrick WB (1990) The roots and mycorrhizae of herbaceous woodland plants. I. Quantitative aspects of morphology. *New Phytologist*, **114**, 457–468.

Dittmer HJ (1937) A quantitative study of the roots and root hairs of a winter rye plant (*Secale cereale*). *American Journal of Botany*, **24**, 417.

Dixon HH and Joly J (1894) On the ascent of sap. *Philosophical Transactions of the Royal Society London Series B*, **186**, 563–576.

Marschner H (1995) *Mineral Nutrition of Higher Plants*, Academic Press, London.

Madson C (1996) Trees born of fire and ice, *National Wildlife*, **34** (6), 28–35.

Mitton JB and Grant MC (1996) Genetic variation and the natural history of quaking. *BioScience*, **46**, 25–31.

Rougier M and Chaboud A (1985) Mucilages secreted by roots and their biological function. *Israel Journal of Botany*, **34**, 129–146.

Russell RS (1977) *Plant Root Systems: Their Function and Interaction with the Soil*, McGraw-Hill, London.

Tyree MT (1997) The cohesion–tension theory of sap ascent: current controversies. *Journal of Experimental Botany*, **48**, 1753–1765.

Wherry ET (1972) Box-huckleberry as the oldest living protoplasm, *Castanea*, **37**, 94–95.

5

Mineral nutrition of plants

Through photosynthesis, plants can potentially make all compounds containing just carbon, hydrogen and oxygen. But plants cannot grow on air and water alone – there are numerous processes which require other elements. For instance, to make DNA plants require nitrogen and phosphorus, and to make proteins plants also require sulphur. A small minority of plants are remarkably effective at taking up these through their leaves and do not rely on a root system for uptake (e.g. *Tillandsia* species); however, most plants rely on their root systems to harvest nutrients. There are estimated to be around 90 different elements frequently found in plants, but of these, so far only around 26 have been shown to be essential. In this chapter, the mechanisms used by plants to take up nutrients from the soil will be studied, and the main elements and their uses in the plant will be examined.

Soil structure and mineral ions

Soil is made up of four layers; surface litter, topsoil, subsoil and bedrock (Figure 5.1). These vary in thickness and extent, depending on soil type and region in the world. The topsoil, on average, makes up the top 30 cm of the soil. It contains around 15–20% organic materials and a mixture of sand, silt and clay. It is in this region that most plants grow their roots and where most of the soil nutrients are. The topsoil is readily subjected to weathering erosion once the plant life on the surface is removed (e.g. through agriculture). It is estimated that over 36 billion tons of soil are lost every year as a result of human intervention. On average it takes 100 years to make 1 cm of topsoil, so it will take thousands of years to replace eroded soils and considerable efforts are being made in some countries to try to mitigate these losses.

Soil acts as a colloidal suspension. Solid soil particles known as micelles are present, suspended in water. These micelles, and particularly those in clay, act as cation-exchangers. Micelles readily develop negative charges (Figure 5.2). These interact with nutrients in the soil and bind cations. The cations have the following order of affinity for the soil particles (highest to lowest):

$$Al^{3+} > H^+ > Ca^{2+} > Mg^{2+} > K^+ > NH_4^+ > Na^+$$

This means that, as nutrients are added through weathering, decay or fertilizers, the cations will bind to the soil particles and are held. In contrast, the soil possesses little capacity to bind negatively charged anions and these are rapidly leached out of the soil by rainfall. Hence, ions such as NO_3^-, SO_4^{2-} and Cl^- are readily lost from the soil and drain into watercourses. In agricultural practice, twice as much nitrogen is applied to crops as is required by the plants to make up for the leaching of nitrates out of the soil. This contributes to the large quantity of nitrates polluting rivers, leading to eutrophication. Too high a concentration of nutrients in soils or bodies of water can kill certain sensitive plants; for example, carnivorous plants are extremely sensitive to soil nutrients and their life strategy of living in nutrient-poor soils leads to them being highly susceptible to nutrient poisoning (see Chapter 8).

The pH of the soil also influences the solubility of mineral ions. Figure 5.3 shows how the solubility of a range of mineral ions important for plants is influenced. For any particular soil pH, there will always be at least one ion which has limited solubility. Soils with an acid pH of around 5.5 tend to be optimal for ion solubility. Mineral ions that are present in the soil at low concentrations and have low solubilities also exhibit low mobility in the soil, and there is little scope for diffusion of the ions. Consequently, the root system must forage for these reserves by growing towards them.

As well as being beneficial, mineral ions can also be toxic to plants. In some instances the pH can make mineral ions so soluble that they rise to a toxic concentration. This can take the form of the ion itself being toxic or it could

Physiology and Behaviour of Plants   Peter Scott
© 2008 John Wiley & Sons, Ltd

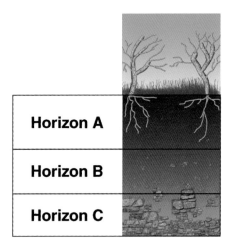

Figure 5.1 Soil profile. Soil can be simply divided into three horizons. The upper horizon, known as horizon A, is that in which most of the roots of plants and soil fauna are found and is rich in organic material. In horizon B some roots are still present but this region mainly contains leached minerals and clay and contains little organic material. The final horizon is horizon C; in this area there is mainly the parent material of the soil (rocks).

interact with other ions, causing them to precipitate. One example of this is high Al^{3+} concentrations in the soil, which can cause soluble phosphates to precipitate out of solution and become unavailable to the root system. A low soil pH can also lead to soluble nutrients leaching from the soil. This has become a particular problem with acid rain, which has tended to cause soil pHs to fall, and thus permitted a fall in the nutrient levels in the soil.

General ion uptake

Roots penetrate large volumes of soil through the production of root hairs, which develop in intimate contact with the soil particles. The root hairs release H^+ ions, using an ATPase proton pump, which has several effects. First, this lowers the pH around the rhizosphere (to around pH 5.5) and moves the soil pH to nearer the optimum for ion solubility, enabling mineral ions to be solubilized. Second, the presence of high concentration of H^+ ions around the soil particles favours the exchange of these ions for mineral cations, such as K^+, Ca^{2+} and NH_4^+, that are bound to the particles (Figure 5.2) and, having been released, are taken up by the root hairs. Third, the presence of a low pH outside of the root allows cotransporters to function, pumping the released ions into the root cells (Figure 5.4). Fourth, pumping the cation H^+ out of the root cells will generate an negative electrical charge within the cell, which will favour the movement of cations from the soil through the plasma membrane into the root cells. The presence of

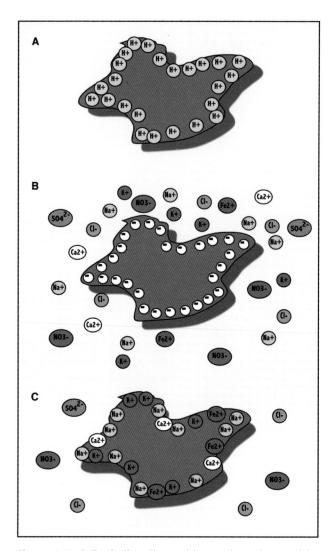

Figure 5.2 Soil micelles. Clay and humus (organic material which has broken down as far as it can) form a complex known as a clay–humus micelle (A). These micelles can be in suspension in water in the soil and they acquire a negative charge in this state (B). This negative charge allows the soil to bind onto any positive charges in the soil, such as metal and ammonium ions (C). Negative ions are not bound and are therefore susceptible to leaching from the soil.

the ATPase proton pump provides the energy for much of the active mineral ion transport into root cells. This is known as secondary active transport, since it uses the energy associated with the H^+ gradient, rather than that from ATP hydrolysis. Although the process of making nutrients around the roots soluble is, to a large extent, indiscriminate (all minerals are affected in an unfocused manner), control of ion uptake into the roots is exerted by regulation of the transporters and cotransporters themselves. Most ions have specific transporters to shuttle them across the plasma membrane (Figure 5.5).

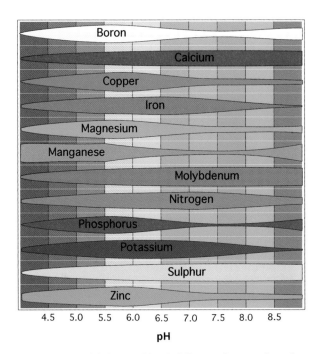

Figure 5.3 Solubility profile of different elements in soil at different pHs. The solubility of different elements or ions varies, depending on the pH of the soil; hence their availability to the roots of plants also varies. This is shown in the figure, where the wider the bar, the more soluble the mineral.

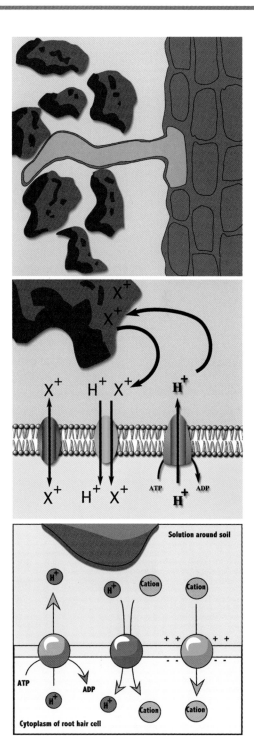

Figure 5.5 Cation entry into the roots. The cations released from soil micelles can be taken up by the root in two different ways. The cation can be taken up by passive cation transporters, which rely on a high concentration of the cation outside the root, in the soil. Alternatively, the root can take up the cations actively, using the H$^+$ gradient established by the H$^+$/ATPase pump. Thus, an H$^+$ ion enters for every cation taken up by the cell.

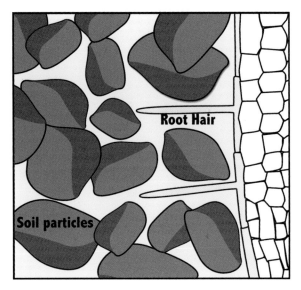

Figure 5.4 Intimacy of roots and the soil. Primary and secondary roots are not very efficient at optimizing the volume of soil that is accessible to a plant. However, the development of numerous root hairs being the growing tip of the root is very effective. Root hairs can make up more than 90% of the root length.

Mineral ions can also be taken up by diffusion. Passive transporters can move the ions across the plasma membrane (facilitated diffusion). The energy to drive this process comes from the presence of a concentration gradient between the inside of the root and the soil, or from the presence of an electrochemical gradient (Figure 5.5).

Once in the root cells, ions move by diffusion between cells through plasmodesmata. They cross the Casparian strip and the endodermis of the root and are then pumped into the apoplastic space between the cells of the stele and the vascular tissue. The ions are then moved in the xylem tissue to other regions of the plant.

This section has covered the general mechanics of movement of mineral ions from the soil into the plant. However, there are specific features that need to be addressed for certain mineral ions, namely phosphorus, nitrogen, sulphur and iron, which, will be covered in the next sections.

Phosphorus uptake and assimilation

Phosphorus is particularly important for plant growth and in most instances its restricted availability limits growth. Since phosphorus is present in the soil as phosphates, when it is soluble it is likely to be leached from the soil through weathering. But phosphates present another problem in the soil: soluble phosphates react very easily with Ca^{2+} and form calcium hydrogen phosphate, which is insoluble. This is a particular problem in alkaline soils, where calcium is especially soluble. The reaction renders most of the phosphates added to the soil insoluble very quickly and only around 20% of phosphates added as fertilizers end up being absorbed by the plant, the remainder being leached out or becoming insoluble in the soil. Because of the reaction between phosphate and Ca^{2+}, phosphates have very low soil mobility, therefore any phosphates arriving in the soil will remain where they fall. This also means that once phosphates are added to the soil they take a considerable period of time to be completely removed. This causes problems when attempts are made to restore agricultural land to its original form, because phosphates will persist and cause difficulties with invasive plant species. For instance, for restoration projects involving the green winged orchid (*Anacamptis morio*), when the orchid is reintroduced onto fertilized old agricultural land, the presence of high nitrogen in the soil causes grass species to outcompete the orchid. In addition, the presence of high phosphates is toxic to the plant and rapidly results in the plant declining in a particular habitat. Leached phosphates from farmland are another problem for watercourses, because high levels of phosphates promote algal blooms and cause eutrophication of the water systems. The cycle of phosphorus in the natural environment is shown in Figure 5.6.

Plants have evolved ways to obtain phosphates from the soil (see Figure 5.7A–C). The pumping of H^+ ions into the soil favours the reaction of calcium phosphate

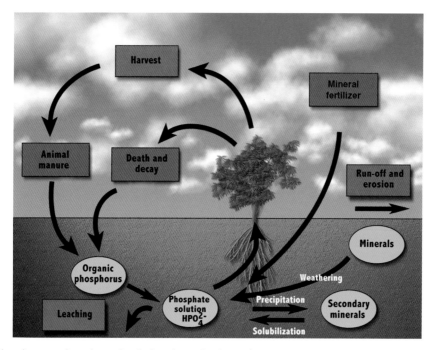

Figure 5.6 The phorphorus cycle – the cycle of phosphorus in the environment.

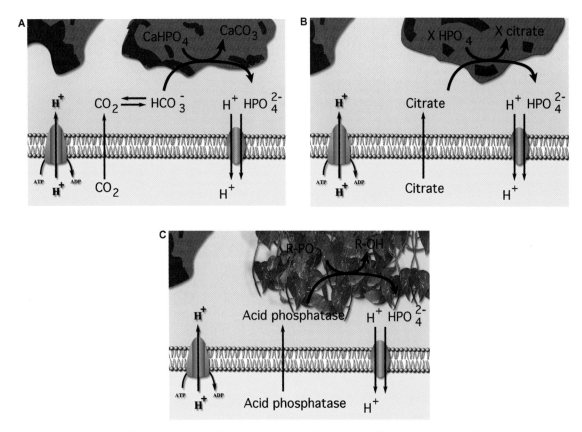

Figure 5.7 The acquisition of phosphate. Phosphate ions are needed for plant growth and are often in limited supply for most plants. There is usually sufficient phosphate in the soil but most of the compound phosphate forms in soils are insoluble. Therefore, most phosphate is bound onto soil particles and is not accessible to a plant. Plants have evolved three strategies to obtain this phosphate. First (A), plants liberate carbon dioxide from the roots, which dissolves and forms the bicarbonate ion. The bicarbonate ion reacts with insoluble calcium phosphates in the soil to release the phosphate ions. The root then uses active transport to take up the released ions. Second (B), the plant releases citrate ions, which again react with insoluble calcium phosphates, liberating free phosphate ions which can then betaken up by the root. Third (C), the root can secrete the enzyme acid phosphatase. This reacts with organic phosphorus-containing compounds in the soil and acts to release free phosphate ions, which are taken up as described earlier.

with dissolved CO_2 (generated from root respiration) to form calcium carbonate. The hydrogen phosphate ion is then released and can be transported into the plant. There is also good evidence that carboxylic acids, such as malic and citric acids, are released by roots to make phosphate reserves in the soil soluble. These carboxylic acids chelate with metal ions, such as Al^{3+}, Ca^{2+} and Fe^{2+}, to release bound phosphates. In addition, there is growing evidence that the hydrolysis of organic phosphate esters is enhanced in the presence of these acids. In some phosphate-deficient plants, there is evidence that they secrete the enzymes acid phosphatase and phytase into the soil (Vance *et al.*, 2003). These enzymes act on insoluble phosphate esters, usually in organic forms, to release the soluble phosphate ions. This is a very costly process because it involves the release of nitrogen in the form of a protein, and would only be used under circum-

stances where nitrogen supplies far outstripped those of phosphorus. The roots take up phosphate into the cells by a H^+-hydrogen phosphate cotransporter. The molecule is then rapidly incorporated into organic molecules or is transported around the plant to sites where it is required.

Phosphorus-deficient plants respond by increasing root growth over shoot growth. The plant attempts to increase phosphate uptake by growing longer roots, and more lateral roots with longer root hairs, which allows the plant to exploit a greater volume of soil. The root cells tend to secrete more phosphatase enzymes, and have a greater number of phosphate transporters. A plant rapidly exploits greater availability of phosphate in any volume of soil by the promotion of root growth into the region and the formation of a dense group of lateral roots, often referred to as cluster roots. The signalling pathway which controls this process is still poorly understood, but it is believed to

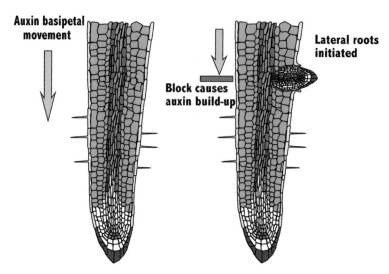

Figure 5.8   Stimulation of lateral root formation. Auxins in the plant exercise a lot of control of the root architecture. There is a gradient of auxin concentration down the roots, with a high concentration of auxins being nearest the shoot. When the root system encounters an area of soil with a high concentration of phosphorus or nitrogen, the transport of auxin in the roots is inhibited. This leads to a local build-up of auxin concentration, which stimulates cells in the root stele to begin to form a lateral root. This new root can than take advantage of the higher concentration of phosphorus or nitrogen.

involve auxin transport down the root. The high concentration of phosphate may inhibit this transport and result in a build-up of auxin in that region of the root. Such a build-up triggers the development of lateral roots (Figure 5.8). Ethylene has also been implicated in controlling this response. The application of ethylene promotes lateral root formation and the production of numerous, dense, root hairs. Cytokinins may also be involved and enzymes that degrade them increase as lateral roots and root hair initiation is induced.

The availability of phosphorus is thought to limit agricultural output in over 30% of farmland, making phosphorus one of the elements most limiting plant growth. Phosphorus has a number of routes into the soil and these are summarized in the phosphorus cycle, shown in Figure 5.6. For farmland, the biggest input of phosphorus is through fertilizers in the form of soluble phosphates, which are derived from mineral reserves around the world, with Morocco being the major producer of high-grade phosphate for fertilizers. However, phosphate from such sources is not renewable and it is estimated that all of the cheap forms of phosphorus will be exhausted soon after 2050. As a consequence, phosphorus limitation of plant growth in agriculture is expected to become a bigger problem in the future.

Nitrogen uptake and assimilation

Nitrogen is a limiting nutrient for the optimal growth of most plants and, given the central role of nitrogen in

protein synthesis, it is vital that the plant takes up as much nitrogen as possible. The atmosphere contains an abundance of nitrogen gas (N_2) and approximately 78% of the air around us is in this form. However, the bond between the nitrogen atoms is hard to break and only a few plants and bacteria can incorporate this into an organic form. Around 10% of the nitrogen fixed in the soil derives from weathering (electric storms) and photochemical reactions between ozone and nitric oxide. Plants and bacteria fix the remainder. This highlights the crucial role of biological fixation of nitrogen in the global nitrogen cycle (Figure 5.9). The competition between plants and bacteria for available nitrogen reserves in the soil is intense, and plants have evolved numerous mechanisms to facilitate nitrogen uptake. Plants respond to high nitrogen in the soil by growing towards the volume of soil containing the high concentration and by producing lateral roots in this region to maximize uptake.

Plants take up nitrogen from the soil or from the air. In the air, nitrogen gas (N_2), through symbiotic N_2 fixation, is converted directly to ammonium ions, using energy in the form of ATP (see later, in nitrogen fixation and root nodules). In the soil, nitrogen is present as a result of decaying organic materials, dissolving of minerals in the soil or through application of fertilizers. The cycle for nitrogen in the natural environment is shown in Figure 5.9. Nitrogen is obtained from the soil in the form of nitrate ions, ammonium ions or, to a lesser extent, amino acids, but nitrate is the dominant form. The presence of ammonium ions in the soil is very advantageous, since these

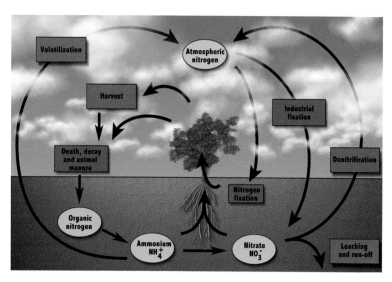

Figure 5.9 The nitrogen cycle – the cycle of nitrogen in the environment. Bacteria form an important role in the cycle by making nitrogen available to a range of higher organisms.

positive ions bind to soil particles and are not readily leached from the soil. However, these ions are especially short-lived in the soil, since bacteria readily convert these ions to nitrate. The mechanisms for the uptake of nitrate and ammonium ions into plant cells is summarized in Figure 5.10.

Once in the plant, nitrate can be stored or converted into ammonium ions. The storage of nitrate could in itself act as a defence mechanism in some plant organs. High

nitrate in mammals is reduced to nitrite, which can act to damage haemoglobin function and can form carcinogenic nitrosamines. High concentrations of nitrate in reserve organs of plants can act as deterrents to herbivorous animals. The pathway for the production of ammonium ions (NH_4^+) are summarized in Figure 5.11. Both the intermediate nitrite and the product NH_4^+ are toxic and need to be converted to other forms immediately. The nitrite ion is converted into NH_4^+ in the plastid of roots or leaves to prevent its build-up in plant tissues. The place in the plant where the process of conversion of nitrate to NH_4^+ occurs varies, depending on the availability of

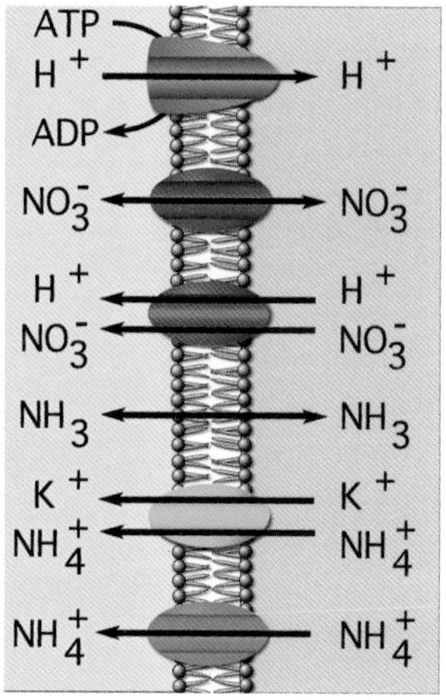

Figure 5.10 Uptake of nitrogen by roots. There are numerous different forms of nitrogen-containing compounds in soils. Therefore, uptake of nitrogen into roots is complicated. Most nitrogen enters the roots as a nitrate ion (passive and active transporters are present in roots), or as ammonia (free diffusion) or as ammonium ions (passive or K^+-linked transport).

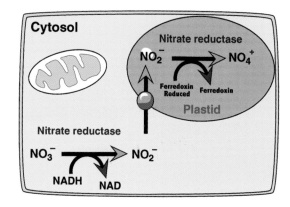

Figure 5.11 Assimilation of nitrate. Nitrogen is assimilated into amino acids and then proteins. However, in order to make amino acids, the plant needs to convert nitrates into nitrites and then into ammonium ions. The scheme for this pathway in plants is shown.

nitrate. The process requires reductase enzymes and a plentiful supply of NADPH is required. If availability of nitrate is low, the roots have sufficient capacity to handle the fluxes through the nitrate reduction pathway. However, when there is a high flux through the pathway, the leaves tend to take over this role, as there is a great deal of NADPH available indirectly from photosynthesis. For this to occur, the nitrate is transported through the xylem to the leaves.

As mentioned earlier, NH_4^+ ions are toxic. They have a tendency to form ammonia (NH_3), which is free to diffuse across membranes. If a structure bound by a membrane contains a low pH, then the ammonia will be reconverted back to the NH_4^+ ion and consequently become trapped in that compartment. This has the net effect of altering the pH in that compartment and causes a dissipation of proton gradients, which can impair the function of the vacuole and mitochondria in plants. To avoid this, plants rapidly incorporate the NH_4^+ ions into amino acids and do not let concentrations build up. Such a release of NH_3 occurs during the process of photorespiration (see Chapter 2).

Amino acid synthesis

Amino acids are made by the incorporation of the NH_4^+ ion to glutamate to make glutamine, using the enzyme glutamine synthetase (see pathway in Figure 5.12). This reaction can occur in the leaves or the roots, depending on the extent of nitrogen flux through the nitrate reductase pathway (see earlier). Glutamine and 2-oxoglutarate are then converted to glutamate by glutamate synthase. These reactions occur in the plastid of the root or leaf cells. Once the nitrogen has been assimilated into an organic form, other amino acids can be produced through transamination reactions, where the $-NH_2$ group is transferred between molecules (Figure 5.13). Amino acids can then be moved

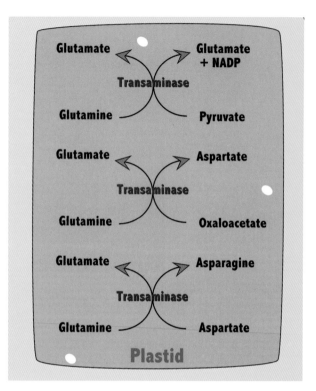

Figure 5.13 The synthesis of amino acids. Once glutamine has been synthesized, a whole range of other amino acids can be made. By removing the $-NH_2$ group from glutamine, it can be transferred through transaminase reaction to other carbon skeletons to make the other amino acids used in protein synthesis.

around the plant through the phloem or, to a lesser extent, the xylem.

Symbiotic nitrogen fixation

There are many different types of bacteria that can fix nitrogen in the soil; most of these are free-living. The only known photosynthetic organisms that can fix nitrogen are the cyanobacteria. Plants must use symbiotic associations with bacteria to achieve this feat but the list of plants which are capable of this is not long. The group of plants that made this relationship most successful are members of the Fabaceae (the pea family). The relationship between the bacteria and the host plant is very specific and a summary of some of the known symbiotic relationships is shown in Table 5.1.

The symbiotic relationship between the plant and the bacterium usually results in the formation of a swollen structure in the root system, called a nodule. There are three main genera of bacteria that form nodules with the

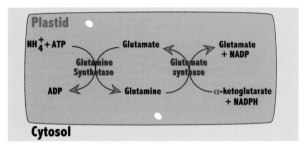

Figure 5.12 Glutamine and glutamate synthesis. The pathway for the incorporation of ammonium ions into glutamate and glutamine is shown.

Table 5.1 Known nitrogen-fixing symbiotic relationships between soil bacteria and members of the family Fabaceae.

Host plant	Bacterium species
Soybean (*Glycine max*)	*Bradyrhizobium japonicum*
Alfalfa (*Medicago* spp.)	*Rhizobium meliloti*
Clover (*Meliloti* spp.)	
Bird's foot trefoil (*Lotus*)	*Rhizobium loti*
Bean (*Phaseolus*)	*Rhizobium leguminosarum*
	biovarieties
Broad bean (*Vicia*)	
Clover (*Trifolium*)	
Lentil (*Lens*)	

The relationship between the bacterial species and the host plant is very specific, and if the correct bacteria are not present in the soil, symbiosis cannot be established with other non-compatible bacterial species. In most cases the bacteria are widespread and there is some speculation that compatible bacteria are carried by seed to new sites for future reinfection.

Fabaceae; these are *Azorhizobium*, *Bradyrhizobium* and *Rhizobium*, referred to under a collective name, rhizobia, in most instances. The only other plants known to form root nodules with rhizobia for nitrogen fixation are of the genus *Parasponia* (family Ulmaceae). Nodules are also known to occur in a few other plant species, such as alder (*Ulnus*) and bayberry (*Myrica*), but the association is with a bacterium of the genus *Frankia* (Figure 5.14).

The relationship is established between the rhizobia and the plant through the root system encountering the

Figure 5.14 Root nodules in the Fabaceae. A root nodule is shown on the root of a pea plant (*Pisum sativum*). In these nodules a symbiotic bacterium (*Rhizobium* sp.) lives within the root cells and fixes nitrogen into an organic form that the plant can use.

bacterium. Once the rhizobia encounter a root hair of a compatible member of the Fabaceae, the root hair extends and curls around the bacteria (Figure 5.15). The root hair cell wall degrades in the localized region where the bacteria are. Vesicles generated by the root hair cell fuse to make a thread, which the rhizobia use to penetrate the root cells and ultimately enter the apoplastic space in the root cortex. Once there, the bacteria spread from cell to cell through further infection threads and then enter the cells via vesicles, where they form structures known as bacteroids. These bacteroids are made up from the bacteria bound by a membrane known as the peribacteroid membrane. The inner cortical cells of the root begin to divide to form the primordium of the nodule. Ultimately the nodule grows larger, forming vascular tissue to supply the region with nutrients. A layer of root cells develops to maintain a low concentration of oxygen in the root nodule interior; here the nitrogen fixation occurs.

The fixation occurs with the rhizobium enzyme nitrogenase converting gaseous nitrogen into ammonia. This enzyme cannot function in the presence of oxygen and requires anaerobic conditions. In the cytoplasm of the plant cells within the nodules, there is a red pigment that bears a remarkable similarity to haemoglobin. This pigment, known as leghaemoglobin, is not present in the bacterium or the plant on their own and is generated as a result of the symbiosis. The plant produces the globin part of the molecule in response to the bacterial infection and the rhizobia produce the haem section. Leghaemoglobin binds onto oxygen with very high affinity and this maintains an anaerobic environment for the nitrogenase enzyme to function (Figure 5.16) (Denison and Harter, 1995). It also releases oxygen to fuel the electron transport chains in the bacterium in a similar manner to haemoglobin in the human bloodstream.

Iron uptake and assimilation

Iron is abundant in the soil but it is mainly in the Fe(III) form as an oxide. This has very limited solubility in most soils, particularly if they are neutral to alkaline (Eide *et al.*, 1996). Under favourable conditions, the roots take up Fe ions directly from the soil. However, if the soil is deficient in soluble iron, plants have evolved two strategies to extract iron (Figure 5.16).

Most plants when deficient in iron use Strategy I, which involves a highly active ATPase proton pump to lower the pH of the rhizosphere and the release of phenolics to act as reducing agents (Ma *et al.*, 1999). This helps to increase the solubility of Fe(III) salts in the

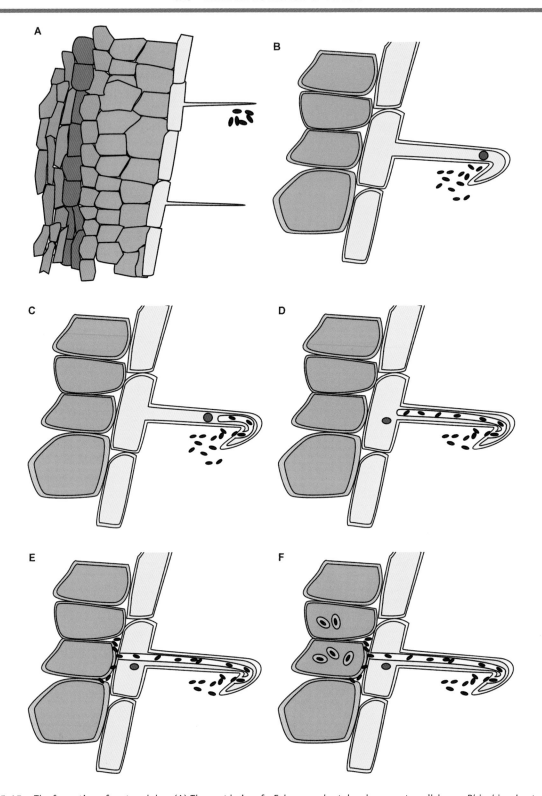

Figure 5.15 The formation of root nodules. (A) The root hairs of a Fabaceae plant develop near to soil-borne *Rhizobium* bacteria. The root hair curves near the *Rhizobium* (B) and forms a hollow channel in the centre of the hair (C). The bacteria enter the root hair channel and enter the root (C–E). Once in the root, the bacteria enter invaginations of the root cortex cells and appear like cellular organelles (F). The infected cells multiply to produce the nodule, in which nitrogen fixation occurs, i.e. atmospheric nitrogen is taken and converted into ammonium ions.

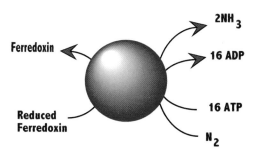

Figure 5.16 The nitrogenase reaction. The *Rhizobium* bacteria contain the enzyme nitrogenase, which catalyses the reaction shown. The energy for the process is gained from the plant.

soil (Figure 5.17). Soluble Fe^{3+} usually forms complexes with anions in the soil solution and these interact with a Fe^{3+} chelate reductase (ferric reductase oxidase) in the plasma membrane of roots, which converts Fe^{3+} ions into Fe^{2+} ions. The Fe^{2+} ions are much more soluble than Fe^{3+} ions, so these are then taken up into the root system by a Fe(II)-transporter.

In some grass species (e.g. wheat and barley), a second strategy (Strategy II) is used when the plant is iron-deficient. The roots of the grass secrete a phytosiderophore that chelates with Fe^{3+} ions to make a soluble complex.

The root then takes up this complex by specific transporters. Seven different phytosiderophores have been identified in grass species, the main one being mugineic acid.

Once in the plant tissue, iron ions are converted to Fe^{2+} and then transported in the xylem chelated with citrate ions. One of the key roles of Fe^{2+} in plants is in the formation of the cytochrome molecules. This incorporation of the ion is achieved by ferrochelatase. Excess Fe^{2+} is not left as a free ion, since there is evidence that this can react with oxygen to generate free radicals. The Fe^{2+} is stored in an iron–protein complex called phytoferritin in the chloroplast (van der Mark *et al.*, 1983).

Sulphur uptake and assimilation

The element sulphur is taken up from the soil by roots in the form of the sulphate ion. This ion is readily replenished in the soil through weathering of rocks. As sulphate is an anion, it is easily leached from the soil. Roots actively take up the sulphate ion on an H^+–sulphate cotransporter. Once in the plant tissues, sulphate is transported to the leaves, where it is converted into cysteine (see the pathway in Figure 5.18). The leaves are mainly used in this process

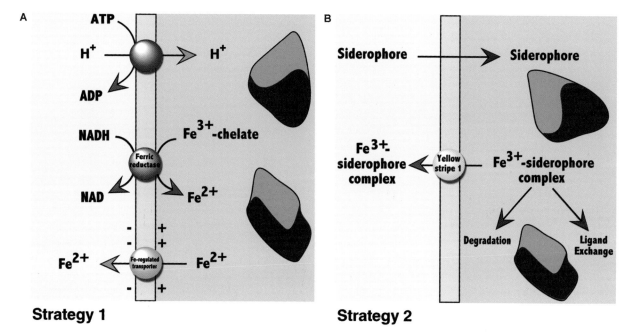

Figure 5.17 Iron uptake from the soil. Most iron in soil is in the form of Fe^{3+} salts, which are generally insoluble. One strategy plants use to take up iron (A) is to convert the chelated Fe^{3+} salts to Fe^{2+} salts, using an enzyme ferric reductase. The Fe^{2+} ion can then be taken up by membrane transporters energized by the membrane potential. A second strategy (B) involves the secretion of a siderophore into the soil by the root. This reacts with Fe^{3+} and produces a complex that can be transported into the root cells.

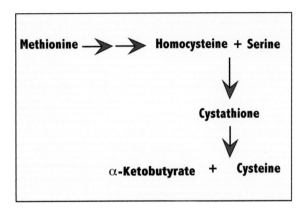

Figure 5.18   Pathway for the synthesis of cysteine in plants.

because of the high ATP and reduced ferredoxin requirement for the pathway, both of which come straight from photosynthesis. Cysteine is then used for the synthesis of methionine and other sulphur-containing compounds.

Function and effects of deficiencies of mineral ions on plants

Mineral ions can be divided up into major elements (essential nutrients needed in large quantities) and trace elements (essential nutrients needed only in very small quantities). However, the symptoms of plants deficient in a mineral ion vary, depending on the mobility of the ion, and the mineral requirements of a plant can also be subdivided into mobile and immobile elements. If a particular element is mobile, then a plant experiencing deficiency in it will usually mobilize the ion from the old leaves to the new ones; if this is the case, the old leaves will be the first tissues to exhibit symptoms of the deficiency. However, if the element is immobile, then mobilization is not an option and the new developing leaves are the first to exhibit the deficiency symptoms.

Major elements

For optimal growth of crop and horticulturally important plants, fertilizers are used. These contain three major components, nitrogen, phosphorus and potassium, which are three essential major elements required by plants.

1. *Nitrogen.* This is available in a number of forms: as a nitrate, nitrite or ammonium salts. Nitrate and nitrite are easily leached from the soil. However, ammonium ions bind to soil particles and persist for longer in the soil, but soil bacteria act on the ion and convert it to nitrate.

Nitrogen is essential for the production of plant proteins, chlorophyll and nucleic acids. If deficient in nitrogen, the plant will develop chlorosis (yellowish leaves) in the older leaves as nitrogen is recycled from them to enable new leaves to grow. In extreme cases, the older leaves will fall off the affected plant. In nitrogen-deficient leaves, they frequently accumulate pigments other than chlorophyll, such as anthocyanins, which give the plant a purple coloration.

2. *Phosphorus.* This is obtained as a soluble phosphate or hydrogen phosphate salt in the soil. Phosphorus is essential for the energy-generating systems in plant cells, such as respiration and photosynthesis, and is needed for ATP and NADP production and nucleic acid formation. If deficient in phosphorus, the leaves tend to be dark green and growth of the plant is severely stunted. Sometimes spots of dead tissue arise on the leaves, often called necrotic spots. Plants frequently accumulate anthocyanin pigments in their leaves, petioles and stem as a result of phosphate deficiency. The symptoms of phosphorus deficiency in plants are easily confused with nitrogen deficiency.

3. *Potassium.* This element is obtained directly from the soil as a cation bound to soil particles. Although potassium is not an essential component in any of the plant cell structures, it is required as a cation for enzyme function and thus plays an important role in many cellular processes. If deficient in potassium, plants develop yellow patches on the leaves, followed by necrotic patches at the leaf tip and margins. The leaves also curl upwards and plant growth is severely restricted.

Other important major elements

1. *Iron.* This is obtained as a sulphite salt. Iron is important in metabolic processes in the plant cell, such as photosynthesis and respiration. A deficiency in iron causes chlorosis in the young leaves, with the other leaves frequently unaltered. Iron cannot be withdrawn from mature leaves to support growth of new leaves. The veins of the leaves maintain their chlorophyll, whilst the regions around lose theirs, giving the plant a characteristic appearance of iron deficiency.

2. *Calcium.* This element is obtained as a cation in the soil. Calcium is needed in a range of cellular processes, such as cell signalling and in the formation of the pectates in the cell wall. A deficiency in calcium causes the death of the apical meristem and a cessation of plant growth.

The roots also remain short and fat. The younger leaves begin to lose their chlorophyll at the leaf margins. Any expanding leaves are frequently deformed. Calcium exhibits low mobility in the plant once a tissue has developed, so developing tissues are affected more dramatically than older tissues.

3. *Magnesium*. This element is obtained as a cation bound to soil particles. Magnesium is essential for the formation of chlorophyll and as an activator of enzymic processes during photosynthesis. A deficiency in magnesium causes chlorosis of the regions of the leaf between the veins. The chlorosis begins on the oldest leaves, as magnesium is mobilized to allow normal development of the new leaves, but as the deficiency worsens the younger ones are also affected. This chlorosis is often accompanied by anthocyanin accumulation.

4. *Sulphur*. This element is obtained from sulphates in the soil, which are released as a result of decay of organic materials or the weathering of rocks. Sulphur is required in amino acids such as methionine and cysteine and is essential in some of the components of the respiratory pathways. A deficiency in sulphur produces chlorotic leaves with anthocyanin accumulation. However, unlike nitrogen deficiency, the chlorosis develops in the youngest leaves first, which reflects the low mobility of sulphur in the plant, not being easily mobilized from developed tissues.

Minor elements

Alongside the major elements, plants also require a range of elements in very small concentrations:

1. *Boron*. This is obtained as a borate salt. Boron is required for a range of processes, such as cellular transport, nitrogen metabolism and cell division. Its precise function is unclear but boron-deficient plants exhibit death of the meristems and growing tissues. Apical dominance of the stem is lost and boron-deficient plants are frequently highly branched. The meristems for these new branches are also short-lived. Another feature of boron deficiency is the death of the base of developing leaves.

2. *Chlorine*. This is obtained from the soil as a chloride salt. Chlorine is required for photosynthesis and plant growth in minute quantities. A deficiency in chlorine is probably very rare in nature, since it is so abundant. In experiments designed to reveal symptoms, a deficiency in chlorine causes wilting of the tips of leaves with chlorosis and death of leaf tissue.

3. *Cobalt*. This is obtained from the soil as the chloride salt. Cobalt is needed in vitamin B_{12} synthesis.

4. *Copper*. This is obtained from the soil as the sulphate salt but it is only required in small quantities, as copper can be very toxic to plants. Copper is required in the energy-conversion processes of the plant cell. Initially, leaves formed by copper-deficient plants are dark green. Necrotic spots appear on the leaves near the tips but these gradually extend across the leaves.

5. *Iodine*. This is obtained as an iodide salt (KI). A deficiency in iodine gives no consistent symptoms.

6. *Manganese*. This is present in the soil as a sulphate salt. Manganese is essential for the functioning and integrity of the chloroplast membrane. It is also involved in the enzyme complex used in the photolysis of the water molecule by photosystem II. Manganese deficiency is characterized by chlorosis and interveinal necrosis on leaves. The marks appear first on young leaves and then the older ones.

7. *Molybdenum*. This is obtained as a molybdate salt. Molybdenum is required for the conversion of nitrate to ammonium for amino acid synthesis. A deficiency in molybdenum begins with mottled chlorosis, with the veins retaining their chlorophyll on the oldest leaves. This is followed by necrosis and curling upward of the leaves.

8. *Zinc*. This is supplied as a sulphate salt. Zinc is involved in the formation of chlorophyll and auxins. The first signs of zinc deficiency are the formation of chlorotic regions between the veins of the older leaves, beginning at the tips and the leaf edges. The plant develops shorter leaves and internode length is reduced, giving a shorter plant. New leaves are also distorted.

References

Denison RF and Harter BL (1995) Nitrate effects on nodule oxygen permeability and leghemoglobin. *Plant Physiology*, **107**, 1355–1364.

Eide D, Broderius M, Fett J and Guerinot ML (1996) A novel iron-regulated metal transporter from plants identified by functional expression in yeast. Proceedings of the National Academy of Science of the United States of America, **93**, 5624–5628.

Ma JF, Taketa S, Chang YC, Takeda K and Matsumoto H (1999) Biosynthesis of phytosiderophores in several Triticeae species with different genomes. *Journal of Experimental Botany*, **50**, 723–726.

van der Mark F, van den Briel W and Huisman HG (1983) Phytoferritin is synthesized in vitro as a high-molecular-weight precursor. Studies on the synthesis and the uptake in vitro of the precursors of ferritin and ferredoxin by intact chloroplasts. *Biochemical Journal*, **214**, 943–950.

Vance CP, Uhde-Stone C and Allan DL (2003) Phosphorus acquisition and use: critical adaptaions by plants for securing a nonrenewable resource. *New Phytologist*, **157**, 423–445.

6

Mycorrhizal associations and saprophytic nutrition

In the natural environment, the growth of plants is rarely limited by the availability of light. Most plants are more than capable of synthesizing their carbohydrate needs. What often does limit growth, however, is the availability of minerals in the soil. The two most important minerals whose deficiency limits plant growth are nitrogen and phosphorus, which is why these are the major components of fertilizers in agriculture. Compounds containing phosphorus need to be soluble for uptake and obtaining these is a particular problem for plants because they are usually insoluble in the soil, so the root system of a plant must be very close to a phosphate source before a plant can even begin to mobilize these reserves. For a plant, this would require an extensive root system to increase the accessible soil volume. How plants actually take up these nutrients from the soil is discussed in Chapter 5, but for a plant to maximize its capacity to harvest phosphates from the soil, it would require a huge root network. In contrast to this, at the same time, there are large numbers of different fungi species in the soil that form extensive networks of thread-like cells called hyphae. These hyphae grow rapidly and are very effective at accessing large soil volumes and mobilizing insoluble phosphates. The hyphae are also active in the decay processes in the soil and thus recycle nutrients back into organic forms. However, what limits the growth of the fungus is the availability of carbohydrate. In the soil there are few sources of free carbohydrate and fungi have evolved complex metabolic pathways to recycle carbohydrates out of cellulose and lignin in decaying material. Hence, there is a huge advantage to both plants and fungi if the ability of fungi to obtain mineral ions from the soil was combined with the ability of plants to synthesize carbohydrate. It is this combination that will be investigated in this chapter. It has proved so successful that the majority of plant species are involved in this symbiotic process. The term used to describe this symbiosis is a *mycorrhizal association*. This term is derived from the Greek *mycos* (fungus) and *rhiza* (root) and literally means 'fungus root'.

In agricultural plant species, high yields are achieved through the application of large excesses of nutrient fertilizers, and mycorrhizal associations are uncommon and have little impact upon plant growth. In fact, mycorrhizal associations can break down under such conditions, as the plant no longer requires the fungus for nutrient acquisition. In contrast, for wild species mycorrhizal associations are very important and there is good evidence that over 80% of higher plant species form mycorrhizal associations (Harley and Smith, 1983). For so many plant species to form this relationship, it has to be of great importance to both the fungal and higher plant species. This association leads to plants and fungi being better able to compete with non-mycorrhizal species and to withstand environmental stresses.

Mycorrhizal associations

Mycorrhizal associations are highly evolved intimate relationships between the roots of a plant and fungi in the soil. Some plants will only form a mycorrhizal association with a specific species of fungus, whilst others will form such relationships with a broad range of fungi. The fungi involved in mycorrhizal associations come from all three of the sexually reproducing groups of fungi, the ascomycetes, basidiomycetes and zygomycetes (Harley and Smith, 1983, Kendrick, 1992; Brundrett, 1991). Summarized in the most basic terms, mycorrhizal associations are symbiotic relationships between plants and fungi in zwhich the fungus obtains carbohydrate products of

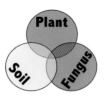

Figure 6.1 Interactions between plant fungi and soil. The roots of plants interact with the soil to obtain water, anchorage and nutrients, but some nutrients can be difficult to obtain in sufficient quantities for optimal growth. Soil fungi obtain nutrients and carbohydrates from the soil through saprophytism, but there is a very limited amount of available carbohydrate in soils. Mycorrhizal associations between plants and fungi represent an optimization of these interactions with the soil. The plant gains access to mineral ions in the soil and the fungus is supplied with carbohydrate from the plant. It is therefore not surprising that over 90% of plant species make mycorrhizal associations.

photosynthesis from the plant, and the plant obtains mineral nutrients from the fungus. Thus, there is a three-way association between the plant, the fungus and the soil (Figure 6.1). Seven different types of mycorrhizal associations have been identified and described so far, although there are likely to be many new forms that will be discovered in the future:

1. Vesicular arbuscular mycorrhiza (VAM).

2. Orchidaceous mycorrhiza.

3. Ectomycorrhiza.

4. Ericoid mycorrhiza.

5. Arbutoid mycorrhiza.

6. Monotropoid mycorrhiza.

7. Ectendomycorrhiza.

The distinguishing features of the seven known forms of mycorrhizal associations are described in Table 6.1. It is important at this point not to get lost in the definitions of the different associations. Generally, all types of mycorrhizal associations achieve the same end but, as a result of convergent evolution, different species of plants and fungi have formed different structures to achieve the nutrient transfer between the partners.

Of these types of mycorrhizal association, the most frequently found are vesicular arbuscular mycorrhizae (VAM). Over 60% of higher plant species possess VAM associations, with a further 20% of plant species forming the other sorts of mycorrhizal associations (Trappe, 1987). There are a few plant families which do not form mycorrhizal associations, the Brassicaceae being the major one. It is not known why they do not form these associations. The discussion in the remainder of this chapter will centre mainly on VAM associations.

Vesicular arbuscular mycorrhizal associations

VAM associations involve the linking of zygomycete fungi with higher plants. The hyphae of the fungi branch throughout the soil, forming a complex network. There are two main forms of hyphae: thick hyphae, acting as channels to, and support for, thinner hyphae which branch off

Table 6.1 The differences between the various forms of mycorrhizal association.

Type of association	Distinguishing features	Types of fungi involved
Vesicular arbuscular mycorrhiza	Fungi produce hyphae, arbuscules and vesicles in plant root	Zygomycetes
Ectomycorrhiza	Swollen lateral roots form on the plant coated with a mantle of fungal hyphae. The hyphae form a Hartig net around the plant cells in the epidermis or cortex. The hyphae do not enter the plant cells	Mostly basidiomycetes but some ascomycetes and zygomycetes
Ectendo-, arbutoid and monotropoid mycorrhizae	These are similar to ectomycorrhizae, but hyphae do enter cells	Mostly basidiomycetes but some ascomycetes and zygomycetes
Orchidaceaous mycorrhiza	Fungus forms coils of hyphae in the root cells of the plant	Basidiomycetes
Ericoid mycorrhiza	Fungi form hyphal coils in the root hairs of plants in the order Ericales order.	Ascomycetes and some basidiomycetes

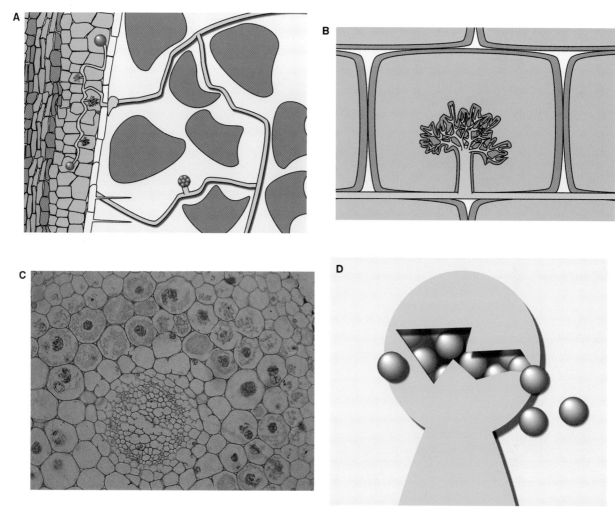

Figure 6.2 Vesicular–arbuscular mycorrhiza (VAM). VAM is the commonest association between fungi and plants. It derives its name from the formation of vesicles (food storage) and arbuscules within the plant tissue (A). Details of the arbuscule structure are shown in (B). The fungal cell never enters the plant cell, but is surrounded by an invagination of the plasma membrane of the root cell. It is thought that mineral ions and carbohydrate are exchanged across the membrane interface between the plant and the fungus. (C) Photograph of a beech tree root section possessing VAM. (D) Spores are periodically released from branches in the fungal hyphae in the soil.

into the soil; and thinner hyphae, thought to be involved in the process of nutrient uptake. The whole hyphal network is invisible to the naked eye and is very difficult to observe. Periodically, the thin hyphae link with roots of a host plant. The fungus at the point of entry forms an appressorium (a swollen section to a hypha). Then, within the root cortex, the hyphae ramify between the cells. The hyphae can actually penetrate cells in some associations but there is always a plasma membrane barrier between the host cell and the hypha. Occasionally within invaginations into cortical root cells, vesicles and arbuscules are formed

(hence the name of this form of mycorrhizal association; Figure 6.2). However, both of these structures are not present in all instances. All mycorrhizal associations where the fungus penetrates the cortical cells of the host root are known as endomycorrhizae.

Arbuscules are branched haustoria that cause invaginations within the plasma membrane of the cortical cells of the host, resulting in a structure that resembles a small tree, hence the name arbuscules', first used by Gallaud 1905. The arbuscule never actually enters the cytoplasm of the root cell, but forms a large surface area of contact between the root and the fungal hyphal system. Arbuscules form rapidly after infection of the plant by the fungus and only last for a couple of days before degrading (Figure 6.2B). It is this large surface area that is believed to be the site of nutrient exchange between the fungus and the plant. The fungus gains the carbohydrate derived from the photosynthetic activities of the plant host, effectively acting as an alternative sink tissue for carbohydrate reserves in the plant. This carbohydrate can then be used for further growth of the fungus within the soil, and for reproduction. Without this carbohydrate supply the growth of the fungus would be severely limited, so the fungal partner greatly benefits from the association. It is unclear whether the plant gains directly from transfer of mineral ions across the arbuscule membrane or whether the dissolution of the arbuscule, at the end of its life, releases nutrients into the apoplastic space within the root, which are then taken up by the root cells. However, what is clear is that the plant host greatly benefits from the extra nutrients.

Sporadically the hyphal swellings within the cortex of the host plant result in the formation of vesicles. These vesicles form either in the apoplastic space between root cells or between invaginations of the plasma membrane and the cell wall of the host cell. These structures are composed of cytoplasm of the fungus and contain lipid stores. In many ways these vesicles resemble spores and there is evidence that they can act as reproductive structures for certain fungus species. These vesicles help to maintain the sink strength of the hyphal network. By storing the imported sucrose as a lipid in the vesicles, the fungus can maintain a low concentration of sucrose or other carbohydrates in the hyphal system and thereby maintain a steep concentration gradient between the hyphae and the root cells.

In a mycorrhizal association, the hyphal network throughout the soil offers the host plant a huge root system. As a result of their microscopic nature, it is not possible to trace hyphae over long distances and the actual length of any hyphal network of individual fungus cannot be measured with any degree of accuracy. This is further complicated by several hyphal systems usually being present in the soil at the same time. In addition, hyphal systems of neighbouring fungi of the same species can link together, forming a truly gigantic network. Thus, the thick hyphae could stretch for long distances through the soil, with average distances ranging from a couple of centimetres to several metres. The thick hyphae act as the equivalent of a runner system which branches through the soil, seeking areas of high nutrient concentration. Where these are found, the hyphal system branches intensely, producing numerous thin hyphae (St John *et al.*, 1983). This is similar to the response of the root system of a plant to areas of soil containing a high mineral concentration. These thinner

hyphal systems branch further to produce spores, accessory vesicles and branched absorptive structures. These finer branched hyphae carry out most of the absorption of nutrients from the soil. The branched absorptive structures are the soil-borne equivalent of the arbuscules formed in the host cells of the plant. The fine hyphal structure in these regions provides a large surface area to support the uptake of nutrients from the soil.

Orchidaceous mycorrhizal associations

Orchidaceous mycorrhizal associations take the relationship between the fungus and the plant to a different level. Most orchids require a mycorrhizal association with a fungus in order to germinate (particularly terrestrial orchids; Figure 6.3). The seeds produced by orchids are very small (usually less that 100 mm) and resemble fine particles of dust. The seed is made up mainly of the embryo and a twisted testa, which acts to protect the embryo. Most orchids produce hundreds of thousands of these small seeds each season.

Terrestrial orchid seeds are dispersed by air currents until they land on the soil, where weathering will slowly take the seeds below the soil surface. If conditions are favourable and a compatible fungus is encountered in the soil, the seed germinates. The seed swells, breaks the testa and forms a small protocorm (a globe-like structure bearing an immature colourless shoot at the upper side and radiating root hairs at the lower side). The fungal hyphae penetrate the root hairs and the fungus supplies the protocorm with mineral ions and carbohydrates (Figure 6.4). Thus, in this growth phase the plant acts as a parasite of the fungus, as there is no obvious benefit to the fungal partner at this stage of the relationship. Growth of the protocorm can take a long time, but for a fast-growing orchid species, such as a member of the genus *Dactylorhiza* (spotted or marsh orchid; Figure 6.5) the protocorm can be mature within around 6 months. At maturity the protocorm swells to approximately 10 mm × 4 mm and the shoot tip turns green. In this stage the orchid can remain dormant over the winter, with the onset of spring triggering the formation of thick roots, which appear from the top of the protocorm, and the bursting and growth of the bud. At this stage there is evidence that the relationship between the orchid and the fungus can break down in the spotted orchids and they cease to be mycorrhizal. However, many species of orchid remain in a mycorrhizal association throughout their lives. Whether other mycorrhizal associations periodically break down has not been researched.

One unusual feature of the orchidaceous mycorrhizal associations is that orchids can frequently fail to have any

Figure 6.3 Orchidaceous mycorrhiza. (A) An orchid seed is shown in the soil, with fungal hyphae linking the seed with the soil matrix and neighbouring plants. Most orchid seeds do not have sufficient nutrient reserves within them to fuel germination. In order to germinate, they wait for specific fungi in the soil to link up with them. They then parasitize the fungus and use its carbohydrate reserves to fuel their germination. The first structure formed is called a protocorm (B, C). This accumulates carbohydrate reserves for growth of the parent plant.

surface growth for up to several years (Figure 6.6). Plants such as the military orchid (*Orchis militaris*) have frequently been reported as having remained dormant after flowering the previous year. For woodland-dwelling species, such as the red helleborine (*Cephalanthera rubra*), not only can the plant remain dormant for several years without any apparent surface growth, but also the plant can grow in areas of very low light intensity. Red helleborines are found in beech woods on chalk. Only through the mycorrhizal association can the red helleborine gain sufficient carbohydrate to grow under such poor conditions. Under such circumstances, the orchid is benefiting from a dual association between the fungus and the orchid and the fungus and neighbouring beech trees. Thus, in many respects the helleborine is acting as an indirect parasite of the trees.

Orchidaceous mycorrhizal associations appear to have great benefits for the plants. The plants can produce huge quantities of seeds with little investment of carbohydrate by the parent plant; the seeds, being small and light, are also dispersed widely by the wind. Then the seeds, on being deposited in suitable environment with a compatible mycorrhizal fungus, can germinate and grow. On forming the mycorrhizal association, the protocorm then has the

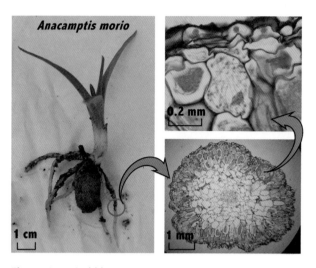

Figure 6.4 Orchidaceous mycorrhiza in *Anacamptis morio*. A mature *A. morio* plant is shown uprooted. Viewed under a microscope, the mycorrhizal association can be seen. The fungus forms coils of hyphae in the outer root cells. These structures are short-lived and degrade rapidly, new ones being constantly formed to replace the lost ones. It is thought that the plant gains nutrients as the coils are degraded. No benefit has ever been identified for the fungus in this relationship.

benefits of a gigantic hyphal network through the soil, which gives it access not only to soil nutrients but also to nutrients in dead and decaying organic matter and from neighbouring plant species linked to the mycorrhizal fungus. Yet, despite these benefits, most orchid species are rare. The explanations that could account for this apparent paradox are as follows. First, viable seed set in many orchid species is poor and inefficient. Second, many orchids have a very restricted range of fungal species which are suitable for forming a mycorrhizal association. While species such as the spotted orchids tend to have a broad range of possible fungus species with which they can form mycorrhizal associations, and consequently are relatively common, species such as the bird's nest orchid appear to have a very strict requirement for fungi that are also mycorrhizal with beech trees (Figure 6.7). Thus, many orchid species are restricted by the frequency of the certain fungus species and their host trees. Third, the mycorrhizal association can be a very difficult balance between the growth of the orchid and growth of the fungus. Many of the mycorrhizal fungus species are known plant pathogens, e. g. *Rhizoctania* spp. However, these same fungus species form mycorrhizal associations with the spotted orchids. In

Figure 6.5   The life cycle of a bee orchid (*Ophrys apifera*). The seeds were sown (A) on a asymbiotic culture medium and photographed as they developed into protocorms. The protocorms are shown 1 (B) and 3 (C) months after sowing the seeds. A mature plant is shown in (D), which will be 3 or more years old.

Figure 6.6 Orchid survival. (A) Whole plant and (B) inflorescence detail of a red helleborine (*Cephalanthera rubra*). This plant can grow in semi- to deep shade as a result of the mycorrhizal association. It is found mainly in beech woods, where the mycorrhizal association can allow the plant to grow with minimal photosynthesis of its own. (C) Whole plant and (D) inflorescence detail of a military orchid (*Orchis militaris*), which has been observed to go dormant for periods of up to 3 years in the soil, where no apparent surface growth of the orchid is seen. Again, the mycorrhizal association allows the plant to do this, as it is not entirely reliant on photosynthesis for its energy requirements.

tissue culture, if nutrient levels are too high in the medium, the fungus will thrive and kill the protocorm during development. If the nutrient concentrations are too low, then the protocorm growth is very slow. Hence, in the natural environment the conditions would have to be very specific for the mycorrhizal association to establish and produce the benefits for the orchid. Curiously, in cultivation, once out of the protocorm stage most photosynthetic orchids grow quite well in a wide range of soils without the presence of the fungus. Even very rare species, such as the red helleborine (*C. rubra*) and the lady's slipper orchid (*Cypripedium calceolus*), are quite vigorous in cultivation and can be propagated rapidly by rhizome division. However, in the natural environment these species are very rare. Thus, the early growth stage for orchids is the critical period for their survival.

The fungi involved in orchidaceous mycorrhizal associations are basidiomycetes, related to those species involved in plant pathogenesis (*Rhizoctania* and *Armillaria*) and in the decay of wood (*Fomes* and *Marasmius*). In terms of structure, the infection of the plant occurs through the root hairs. The fungus grows into the plant cell, causing an invagination of the cell membrane. The hyphae then form coils within the cells. Like the arbuscules, these coils are maintained for a few days, after which they disintegrate. Upon this degeneration, the orchid absorbs the nutrients from the fungus. The plant is believed to gain both carbon and vitamins from the fungus. The host plant initiates the disintegration of the hyphal coils, and if this process does not occur the fungus can become parasitic and kill the orchid. One factor orchids are known to produce in order to limit fungal growth is the fungicide orchinol

Figure 6.7 Mycotrophic orchids. The mycorrhizal association in some orchids has permitted the evolution of some plants which have completely lost their ability to photosynthesize. (A) Whole plant and (B) inflorescence detail of the bird's nest orchid (*Neottia nidus-avis*). This plant lacks chlorophyll completely from its small brown scale leaves and is wholly reliant on the fungus for survival.

(dehydroxyphenanthrin; Broering and Morrow, 1999), so the orchid also possesses effective means of influencing the growth of the infecting fungus. Consequently, with orchidaceous mycorrhiza there is a very delicate balance in this relationship which is easily tipped in either direction.

Ectomycorrhiza

The fungi involved in ectomycorrhizal associations are mainly basidiomycetes, but several ascomycete species are also involved in some interactions (Bending and Read, 1995; Figure 6.8). Ectomycorrhizal associations are recognized by the presence of fungal hyphae between the cells in the root cortex, forming a structure that resembles a net known as the Hartig net (named after a famous forest botanist) (Figure 6.9). It is across the Hartig net that nutrient exchange between the plant and the fungus is thought to occur. In addition, in many instances the fungus can completely envelop the finer feeding roots in a sheath of tissue. There is some evidence that the sheathing of the roots by the fungus can protect the plant from pathogen fungal attack by other fungi, such as that by the honey fungus (*Armillaria mellea*). This sheath influences the morphology of the root system, causing root bifurcation (forked into two). Hyphae are linked with the sheath and spread out into the soil. In some instances the hyphae aggregate to form rhizomorphs, which can be tube-like structures involved in long-distance transport of mineral ions and water. In this association there is evidence that the fungus gains carbohydrate from the plant in the form of glucose and vitamins.

Ectomycorrhizal associations are found between fungi and trees and shrubs, especially members of the families Betulaceae, Cupressaceae, Dipterocarpaceae, Fagaceae, Myrtaceae, Pinaceae and Salicaceae. The host–fungus relationship can be very specific, e.g. the species *Boletus elegans* only forms an association with larch species. However, the fly agaric fungus (*Amanita muscaria*) forms a broad ectomycorrhizal relationship with many tree species, including birch, pine, oak, beech and spruce.

Ericaceous mycorrhiza

Ericaceous mycorrhizal associations form between fungi and plants of the order Ericales. The hyphae of the fungi penetrate the cortical cells of the host plant root but no arbuscules are formed. This distinguishes this group from the ectomycorrhizae, as the latter do not form coils within the root cortical cells. This group is subdivided into three main forms of this association, ericoid, arbutoid and monotropoid.

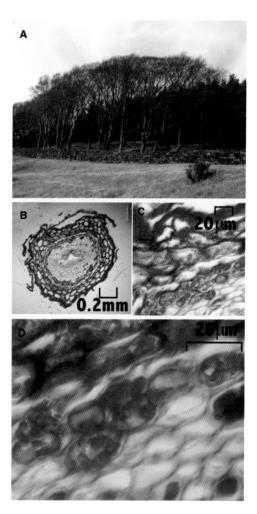

Figure 6.8 Ectomycorrhiza. (A) Typical habitat of beech tress, which form ectomycorrhizal links with fungi. In ectomycorrhiza the fungal hyphae form a sheath around the roots (B). The hyphae then form a net-like structure around the plant cells (C) and it is across this that the nutrient and carbohydrate exchange occurs. Magnified detail of the association is seen in (D).

Ericoid mycorrhiza

Species of the families Ericaceae and Monotropaceae are always associated with mycorrhiza and are regarded as obligate mycotrophs. These species can be grown without a mycorrhizal association, but in fertile soils growth is severely reduced. It is thought that the roots of species such as heathers (*Calluna*), blueberries (*Vaccinium*) and azaleas (*Rhododendron*) produce a secretion that reacts with nutrients in the soil and causes the build-up of toxins. The fungus detoxifies the soil environment and allows the plant to grow. Thus, ericoid mycorrhizae can grow effectively on nutrient-poor or -rich soils with a mycorrhizal partner.

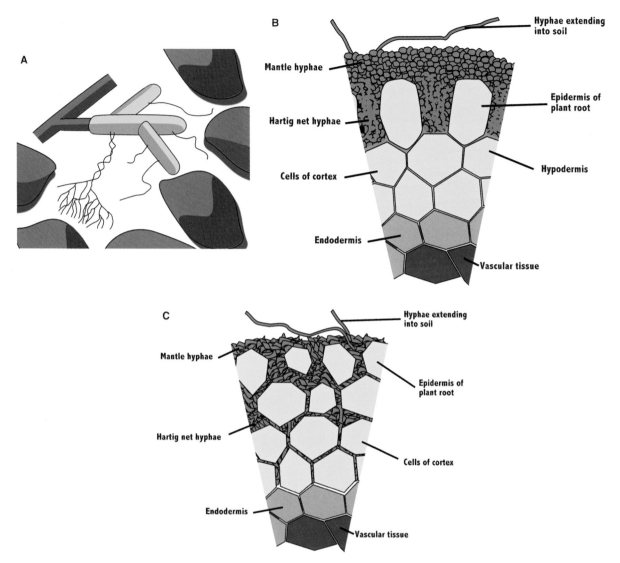

Figure 6.9 Schematic figure of a typical root infected with an ectomycorrhiza (A). For some ectomycorrhiza the sheath just encircles the root without penetrating into the root cortex (B). However, in some other associations the cortex is penetrated up to the endodermis, as shown in (C).

The fungi involved in this association are ascomycetes of the genus *Hymenoscyphus*. The fungal partner in an ericoid mycorrhiza also secretes toxins that reduce the growth of other mycorrhizal fungi. Thus, many tree species which depend upon mycorrhizal associations for growth often have great difficulty establishing in heathland. Only some *Amanita* and *Boletus* species are unaffected, which leads to birch trees being common colonizers of such habitats (Figure 6.10).

In ericoid mycorrhizal associations, the fungus forms a sheath around the root in a similar fashion to the ectomycorrhizae, but in this instance the association is much looser. The fungal hyphae extensively infect the epidermal and cortex cells. The hyphae penetrate the root cells and form coils, which degenerate as a result of the host activity and are longer-lived than similar structures in other mycorrhizal associations. However, during the degeneration process the plant cells also die. No Hartig net is formed in this association.

Arbutoid mycorrhiza

During this infection, a sheath of hyphae forms around the root and a Hartig net is present intercellularly around the

Figure 6.10   Moorland (A) with a birch tree (*Betula pendula*) (B) and a *Boletus edulis* fungus (C). The plant family Ericaceae form ericoid mycorrhizal associations with soil fungi such as *Boletus edulis* and *Amanita* spp. These fungi help to exclude other plant species from the moorland habitat. *Betula* spp., however, form mycorrhizal associations with the fungi and can colonize moorland, as shown.

outer layer of cortical cells. The morphology of the root system is similar to that described for ectomycorrhizae. The association forms between a basidiomycete fungus and a species of *Arbutus* (madrone species), *Arctostaphylos* (bearberry) and Pyrolaceae (e.g. *Pyrola asarifolia*). The fungi involved in the mycorrhizal association are basidiomycetes and can also form ectomycorrhizal associations with nearby tree species. The hyphae penetrate the root cells and form coils, these coils degenerate as a result of host action, but the host cell does not die.

Monotropoid mycorrhiza

In this mycorrhizal association the fungus infects species of the family Monotropaceae (a close relative of the Ericaceae), e.g. Indian pipe (*Hypopitys uniflora*, formerly *Monotropa*) and pinedrops (*Pterospora andromedea*) (Figure 6.11). In this infection a sheath around the root is formed and a Hartig net, in a similar fashion to the ectomycorrhiza. However, this form of mycorrhizal association is distinguished from the others by the lack of any

Figure 6.11   (A, B) Monotropoid mycorrhiza; pictures taken in Glacier National Park, Montana, USA. (C, D) An example of an Indian pipe plant (*Hypopitys uniflora*, formerly *Monotropa*), which forms a monotropoid mycorrhizal association.

hyphal penetration of the walls of cells in the root cortex. The mycorrhizal fungus also forms a mycorrhizal association with nearby tree species and thus there is a flow of carbohydrate from the tree to the Monotropaceae species. The fungi involved in this association are basidiomycete species and are discussed further in this chapter in the section on saprophytic plants.

Ectendomycorrhiza

Although this form of mycorrhizal association is very similar to the endomycorrhizal associations, the hyphal penetration of the root cortex cells is much more extensive and hence has been placed in a category of its own. It is mainly restricted to species of the genera *Picea* (spruce) and *Pinus* (pines), and to some extent *Larix* (larches). The fungi involved in the partnership are ascomycetes (mainly of two taxa, *Wilcoxina mikaloe* and *W. rehmii*). However, these species rarely form distinctive fruiting bodies in the soil and their presence can be difficult to identify. The fungal infection of the root system begins with the formation of the Hartig net, which develops behind the root meristem. As the root matures, the hyphal penetration of the root cortex cells becomes extensive and the hyphae form intracellular coils. These coils do not degenerate, as happens in many other mycorrhizal associations. Ectendomycorrhizal infection of the host leads to a short stunted root system of the host, covered with a network of branched hyphae. The term ectendomycorrhizal' should really be viewed as a description of a form of mycorrhizal association that combines the features of the ecto- and endomycorrhiza mycorrhizae.

Development of the mycorrhizal association

In a mycorrhizal association, both the participating fungus and the host plant can grow in isolation from one another, but the union is greater than the sum of the parts and the plant and the fungus thrive on the association. The relationship begins with the fungus encountering the root system of a compatible plant. The fungus then forms a swollen structure, known as an appressorium, against a junction between two root epidermal cells. In this region there is evidence that there is some digestion of the root surface by enzymes released by the fungus. Pressure is then thought to build up in the appressorium, which causes a rupture to occur in the root surface, and the fungal hypha gains entry into the root cortex through the epithelium. Whether a mycorrhizal association will develop depends on numerous factors. In some instances the health of the plant can be a factor; in some cases a healthy plant will resist the fungus and will prevent the establishment of the infection. However, in most instances the fungus will be able to extend its hyphae through the root cortex. Then, depending on the type of mycorrhizal association, different structures will develop within the cortical cells (these structures have been discussed earlier under the individual sections on the types of mycorrhizal associations). The hyphae never leave the root system, but a mycorrhizal association is not a static relationship and the fungal hyphal system is constantly growing and being degraded.

The role of the mycorrhizal association

This role has been touched on in earlier sections of this chapter, but it requires further consideration to appreciate fully the advantages that result from this symbiotic relationship. The fungus benefits from the increased carbohydrate supply originating from the host plant. It only requires a little knowledge of mycology to realize that certain fungi are only found in the vicinity of specific plant species (usually trees). It would be true to say that the presence of most fungi would go unnoticed without the production of fruiting bodies, such as the basidiocarps and ascocarps produced by the basidiomycetes and ascomycetes, respectively (usually termed toadstools). For example, the fly agaric fungus (*Amanita muscaria*) is usually found under pine or birch trees. It will be these trees that are providing the carbohydrate for the fungus to produce these fruiting bodies.

In many respects, for the plant, the mycorrhizal association can be seen as an extension of the root system; that is to say, it allows the plant to penetrate soil volume that is beyond the direct influence of the root system (Figure 6.12). But there is a growing body of evidence that suggests that the benefits to the plant of the mycorrhizal association extend beyond this. The hyphal system can grow faster than the root system of a plant; thus, a fungus can react more rapidly to available mineral resources in the soil and pass this on to the host plant (Harley and Smith, 1983). In addition, the thinner hyphae ramify through dead and decaying material in the soil (Mosse, 1959; Nicolson, 1959). In much of this decaying material there is little soluble material available (e.g. leaf litter in autumn). Thus, the fungus has to mobilize minerals and carbohydrates in the structure through enzymatic action. Again, the host plant can benefit from these nutrients harvested from the decaying material. There is also evidence that fungi can dissolve insoluble nitrogen- and

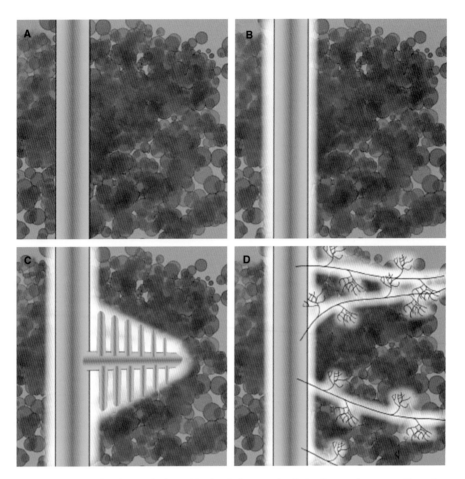

Figure 6.12 Advantages of mycorrhizal associations. The hyphal network of the fungus increases the volume of soil which is accessible to the plant (A, B). A primary root has little impact on the soil, with the rhizosphere being very limited (A). However, if the root is branched or contains numerous root hairs (C), the accessible soil volume is increased. However, a mycorrhizal association can be much more advantageous to the plant (D).

phosphorus-containing compounds through the release of organic acids around the substrate (Griffiths and Caldwell, 1992). Thus, rather than just extending the root system, the fungal hyphal network actually makes nutrients available to the plant that were previously inaccessible (Kahiluoto and Vestberg, 1998).

The presence of the mycorrhizal associations can also offer numerous competitive advantages to the host plant, such as reducing infection by pathogenic fungi. Through the competition between the fungal mycorrhizal partner and other soil fungi, the incidence of infection of the host plant by the pathogen can be decreased. In addition, the depletion of mineral resources by the fungus in the soil can reduce the effective growth of neighbouring non-mycorrhizal plant species. Plants of the same species can be linked to the same fungus, and thus a young plant can benefit from a being linked to a mature plant (Hogberg et al., 1999; Horton *et al.*, 1999). Consequently, saplings or

orchid species growing under the canopy of a forest can benefit from neighbouring trees through an increased carbohydrate supply. This has been demonstrated through the feeding of radioactive CO_2 to the parent plant and its subsequent rapid detection in the sapling. There is evidence that different plant species neighbouring one another can be mycorrhizal with the same fungus, and thus there can be nutrient transfer between different plant species (Simard *et al.*, 1997). This can reduce competition between plant species and so help to stabilize a plant community.

Thus, the benefits of mycorrhizal associations for plants and fungi are very great. Undoubtedly there would not be the diverse range of plant and fungi species there are today without these associations. There is good evidence that modern agricultural practices are upsetting this delicate balance between plants and fungi. Eutrophication of the environment through excessive application of fertilizers and animal waste are changing the balance of fungi in the soil. As a consequence, certain plant species are in dramatic decline as neighbouring species out-compete them, due to increased nutrient availability. In addition, loss of large areas of established ancient forest in the world is probably one of the major contributing factors to the decline of fungal species. Hence, plant species that are co-mycorrhizal with tree species and fungi are also in sharp decline, e.g. species in the family Orchidaceae. Some crop plants do form mycorrhizal associations but these are very limited. The mycorrhizal association benefits from a large hyphal network through the soil. This network is disrupted during ploughing of the soil and as a consequence the effectiveness of the association is restricted. There is a desperate need for the relationship between plants and fungi to be better understood, so that plant and fungi diversity can be preserved and perhaps the advantages of mycorrhizal associations can be harnessed for the benefit of agriculture. There is a danger of ignoring mycorrhizal associations because they cannot be seen, seriously damaging the diversity of life on this planet.

Saprophytic nutrition

A saprophyte is an organism that obtains its nutrition from assimilating organic matter from dead and decaying material. Like a fungus, a true saprophyte is a heterotroph, which means that the plant cannot make its own food (Figure 6.7). As a consequence of their life style, the plants reported to be saprophytic all possess no chlorophyll and live where there is an abundance of decaying plant material (usually on the floor of a forest). There is good evidence that the growth of the plant is dependent upon the presence of decaying material, but there is little evidence that plants release enzymes to facilitate the decay of material in the soil, thereby releasing nutrients that can be utilized by the plant. Plants can release acid phosphatase, which can facilitate the dissolving of esterified phosphorus-containing compounds in decaying matter (Dinkelaker and Marschner, 1992). However, this is a long way from having a plant wholly dependent on saprophytic nutrition. There are references to saprophytic plants in plant identification guides, such as *The Wild Flowers* (Fitter *et al.*, 1991), but all of these examples have mycorrhizal associations with fungi. It is the fungi that possess the saprophytic activity, not the plant, and there are no true saprophytic plants. So-called saprophytic plants are more accurately termed mycotrophic, since they derive much of their nutrition from mycorrhizal associations with soil fungi. Examples of these plants include: from the Orchidaceae;

ghost orchid (*Epipogium aphyllum*), birdsnest orchid (*Neottia nidus-avis*; see Fig. 6.7) and coral root orchid (*Corellorhiza trifida*); from the Monotropaceae, Indian pipe (*Hypopitys uniflora*, formerly *Monotropa*), sugar stick (*Allotropa virgata*) and pinedrops (*Pterospora andromedea*).

All of these plant species possess mycorrhizal associations with fungi in the soil. These fungi derive their nutrients either from further mycorrhizal associations with neighbouring tree species or through the assimilation of nutrients from decaying material. The location of most of these species, under the canopy of forests, would suggest that the fungal partner obtains its nutrients from a combination of the two sources. In these examples, the plants are really acting as direct parasites on the fungus and indirect parasites on the tree species, since there is no evidence that the fungus gains from this relationship. Since the fungus initiates the formation of the mycorrhizal association between itself and the plant host, the plant is deceiving the fungus that the relationship will be to its benefit. Plants that use this form of nutrition are usually very rare. A case in point is the example of the ghost orchid, which has been repeatedly declared extinct in the UK, only to reappear after years of apparent dormancy! There are no true saprophytic plants and the plants thought of as saprophytic are more accurately called mycotrophic.

It could be argued that carnivorous plants act as semi-saprophytic plants. Although carnivorous plants can digest animals whilst they are alive, they usually digest the dead material. Thus, much of their mineral nutrition is gained from the digestion of insect carcasses but a vast majority of their carbohydrate comes from photosynthesis. Therefore, in no way could a carnivorous plant support its whole life cycle through carnivory.

The examples presented in this chapter concerning the remarkable benefits of mycorrhizal associations in plants clearly reveal that the impact of this mutually beneficial relationship for plants and fungi is immense. It is a hidden interaction between the two kingdoms, rarely appreciated because of its microscopic nature. However, it supports much of the diversity of plants and fungi that we see today. It is a topic that is usually studied in isolation from plant physiology but it is intimately linked to how plants function and grow in their natural habitats.

References

Bending, GD and Read, DJ (1995) The structure and function of the vegetative mycelium of ectomycorrhizal plants V. Foraging behaviour and translocation of nutrients from exploited litter. *New Phytologist.* **130**, 401–409.

Broering, TJ and Morrow, GW (1999) Oxygenated dihydrophenanthrenes via quinol acetals: a brief synthesis of orchinol. *Synthetic Communications*, **29**, 1135.

Brundrett, MC (1991) Mycorrhizas in natural ecosystems, in *Advances in Ecological Research*, **vol. 21** (eds A. Macfayden, M. Begon and AH Fitter), Academic Press, London, pp.171–313.

Dinkelaker, B and Marschner, H (1992) *in vivo* demonstration of acid phosphatase activity in the rhizosphere of soil-grown plants. *Plant and Soil*, **144**, 199–205.

Dinkelaker, B Romheld, V, *et al.* (1989) Citric acid excretion and precipitation of calcium citrate in the rhizosphere of white lupin (*Lupinus albus* L.). *Plant Cell and Environment*, **12** (3) 285–292.

Fitter, R, Fitter, A and Blamey M (1991) *The Wild Flowers of Britain and Northern Europe*, Collins, London.

Gahoonia, TS and Nielsen, NE (1997) Variation in root hairs of barley cultivars doubled soil phosphorus uptake. *Euphytica*, **98** (3) 177–182.

Gallaud, I (1905) Études sur les mycorrhizes endophytes. *Revue General de Botanique* **17**, 5–48, 66–83, 123–136, 223–239, 313–325, 425–433, 479–500.

Gerke, J and Römer, W *et al.* (1994) The excretion of citric and malic acid by proteoid roots of *Lupinus albus* L.; effects on soil solution concentrations of phosphate, iron, and aluminium in the proteoid rhizosphere in samples of an oxisol and a luvisol. *Zeitschrift für Pflanzenernährung und Bodenkunde*, **157**, 289–294.

Griffiths, RP and Caldwell, BA (1992) Mycorrhizal mat communities in forest soils, in *Mycorrhizas in Ecosystems*, (eds DJ Read, DH Lewis, AH Fitter and IJ Alexander), CAB International, Wallingford, UK, pp.98–105.

Harley, JL and Smith, SE (1983) *Mycorrhizal Symbiosis*, Academic Press, London.

Hogberg, P., Plamboeck, AH, Taylor, AFS. and Fransson PMA. (1999) Natural C-13 abundance reveals trophic status of fungi and host-origin of carbon in mycorrhizal fungi in mixed forests. Proceedings of the National Academy of Science of the United States of America, **96**, 8534–8539.

Horton, TR, Bruns, TD and Parker, VT (1999) Ectomycorrhizal fungi associated with *Arctostaphylos* contribute to *Pseudotsuga menziesii* establishment. *Canadian Journal of Botany*, **77**, 93–102.

Kahiluoto, H and Vestberg, M (1998) The effect of arbuscular mycorrhiza on biomass production and phosphorus uptake from sparingly soluble sources by leek (*Allium porrum* L.). in Finnish field soils. *Biological Agriculture and Horticulture*, **16**, 65–85.

Kendrick, B (1992) *The Fifth Kingdom*, Mycologue Publications, Waterloo, Ontario, Canada.

Mosse, B (1959) Observations on the extra-matrical mycelium of a vesicular–arbuscular endophyte. *Transactions of the British Mycological Society*, **42**, 439–448.

Nicolson, TH (1959) Mycorrhiza in the Gramineae. I. Vesicular–arbuscular endophytes, with special reference to the external phase. *Transactions of the British Mycological Society*, **42**, 421–438.

Simard, SW, Perry, DA, Jones, MD, Myrold, DD, *et al.*(1997) Net transfer of carbon between ectomycorrhizal tree species in the field. *Nature*, **388**, 579–582.

St John, TV, Coleman, DC and Reid, CPP (1983) Growth and spatial distribution of nutrient-absorbing organs: selective exploitation of soil heterogeneity. *Plant and Soil*, **71**, 487–493.

Trappe, JM (1987) Phylogenetic and ecologic aspects of mycotrophy in the angiosperms from an evolutionary standpoint, in *Ecophysiology of VA Mycorrhizal Plants*, (ed. GR Safir), CRC Press, Boca Raton, FL, pp.5–25.

Further reading

Harley, JL and Harley, EL (1987) A check-list of mycorrhiza in the British flora. *New Phytologist*, **105** (suppl) 1–102.

Harvey, AE, Larsen, MJ and Jurgensen, MF (1976) Distribution of ectomycorrhizae in a mature Douglas-fir/larch soil in western Montana. *Forest Science*, **22**, 393–633,

Lewis, JD and Koide, RT (1990) Phosphorus supply, mycorrhizal infection and plant offspring vigour. *Functional Ecology*, **4**, 695–702.

Marschner, H and Dell, B (1994) Nutrient uptake in mycorrhizal symbiosis. *Plant and Soil*, **159**, 89–102.

Smith, SE (1995) Discoveries, discussions and directions in mycorrhizal research, in *Mycorrhiza*, (ed. A. Verma and B. Hock), Springer-Verlag, Berlin, pp.3–24.

Smith, SE and Gianinazzi-Pearson, V (1988) Physiological interactions between symbionts in vesicular–arbuscular mycorrhizal plants. *Annual Review of Plant Physiology and Plant Molecular Biology*, **39**, 221–244.

7

Parasitic plants

Autotrophy is, perhaps, the most remarkable ability of plants: to be able to make carbohydrate from light and inorganic compounds. This carbohydrate can be stored in various plant structures such as leaves, roots, seeds and tubers. The rich source of nutrition is an obvious target food source for a large range of herbivores which, on average, devour an estimated 18% of global plant production annually (Cyr and Pace, 1993). This makes plants the basis for the vast majority of the food chains on the planet. However, there is a small but diverse group of plants, numbering around 4000 different species, which have evolved the ability to prey on other plant species. This group are the parasitic plants. Many of these species have been freed from the usual constraints on plant structure dictated by a photosynthetic life style and have evolved unusual forms. In some species chlorophyll, leaves and roots have been abandoned, allowing the plants to form structures more reminiscent of fungi than plants.

Parasitic plants make up around 1.5% of flowering plant species. Parasitism has arisen in 22 different plant families and it is estimated to have arisen 11 times during evolution (Nickrent *et al.*, 1998). Parasitism can be found between plants in most ecosystems across the world. Species from four of the families of parasitic plants are parasitic to agricultural crops and as a consequence have some impact on crop productivity. The families of angiosperms known to have parasitic members are listed in Table 7.1. Given the number of plant species that use parasitism as a mode of nutrition and their agricultural significance, this is not a group that should be ignored, and it is important to have some understanding of the physiological processes occurring, so this chapter gives further attention to this fascinating group of plants.

Establishing a parasitic relationship

A vast majority of parasitic plants target the root system of the host (Table 7.1, Figure 7.1). However, whether it infects a stem or root, the parasite aims to form a link between its tissue and the host's vascular tissue. The vascular tissue of a plant contains water, organic acids, mineral ions, amino acids and carbohydrates, and so through a link with this tissue the parasite can tap the host for all of the food reserves it requires for growth (Figure 7.2). It is the extent to which the parasite can link up to the host vascular tissue that dictates the degree of parasitism that will be involved between the host and parasite. As would be expected, there is a gradation in the parasitic ability of these plants, with some species being only slightly dependent on parasitism, whereas others are completely dependent upon a parasitic relationship for their survival. Hence, this group has been subdivided into:

1. *Facultative parasitic plants*: plants that can grow without a host plant but usually grow more rapidly in a parasitic relationship (hemiparasites), e.g. *Rhinanthus minor* (yellow rattle).

2. *Obligate parasitic plants*: plants that are usually not autotrophic and can only survive in the parasitic relationship. These can be subdivided into: Species that can photosynthesize (hemiparasites), e.g. *Cuscuta reflexa* (dodder). Species that have lost their chlorophyll and are non-photosynthetic (holoparasites), e.g. *Orobanche* species (broomrapes).

3. *Plants parasitic as a result of mycorrhizal associations*: these are termed mycotrophic relationships (these again can be facultative or obligate mycorrhizal parasites), e.g. ghost orchid (*Epipogium aphyllum*). These have been discussed in detail in Chapter 6 in the section on mycorrhizal associations.

The amount of nutrients obtained from host plants varies, and this variability is a direct result of the effectiveness of the vascular links between the parasite

Physiology and Behaviour of Plants   Peter Scott
© 2008 John Wiley & Sons, Ltd

Table 7.1 Types of parasitic interaction with plants, summarizing the different plant families in which parasitism is represented and the type of association formed.

Plant family	Parasite type		Host tissue association		
	Holoparasite	*Hemiparasite*	*Root*	*Stem*	*Leaf*
Apodanthaceae	√	–	√	√	
Balanophoraceae	√	–	√		
Cuscutaceae	√	√		√	√
Cynomoriaceae	√	–	√		
Cytinaceae	√	–	√		
Hydnoraceae	√	–	√		
Krameriaceae	–	√	√		
Latraeophilaceae	√	–	√		
Lauraceae	–	√		√	
Lennoaceae	√	–	√		
Lophophytaceae	√	–	√		
Loranthaceae	–	√	√		
Misodendraceae	–	√	√		
Mitrastemonaceae	√	–	√		
Olacaceae	–	√	√		
Opiliaceae	–	√	√		
Orobanchaceae	–	√	√		
Rafflesiaceae	√	–	√	√	
Santalaceae	–	√	√		
Sarcophytaceae	√	–	√		
Scrophulariaceae	–	√	√		
Viscaceae	–	√		√	

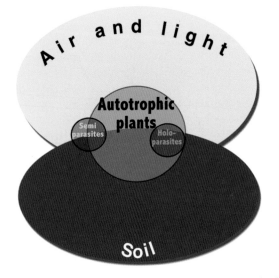

Figure 7.1 Interactions between host plant, parasite and soil. There are a number of different types of parasitic plant. Semi-parasites can grow on their own in some instances, or in association with a host plant. Holoparasites need to be associated with a host but can be linked to the host via a root or shoot.

and the host. If the links are poor, the parasite is likely to be a facultative parasite and will appear no different from an autotrophic plant. However, if the links are very effective, the parasite is more likely to have abandoned the ability to photosynthesize, possess a poorly developed root system and become a holoparasite. Such species are frequently colourless in appearance as a result of that lack of chlorophyll, and it is these plants that resemble fungi rather than plants.

The initial stages of the life cycle of parasitic plants are crucial. These stages are the germination of seeds of the parasite, and the establishment of the infection of the host. The seed of a parasitic plant needs to germinate in the close vicinity of a host plant. Food reserves in the seed are limited, and in many instances autotrophic nutrition cannot be used by the parasite to fuel the early stages of seedling growth. Thus, in many instances the parasite has an absolute requirement for a host plant in very close proximity to the seed. The seeds are able to sense the presence of the host plant by the release of specific compounds from its root system into the soil. For certain parasites with a limited host range, these chemicals can be very specific to a particular host.

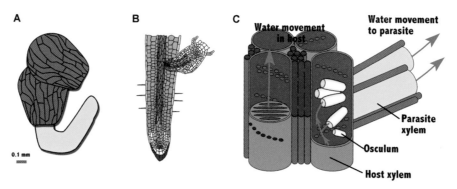

Figure 7.2 Establishment of a host–parasite interface in the roots. The seeds of the parasitic plant *Striga* (A) germinate only when very close to a host plant. The limited root system locks on to the host root (B) and forms a haustorium, which penetrates into the root cortex and through the endodermis. Then the haustorium penetrates the xylem tissues, with oscula to permit water flow into the parasitic plant. No penetration of the phloem is required, since producing phloem tissue in the vicinity of the host phloem tissue will allow the parasite to act as a sink tissue and therefore obtain carbohydrate.

The parasitic plant system that has been studied in the greatest detail is that of the witchweed (*Striga* spp.). Witchweed is a hemiparasite of grass species that use the C4 photosynthetic pathway. In particular, *Striga* parasitizes crop species such as maize and sorghum and is a very damaging pest of Third World agriculture. The parasite tends to inhabit fields of these crops, and seeds from the parasite easily get mixed with those from the host plant during harvesting. Then, once the seeds are sown for the next season, the seeds of the parasite are also sown. Thus, an infestation is perpetuated if strict controls of seed are not exercised. Through this route, the parasite has become very widespread and has even spread to the eastern states of North America through poor seed quality control. The seeds of witchweed need to be within 3–4 mm of a host plant to germinate, since the seed is capable of a maximum of 4 mm growth after germination. To compensate for this, each plant is capable of producing around 500 000 seeds. For dispersal, the seeds are sticky and can adhere to plants, animals and seeds of the crop plant. Witchweed seeds are believed to be able to sense the presence of sesquiterpene (called strigol) and lactones released by the host, and these act as a germination cue (Sugimoto et al., 1998). The parasite can detect minute amounts (picomolar concentrations) of these compounds in the soil, and through the sensing of the concentration gradient of these compounds the parasite can locate the host precisely. In other parasitic plant species, little progress has been made in identifying the factors that trigger seed germination, mainly because of the high sensitivity of the seeds to the compounds and their extremely low concentrations in the soil.

On contact with the host root, the parasite forms a haustorium (plural, haustoria; derived from the Latin *haurire*, to drink). This structure is functionally similar to that formed when fungi infect plants; however, they are structurally very different. In order to form the haustorium, the parasite responds to further signals from the host tissue. There is generally a broader range of compounds that induce haustorium formation than trigger germination. Such signal compounds have been identified as coming from the following groups: cytokinins, flavenoids, hydroxyacids and quinones (Lynn and Chang, 1990). In *Striga asiatica* the haustorium is initiated in response to a single quinone, 2,6-dimethoxy-*p*-benzoquinone (2,6-DMBQ; Figure 7.3). This quinone is found in sorghum roots after

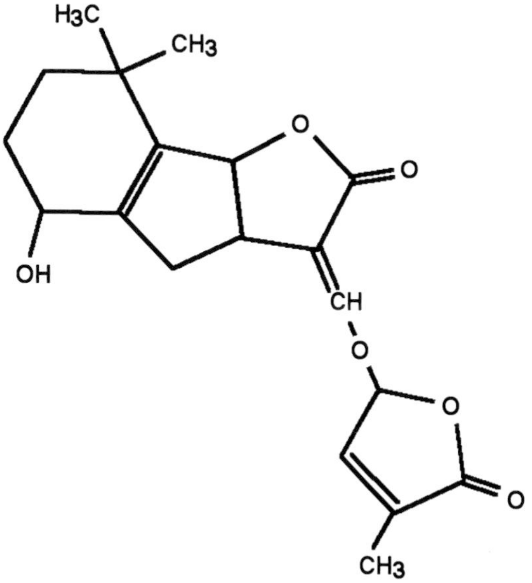

Figure 7.3 Strigol, a signal for parasitic seed germination. The seeds of the parasitic plant *Striga* germinate only when the seeds detect the compound strigol, whose chemical structure is shown here, in the soil. It is released from its host plant but has a very limited range of mobility. Thus, *Striga* seeds will only germinate very close to a host plant; otherwise, they remain dormant.

severe agitation, but not in root exudates (Chang and Lynn, 1986). This suggests that the quinone is produced as a result of a cell wall-degrading enzyme, lactase, produced by the parasite in response to the proximity of the host (Stewart and Press, 1990). The parasite then uses the haustorium to penetrate the host tissue. This is achieved through a combination of enzymatic activity at the surface of the root and a build-up of pressure. The parasite may then penetrate the host cells or, more usually, use partial enzymatic degradation of the cell walls to force entry between the host cells. The parasite then links up with the host vascular tissue.

The plant parasite can in some instances form direct continuous links through the lumen of the host xylem and its own xylem tissue. However, in most cases the parasite vascular tissue abuts that in the host, and transfer of nutrients occurs across this interface or through transfer cells (see Figure 7.2). The nutrients gained by the parasite will depend on whether the parasite mainly links up to the xylem and or the phloem. This topic is best explained by studying the different subdivisions of parasitism.

Facultative parasitic plants

Facultative parasitic plants can be viewed as a half-way house between parasitism and autotrophy. To look at, these plants are generally like any other plant and they do not bear the typical characteristics of parasitic plants. They possess leaves through which they obtain autotrophic nutrition, and a root system through which they can take up water and nutrients. The real difference lies in the development of the root system, as all facultative parasitic plants are root parasites. The root system is not as well developed as in other plants, because periodically the parasites roots join via haustoria to the roots of neighbouring plant species (Figure 7.2). The haustorium forms a connection with the xylem tissue and then allows the parasite to tap nutrients and water in the xylem of the host for its own uses. The haustorial connection is not as well developed as in other types of parasitic association, because the phloem is not tapped.

The mineral ion contents of xylem tissue are not high, so in order for the parasite to redirect as much of the xylem flow as possible from the host, the parasite possesses several adaptations. First, the stomatal apertures of some hemiparasitic plants display some unusual behaviour compared to the host plant. The stomata of the parasite do not close fully at night. Therefore, transpiration rates are maintained at a high rate at night and this guarantees a greater flow of xylem contents to the parasite (Press *et al.*,

1999). Second, the water potential within the parasite tends to be lower than that in the host plant. This again will favour high flow rates from the host. Consequently, the parasite effectively competes with the host for the contents of the xylem tissue (Ehleringer and Marshall, 1995).

Species that exhibit this form of parasitism include yellow rattle (*Rhinanthus minor*), red bartsia (*Odontites verna*), Indian paintbrush (*Castilleja indivisa*) and eyebright (*Euphrasia officinalis*) (Figures 7.4–7.8). Most of these species tend to be parasites of a wide range of grass species. There is some evidence that they are selective about the hosts they parasitize, and they seek out the best host to sustain high growth rates. They can grow independently of a host plant, but under such conditions they tend to have low chlorophyll amounts in the leaves and consequently they grow poorly. The independent plants also possess very limited root systems and are very sensitive to drought stress. The Indian paintbrush plant from North America, known to parasitize the Texas bluebonnet (*Lupinus texensis*), is an interesting example (Figure 7.8). Through this interaction, the parasite gains alkaloids

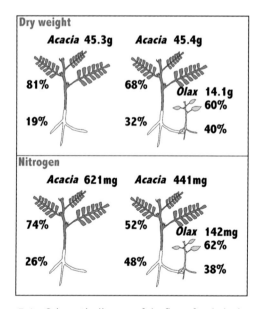

Figure 7.4 Schematic diagram of the flow of carbohydrate and nutrients from host to parasite. The parasitic relationship between *Olax phyllanthi* (Olacaceae) and *Acacia littorea* is probably the best-studied host–parasite relationship. This figure is a simplified version of that presented in Tennakoon *et al.* (1997). (A) The total weight of the *Acacia* plant is not affected by the parasite, in fact it could be considered to be more productive. The only major difference is that more of the photosynthetically fixed carbon is accumulated in the host root system than would be seen in normal, unparasitized plants. However, nitrogen availability (B) to the aerial tissue of the host is severely reduced because it is redirected to the parasite.

Figure 7.5 Semiparasites– the manipulators of the landscape. There are numerous hemiparasitic plants in Northern Europe; those pictured above belong to the family Scrophulariaceae. (A) Red bartsia (*Odontites verna*); (B) Common eyebright (*Euphrasia officinalis*); and (C) sward inhabited by red bartsia and eyebright. These plants act to suppress the vigorous growth of grasses and thus allow other species to thrive. Hence, hemiparasites have a dramatic effect upon the flora that grow in a particular habitat.

Figure 7.6 Growth of the annual semiparasite yellow rattle (*Rhinanthus minor*). (A) Unlike the seeds of the Orobanchaceae, the seeds of yellow rattle are large and contain food reserves sufficient for growth of the seedling, even in the absence of a host. The seeds of yellow rattle require a considerable period of cold treatment before they germinate (around 3 months). (B) Once germinated, the root system quickly develops and the seed coat splits, revealing the cotyledons. The root system then develops secondary roots, which will be used to attach to host roots. If these fail to attach, in most cases the seedling will die, even if grown under very favourable conditions. (C) Experimental pot plant, showing two parasitic yellow rattle plants and a host grass. (D) Removing the soil substrate reveals the limited extent of the yellow rattle root system. (E) Close-up of the haustorial attachment points between the parasite and the host. (F) Yellow rattle in flower.

Figure 7.7 The spread of semiparasites. As *Rhinanthus* sheds seeds over generations, large areas of grassland can be parasitized and in many instances killed. (A) Grassland that has been untouched by *Rhinanthus*. (B) A very dense colony of *Rhinanthus*, in which many individual plants will be growing without a host (C). Such colonies tend to thrive on their margins (where parasitism can occur) but are stunted in the centre, where there are few hosts. In this instance, the specimens in the centre all died due to drought stress over the summer period. *Rhinanthus* forms a poor root system that cannot survive prolonged drought if no host plant can be found to parasitize.

Figure 7.8 Indian paintbrush (*Castilleja indivisa*), a hemiparasitic plant of mountain meadows in North America. As in Europe, in North America there are numerous hemiparasitic plants, mainly represented in the family Scrophulariaceae. (A) Mountain meadow in Glacier National Park, Montana. The red flowering plants are the Indian paintbrush plant, shown close up (B).

(defensive compounds; see Chapter 16) from the bluebonnet that accumulate in the tissues. Plants that accumulate alkaloids exhibit an enhanced ability to resist herbivory, attract pollinators and improve seed set. Thus, the acquisition of toxic defensive compounds from the host could be of great significance to the survival and success of the parasite.

The parasitic relationship between *Olax phyllanthi* (Olacaceae) and *Acacia littorea* is probably the best-studied host–parasite relationship and is illustrated in Figure 7.4 (based on the work of Tennakoon et al., 1997), which shows how carbon and nitrogen flows are influenced in the plants as a result of the parasitism. To summarize, the photosynthetic rate in the host is generally unaffected, but more of the photosynthetically fixed carbon is accumulated in the host root system than would be seen in normal, unparasitized plants. Nitrogen availability

to the aerial tissue of the host is severely reduced because it is redirected to the parasite.

Obligate parasitic plants – hemiparasites

This group of parasitic plants includes some very interesting plant species. Usually plants in this group can photosynthesize for themselves but have an obligate dependency on a host plant at some stage in their life cycle. In addition, this group includes some of the stem parasites, such as the mistletoes (Viscaceae) (Figures 7.9, 7.10), dodder (Cuscutaceae) and witchweed (Scrophulariaceae). These three families of plants contain crop parasites and thus are of agricultural importance.

The mistletoes are probably one of the most well-known species of parasitic plants. Most readers will be familiar

Figure 7.9 Mistletoe (*Viscum obscurum*) on a tree in the Sewula Gorge, KwaZulu Natal, South Africa. Here are shown several images of the same mistletoe infection on a tree in South Africa. The name 'mistletoe' is derived from the Anglo-Saxon word *mistel*, which means dung, and *tan*, which means twig. The name perfectly describes the growth habit of the plant. Birds eat the berries produced by mistletoes in the winter and some are deposited on branches of suitable host trees. These then germinate (A) and the seed rapidly inserts an haustorium into cracks in the bark, thus establishing the infestation. The seed case eventually breaks, revealing the leaves, and the mistletoe plant grows and after several years reaches maturity. (B) Small plants; (C) a mature berry bearing infestation.

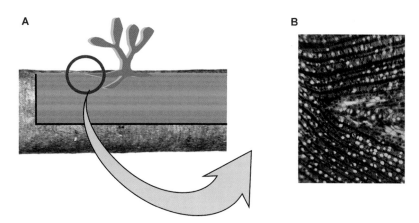

Figure 7.10 The interface between a host and mistletoe plant. The mistletoe produces haustoria, which enter the host bark and penetrate the vascular tissue (A). This permits links between the xylem and the phloem of the two plants to be established (B). Mistletoes are capable of limited photosynthesis but rely on their host for most carbohydrate.

with the European mistletoe, *Viscum album*, which has been used for centuries as a druidic symbol and a piece of greenery to kiss under at Christmas time! European mistletoe parasitizes a large range of tree species but is most frequently found on poplars and apple trees. *V. album* rarely does great damage to the host unless the infestation is particularly prolific. However, the dwarf mistletoes (*Arceuthobium* spp.) are major pests of coniferous forestry in the Northern Hemisphere. Their seeds are spread by explosive ejection from the plant and they stick to neighbouring plants and passing animals. The seeds can be propelled up to 15 m from their source. They then germinate and send a root-like structure (known as a sinker-root) through the bark to form a haustorium. At this point in its development, the mistletoe acts as an endophyte, with all of the parasite tissue being within the host. Little can be seen on the surface of the tree for several years, as the mistletoe spreads under the bark and produces branch or stem swelling. At this point the structure of the tree can be severely deformed by the parasite. In many instances, masses of multiple branches, known as witches' brooms, can form near the infestation. The parasite can severely restrict tree growth as it absorbs water and nutrients from its host. The points of infestation also act as sites for fungal infection and disease in the host trees. Small trees are usually killed by the infestations. Consequently, in terms of silviculture, the trees are stunted, diseased and the timber yield is of poor quality. Dwarf mistletoes cause more damage to coniferous forests in North America than any other disease or pest. Natural control of the dwarf mistletoe is through wildfires destroying the host and parasite. However, with the advent of intense forestry practice, wildfires are strictly controlled and therefore few barriers are presented to stop the spread of an infestation.

The family Loranthaceae of flowering plants follows a similar strategy to the mistletoes in tropical regions of the world. One interesting example of this family, *Dendropemon emarginatus*, is one of the few examples of a plant hyperparasite. This plant parasitizes *Ximenia americana*, a root parasite in the family Olacaceae.

The Cuscutaceae are further examples of stem parasites. There are almost 150 different species of *Cuscuta* but the most well-known example of this family is the common dodder (*Cuscuta epithymum*; Figure 7.11). In many species of *Cuscuta* the plant contains some chlorophyll and is capable of limited photosynthesis, but dodders cannot grow in the absence of a host plant. Seeds of dodder fall to the ground and can germinate if a suitable host is nearby. The seed reserves are limited and once a seed germinates it has sufficient reserves to support it for 6 days. If during this time no parasitic link with a host plant is established, then the seedling will die. In the absence of a host, the seeds can remain dormant for up to 5 years. The seedlings are touch-sensitive and sway from side to side in an effort to locate the host plant. Once contact is established, the seedling entwines the stem. If the plant is a suitable host, the seedling produces adventitious roots that penetrate the stem and form a haustorium. The contact with the ground then tends to shrivel and the threads of the dodder spread upwards. The plant only forms small scale-like leaves and is only evident as a thread-like mass covering the host plant. The dodder is almost completely dependent upon the host for water, nutrients and carbohydrate. The parasite forms links to both the xylem and the phloem tissue of the host (Wolswinkel, 1974). One unusual feature of dodder plants is that they form numerous haustorial connections with the host plant. Sometimes these connections can be in the leaves, but this is not the usual means of connection

Figure 7.11   Infestation of plants by dodder (*Cuscuta* spp.). (A) *Cuscuta campestris* parasitizing some Asteraceae species in Portugal. (B–D) *Cuscuta planiflora* on a range of plant species in South Africa. (D) The points of penetration in the leaves of a bay tree.

with the host. The plant is also known to form haustoria on its own stems and is a rare example of a plant autoparasite.

Dodder has a very broad host range of both wild and cultivated species. In the wild, dodder parasitizes nettles, clover and even bindweed, to name but a few. Alfalfa is one of the major agricultural species infested by dodder. In agriculture, as a result of the seed longevity and the broad host range, it is a difficult parasite to control. The small seeds also mix easily with crop seeds and can be spread through seed stock contamination. Several countries have very strict guidelines with regard to this species and importation of crop seeds. There are several reports of infestation by dodder also resulting in the spread of crop diseases, e.g. Yellows disease in pears, caused by a bacterium that is spread by dodder infestations.

The witchweeds (*Striga* spp.) are also obligate parasites. The witchweeds are devastating parasites of corn, millet, rice and sorghum in the developing world. *Striga* compensates for the lack of a developed root system by penetrating and parasitizing the roots of other plant species. It has been estimated to infest around 66% of the land devoted to cereal agriculture in Africa. Its infestation can result in crop losses of 70% or greater in subsistence farms and is one of the major obstacles to increasing crop production in Africa. There are few methods of combating this pest; it is particularly aggressive in nutrient-poor arid soils with low biodiversity, which are precisely the soils

used in subsistence farming. The seeds can survive for up to 10 years in the soil, hence crop rotation cannot be used to eradicate an infestation. There are two main species which are agricultural pests: *S. asiatica* (South Africa and India) and *S. hermonthica* (East and West Africa); however, a further species, *S. gesneroides*, parasitizes cowpeas and sometimes sweet potato. The seeds of *Striga* germinate and then latch onto the host plants. The plant gains carbohydrates, water and mineral ions from the host and severely reduces the growth rate of the crop. The haustoria form tubular structures, called oscula, that directly penetrate the host xylem and tap the reserves of the host. Once established, the *Striga* plant can photosynthesize for itself, but an estimated 35% of the carbohydrate required for growth comes from the host. The witchweeds primarily feed from the host's xylem vessels, but these vessels do not contain the large concentration of nutrients it requires. Consequently, the parasite has adopted the strategy of transpiring more water from its aerial tissue in order to draw more xylem contents to itself, and hence increase the supply of nutrients. The xylem does not usually contain much carbohydrate; thus, *Striga* probably obtains sugars from the apoplastic spaces in the host root. The parasite is therefore a large drain on the carbohydrate, nutrient and water supply of the host and it is a combination of these factors that severely reduces the growth of the crop and causes the subsequent failure to set seed.

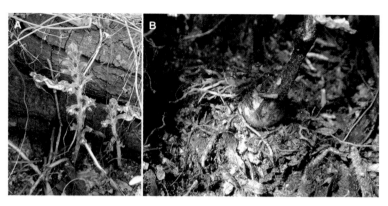

Figure 7.12 *Orobanche purpurea* parasitism of *Acacia* sp. *Orobanche purpurea* is a perennial parasite of a range of plant species, here shown parasitizing an *Acacia* tree (A). The parasite, once linked to a host, forms an underground corm (B), which enlarges until it is large enough to support flowering.

Obligate parasitic plants – holoparasites

The best examples of this division of parasitic plants are the members of the Orobanchaceae. This family is one of the largest groups of parasitic plants. Most of the species contain no chlorophyll and are only evident when an inflorescence is formed. In the UK, the commonest sp ecies is *Orobanche minor* (common or small broomrape) (Figures 7.12–7.14). In a similar fashion to witchweeds,

broomrapes release large quanities of small seeds, which are dispersed by the wind and settle on the soil. Through a process of weathering the seeds get into the soil and, if fortunate, into the close vicinity of a host plant. The broomrapes vary between being very host-specific to having a broad range of possible hosts. For example, the

Figure 7.13 *Orobanche ramosa* parasitism of *Oxalis pes-caprae*, a common invasive weed in many Mediterranean regions of Europe. Several *Orobanche* species have become parasitic on the plant. Here *Orobanche ramosa* is shown (A). The plant forms a matted network of thick fibrous roots around 3–4 cm in length. The plant than attacks roots of some plant species that approach close to this mat. A ribwort plantain (*Plantago lanceolata*) is shown growing close to the *Orobanche* plant (B). The root network of this plant intertwined with the *Orobanche* but no haustorial links were noted between the two plants. The *Oro-banche* was feeding solely off *Oxalis pes-caprae* roots. (C) The haustorial links.

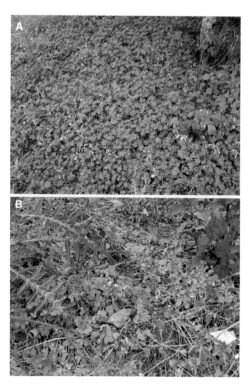

Figure 7.14 The influence of *Orobanche ramosa* parasitism on *Oxalis pes-caprae*, whose colonies (A) can cover large areas of ground in Mediterranean habitats (here shown near Lisbon, Portugal). However, the presence of *Orobanche ramosa* can severely debilitate an *Oxalis pes-caprae* colony (B).

Figure 7.15 *Cistanche phelypaea* parasitism of salt marsh species. *Cistanche phelypaea* is a perennial parasite of a range of woody members of the family Chenopodiaceae.

thistle broomrape (*Orobanche reticulata*) only infests thistle species of the genus *Carduus* and *Cirsium*. This has led this species of broomrape becoming very scarce and confined to one locality in the UK in Yorkshire (Hughes and Headley, 1996). In contrast, species such as *O. minor* have a broader host range of clover species (*Trifolium*), vetches and grasses.

Seeds of the Orobanchaceae have only limited food reserves and for successful infestation of a host, the seeds need to be within 4 mm of a host root system. The chemical triggers of germination and formation of the haustorium have not been studied in detail in the Orobanchaceae. The seed forms a hyaline root (a white translucent structure) that

seeks the host root. On contact with the host, the parasite attaches via a structure known as an appressorium. Then, through a combination of enzymatic digestion of the root cell wall and pressure, the parasite penetrates the host root. There is evidence for the secretion of cellulase, xylanase, pectin methylesterase and polygalacturonase (Ben-Hod *et al.*, 1993; Singh and Singh, 1993). Once the host tissue is penetrated, a tubular link into the host xylem is formed, known as a tubercle, and through this structure the parasite obtains water and nutrients from the host. Shortly after contact with the host vascular tissue, the plant begins to form a shoot. The leaves are always small and scale-like. Although the plants lack chlorophyll, most species have some form of pigmentation and their colour can range from pale pinkish purple to yellow to bright red. The shoot forms a short-lived inflorescence. Most Orobanchaceae are annuals and depend upon seed set for the next generation.

The Orobanchaceae are rarely agricultural pests. The most serious pest is *O. cernua*, which parasitizes sunflowers in Eastern Europe. It is estimated that an infestation of this parasite can cause a fall in productivity of up to 30%. *O. crenata* and *O. aegyptiaca* are also pests of potato, tomato and tobacco crops in North Africa (Linke *et al.*, 1989).

Examples of other plant species that are parasitic on dicotyledonous plants are shown in Figures 7.15–7.18.

Parasitic species other than the dicotyledons

Curiously, all parasitic plants are dicotyledons and no parasitic plants are represented in other plant divisions. There is only one questionable exception, and that is a gymnosperm from the family Podocarpaceae, *Parasitaxus ustus*. This is a rare species, found only on the island of New Caledonia. *Parasitaxus* is a small woody shrub that can reach around 1.5 m in height. The plant has no root system and produces purple scale leaves, which resemble those in the Orobanchaceae. It is believed to parasitize

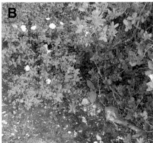

Figure 7.16 Parasitism of garrigue species. (A) Typical habitat on the Serra da Arrabida, Portugal of *Cytinus hypocistis*. This plant is a perennial parasite of the family Rafflesiaceae (related to the largest of the parasitic plants, *Rafflesia arnoldii*, which bears the largest flower of any plant). It is parasitic on *Cistus* species (B) in the Mediterranean garrigue or machia. (C) Detail of the inflorescence.

Figure 7.17 (A) Typical habitat of *Hyobanche sanguinea*, the inkplant (family Orobanchaceae) on the Cape Peninsula Nature Reserve, which forms part of the Fynbos floral kingdom. Over 70% of the species in this region are found nowhere else in the world. The flora in the area is mainly made up of fine-leaved Proteaceae and Ericaceae. The five members of the genus *Hyobanche* possess the most striking appearance of all the holoparasitic plants. (B, C) Details of the inflorescence.

Figure 7.18 *Conopholis americana* holoparasite, photographed in the Blue Ridge Parkway (North Carolina, USA). *C. americana* is a perennial member of the family Orobanchaceae, frequently called cancer root or squawroot. It is a parasitic plant found in hardwood forests (A, B) of Eastern North America and is only found to parasitize the red oak species *Quercus borealis*. It flowers in spring (May–June) and the inflorescence (C) reaches around 25 cm in height. It is usually found under oak or beech trees, which are used as a host. The plant completely lacks chlorophyll and produces scale-like leaves.

Falcatifolium taxoides, a close relative of *Parasitaxus ustus*, and is only found attached to the root system of this plant. However, there is some confusion over whether *Parasitaxus* actually is parasitic. It has been noted to attach to the roots of *F. taxoides* through a system that links the xylem of the partners (Köpke *et al.*, 1981). This link is described as a 'root graft' rather than the usual haustorium formation in other parasitic relationships. To add to the confusion, there is evidence of a vesicular arbuscular mycorrhizal association between a fungus and the two plant species (Woltz *et al.*, 1994). Thus *Parasitaxus* is not a true parasitic plant in the sense that the other dicotyledon species listed earlier are.

Parasitism cannot be ignored as a means of plant nutrition. Although parasitic plants make up only around 1.5% of flowering plant species, several of them are parasites of important crop plants. *Striga*, for example, can cause major crop failures and ultimately starvation in poor countries, making it one plant whose parasitic life

style has a significant effect on food availability and human survival. It is therefore crucial that we understand this fascinating means of nutrition in plants as well as possible, to be able to combat its devastating effects in some crops.

References

Ben-Hod G, Losner D, Joel DM and Mayer AM (1993) Pectin methylesterase in calli and germinating seeds of *Orobanche aegyptiaca*. *Phytochemistry*, **32** (6), 1399–1402.

Chang M and Lynn DG (1986) The haustoria and the chemistry of host recognition in parasitic angiosperms. *Journal Chemical Ecology*, **12**, 561–579.

Cook CE, Whichard LP, Wall ME, Egley GH, *et al.* (1972) Germination stimulants. II. The structure of strigol – a potent seed germination stimulant for witchweed (*Striga lutea* Lour.). *Journal of the American Chemical Society*, **94**, 6198–6199.

Cyr H and Pace ML (1993) Magnitude and patterns of herbivory in aquatic and terrestrial ecosystems. *Nature*, **361**, 148–150.

Ehleringer JR and Marshall JD (1995) Water relations, in *Parasitic Flowering Plants*, (eds MC Press and JD Graves), Chapman and Hall, London, pp.125–140.

Hibberd JM and Jeschke WD (2001) Solute flux into parasitic plants. *Journal of Experimental Botany*, **52**, 2043–2049.

Hughes M and Headley A. (1996) The biology and ecology of the thistle broomrape, *Orobanche reticulata* Wallr. *Naturalist*, **121**, 3–10.

Köpke E., Musselman LJ and De Laubenfels DJ (1981) Studies on the anatomy of *Parasitaxus ustus* and its root connections. *Phytomorphology*, **31**, 85–92.

Linke KH.Sauerborn J and Saxena MC (1989) *Orobanche Field Guide*, ICARDA-FLIP, Aleppo, Syria. p. 42.

Lynn DG and Chang M (1990) Phenolic signals in cohabitation: implications for plant development. Signals mediating symbioses (*Rhizobium*) and parasitism (*Agrobacterium*, witchweed) are discussed. *Annual Review of Plant Physiology and Plant Molecular Biology*, **41**, 497–526.

Nickrent DL, Duff RJ, Colwell AE, Wolfe AD, *et al.* (1998) Molecular phylogenetic and evolutionary studies of parasitic plants, in *Molecular Systematics of Plants II. DNA Sequencing*, (eds DE Soltis, P.S. Soltis and Doyle JJ Kluwer Academic, Boston, MA, pp. 211–241.

Press MC, Scholes JD and Watling JR (1999) Parasitic plants: physiological and ecological interactions with their hosts, in *Physiological Plant Ecology* (eds M.C. Press, JD Scholes and M.G. Barker), Blackwell, London, pp. 175–197.

Singh A and Singh M (1993) Cell wall degrading enzymes in *Orobanche aegyptiaca* and its host *Brassica campestris*. *Physiologia Plantarum*, **89** (1), 177–181.

Stewart GR and Press M.C. (1990) The physiology and biochemistry of parasitic angiosperms. *Annual Review of Plant Physiology and Plant Molecular Biology*, **41**, 127–151.

Sugimoto Y, Wigbert SCM, Thuring JWJF and Zwanenburg B (1998) Synthesis of all eight sterioisomers of the germination stimulant sorghlactone. *Journal of Organic Chemistry*, **63**, 1259–1267.

Tennakoon KU, Pate JS and Fineran BA (1997) Growth and partitioning of C and fixed N in the shrub legume *Acacia littorea* in the presence or absence of the root hemiparasite *Olax phyllanthi*. *Journal of Experimental Botany*, **48**, 1047–1060.

Wolswinkel P (1974) Complete inhibition of setting and growth of fruits of *Vicia faba* L. resulting from the draining of the phloem system by *Cuscuta* species. *Acta Botanica Neerlandica*, **23**, 48–60.

Woltz PR, Stockey A., Gondran M. and Cherrier JF (1994) Interspecific parasitism in the gymnosperms: unpublished data on two endemic New Caledonian Podocarpaceae using scanning electron microscopy. *Acta Botanica Gallica*, **141** (6–7), 731–746.

8

Carnivorous plants

Carnivorous plants are often viewed as being a minor group of unusual but exceptional higher plants; as such they can be regarded as being of very little relevance to the rest of plant physiology. But this view is in danger of dismissing one the most fascinating and highly adapted plant groupings. At the functional level, these plants are no different from other higher plants, but in addition they possess elaborate trapping mechanisms which have evolved from alterations in leaf structure and trichome and glandular appendages. This gives these plants their unusual structural appearance. They also produce enzymes, just like every other higher plant, capable of digesting proteins, phosphate esters, lipids, and carbohydrates. However, what is unusual is that carnivorous plants can secrete these enzymes from glands onto the leaf surface. This behaviour underpins plant carnivory and is the essential characteristic of a carnivorous plant. The nutrients released from animal prey can then be taken up through the leaf surface and assimilated into the plant. The latter characteristic is another common feature of higher plants. Plants can take up soluble nutrients placed onto a leaf surface, provided there are no barriers to solute movement such as the leaf cuticle. This means that if an insect dies on the surface of any leaf or near a root system, and solutes leak out of the carcass through autolysis or fungal and bacterial decay, then these nutrients can be accessible to the plant through the leaf epidermis or the root system. All plants will act as scavengers, given the opportunity. Thus, the distinctions between a carnivorous and a non-carnivorous plant are in reality blurred.

Carnivory: the search for a definition

What is a carnivorous plant? This blurring of a distinction between carnivorous and non-carnivorous plants presents a problem. Few would argue against the strictest definition, that a carnivorous plant needs to be able to:

- Attract animal prey.

- Capture, immobilize or kill prey.

- Digest animal prey.

However, it quickly becomes clear that the last statement, concerning digestion, needs to be further qualified. There are some plants within the carnivorous plant grouping that do not produce their own enzymes to digest prey. These plants appear to use a surrogate organism to carry out the digestion for them, such as a bacterium, fungus, insect or arachnid. These plants have been grouped into a second band of plants, referred to as sub-carnivorous plants. This does not make these plants deficient in any way, since in many environments it may be a positive advantage to form such a symbiotic relationship with other organisms. In addition, it could be a waste of vital nitrogen to make proteins to digest prey, particularly under circumstances where the secreted enzyme was in danger of being washed away, for example. To allow plant carnivory to be lost at this point in the technicalities of a definition would be a mistake. A list of the major genera of carnivorous plants and their ranking is given in Table 8.1.

Why have some plants turned to carnivory?

Almost without exception, the 16 genera of carnivorous plants all inhabit nutrient-poor soils (Figure 8.1). Generally, this takes the form of acidic boggy areas that contain low amounts of nitrogen and phosphorus and high tannin levels. The actual nutrient content of these soils is not low, but the presence of acidic tannic waters prevents bacterial action in the soil and leads to the nutrients being maintained in an insoluble form; thus, the environment is very low in *free*

Physiology and Behaviour of Plants Peter Scott
© 2008 John Wiley & Sons, Ltd

Table 8.1 The extent of carnivory in plant species described as being carnivorous.

Species	Trap mechanism	Sub-carnivorous		Carnivorous	
		Low	Medium/low	Medium/high	High
Albany pitcher plant	Pitfall trap				✓
Tropical pitcher plant	Pitfall trap			✓	✓
North America pitcher plant	Pitfall trap		✓	✓	✓
Cobra lily	Pitfall trap		✓		
Marsh pitcher plant	Pitfall trap		✓		
Bromeliads	Pitfall trap	✓			
Venus fly trap	Spring mantrap				✓
Waterwheel plant	Spring mantrap				✓
Bladderwort	Spring suction trap			✓	✓
Corkscrew plant	Lobster pot trap				✓
Sundew	Sticky flypaper				✓
Butterwort	Sticky flypaper				✓
Portugese dewy pine	Sticky flypaper				✓
Rainbow plant	Sticky flypaper		✓		
Roridula	Sticky flypaper		✓		
Devil's claw	Sticky flypaper	✓			

The terms are defined as follows: High, plant is truly carnivorous, displaying prey attraction, killing and digestion; Medium/high, plants are truly carnivorous but there is evidence of a role for bacterial and fungal involvement in prey digestion; Medium/low, the plant uses a second surrogate organism for prey digestion but does possess a modified leaf structure for prey capture; Low, the plant uses a second surrogate organism for prey digestion but does not possess a substantially modified leaf structure for prey capture. Based on Juniper *et al.* (1989).

nutrients for plants. For any plant to thrive under such conditions, it would need either a very low requirement for nitrogen and phosphorus or a means of obtaining these elements via an alternative route. Carnivorous plants have evolved to fill this niche. Through the intriguing and fantastic evolution of their leaf structure, carnivorous plants have developed traps to lure and capture a variety of organisms. It is through the capture and digestion of these organisms that the nutrient deficit in the soil is compensated for. In the correct environment, these plants can be very successful. However, due to growing nutrient pollution in boggy areas as a result of the expansion of agriculture, the native habitat for these plants is becoming increasingly threatened, and many of these species are becoming very rare indeed.

Worldwide, there are around 600 different species of carnivorous plant. This number may seem large, but it probably reflects more the intense interest in the group as a whole than the absolute necessity to divide the plants into separate species. The trapping mechanisms of these plants can be subdivided into four broad groups:

- Pitfall trap.

- Active spring trap.

- Sticky fly paper.

- Lobster pot trap.

It is beyond the scope of this book to discuss all of the different genera of carnivorous plants in detail. The major genera have been listed in Table 8.1 and individual photographs of specimens are shown in Figure 8.1. Further discussion in the remainder of this chapter will be limited to examples of the first three trapping mechanisms, as the last mechanism is only used by a single genus (*Genlisea*).

Trapping mechanisms

The pitfall trap – North American pitcher plant

The pitfall trap is the most basic of the trapping mechanisms of carnivorous plants. In most instances it can be viewed as a rolled-up leaf, sealed down one side, into which flies tumble. Only in *Nepenthes* and *Cephalotus* is the trap development much more complex. As an illustration of the features of a pitfall trap, the North American pitcher plant (*Sarracenia*) will be used. There are eight different species in this genus (Figure 8.2). The traps can be subdivided into three further

Figure 8.1 Sixteen genera of carnivorous plants: (A) yellow pitcher plant; (B) cobra lily; (C) Venus flytrap; (D) greater bladderwort; (E) pygmy sundew; (F) Albany pitcher plant; (G) marsh pitcher plant; (H) Mexican butterwort; (I) tropical pitcher plant; (J) bromeliad; (K) rainbow plant; (L) water wheel plant; (M) corkscrew plant; (N) devil's claw plant; (O) Portuguese dewy pine; (P) *Roridula*.

groups: trumpet pitchers (*S. rubra*, *S. alata*, *S. oreophila*, *S. flava*, *S. leucophylla* and *S. minor*), reclined pitchers (*S. purpurea*) and lobster pot pitchers (*S. psitticina*). They all use a slightly different means of capturing insects, but many generalizations can be made. For this discussion, *S. rubra* ssp. *alabamensis* will be used to illustrate how pitcher plants catch prey (Figure 8.3).

Figure 8.4 shows a cross-section of a trumpet pitcher trap. The trap can be functionally divided into four zones (Slack, 2000). Zone One defines the regions on the trap that secrete nectar, which serves to attract insects to the pitcher. The nectar can also be associated with a scent that may also function as an attractant. Glands on the leaf surface secrete the nectar. This secretion can contain a

Figure 8.2   The different species of North American pitcher plant (*Sarracenia* spp.) and the cobra lily. (A) Yellow pitcher plant (*S. flava*); (B) purple pitcher plant (*S. purpurea*); (C) pale pitcher plant (*S. alata*); (D) hooded pitcher plant (*S. minor*); (E) white pitcher plant (*S. leucophylla*); (F) green pitcher plant (*S. oreophila*); (G) parrot pitcher plant (*S. pscitticina*); (H) sweet pitcher plant (*S. rubra*); (I) cobra lily (*Darlingtonia californica*).

range of sugars but sucrose is the dominant carbohydrate. In addition, amino acids have also been detected in the secretions in some pitcher plant species (Dress *et al.*, 1997). The secretion can contain between 1 and 3 M sucrose, thus it is usually very sticky (the actual concentration varies greatly, depending upon humidity and the age of the trap). This forms the bait for the trap and the range of components in the nectar has a great bearing on the target prey. The poison coniine has also been detected in the nectar of *S. flava* (Mody *et al.*, 1976). This toxin acts as a neurotoxin or nerve poison, attacking the motor function of the insect. Any insect feeding for a period of time on the traps is likely to become more and more intoxicated and eventually fall into the trap opening. In

addition, a further attractant to insect prey is that the trap lid and opening are often highly coloured and may act to mimic the appearance of a flower.

In Zone Two, the secretion of nectar ceases. The leaf surface here is exceedingly slippery, although under a microscope the leaf surface looks surprisingly rough (Figure 8.4). Regular patterns of cone-shaped shaped cells, which resemble teeth, protrude from the epidermal surface. One needs to understand a little about the physiology of fly feet in order to understand how this surface challenges an insect's footing. Flies possess two forms of grip on their feet, three hooks and two suction cups covered with fine hairs. The insects use these two means to negotiate a range of surfaces, from rough to very smooth. However, as a

Figure 8.3 North American pitcher plant (*Sarracenia rubra* ssp. *alabamensis*): (A) the upright trumpet pitcher trap; (B) fly's eye view of the pitcher opening; (C) dead insect remains at the bottom of the pitcher; the remains usually consist of the chitinous shell of the insect. Here a blowfly larva is seen living scavenging a living within these remains. In this form the larva is unaffected by the digestive juices, but it is unlikely ever to escape the trap.

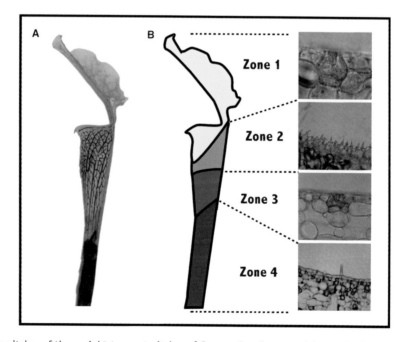

Figure 8.4 Zones on pitcher of the upright trumpet pitcher of *Sarracenia rubra* ssp. *alabamensis*. (A) Cross-section of pitcher. (B) Colour coding of the zones within the pitcher: zone 1, nectar secretion zone; zone 2, slippery surface (d); zone 3, digestive glands secrete enzymes into the pitcher (E); zone 4, the lack of a cuticle makes the lower region of the pitcher ideal for the absorption of the nutrients resulting from prey digestion (F).

result of the dimensions of the conical cells, the suction cups on an insect's feet can no longer grip effectively. As the fly feeds from the nectar on the pitcher, it inevitably hunts for the highest density of the secretion, which is near to Zone Two of the pitcher. Ultimately the insect will move towards the slippery region and, whether through greed or intoxication, it slips into the trumpet of the pitcher and enters either Zone Three or Zone Four.

The mouth of the trumpet tends to flare outwards. It is wide at the top but narrows rapidly. Thus, a falling fly has little opportunity to get its wings into action and escape death. The fly then either jams in Zone Three or falls all the way to Zone Four. This depends on the narrowness of the pitcher and the size of the insect. In Zone Three there are secretory glands that secrete enzymes into the pitcher tube (Figure 8.4). They appear as dew on the side of the pitcher and dribble down into Zone Four. If an insect is trapped in Zone Three, the enzymes will dribble down to it and begin to act on the body. If the insect is still alive, then the narrowness of the pitcher and the slipperiness of the sides ensure that there is no escape. A trapped fly can last up to 3 days in the trap before finally starving or dying as a result of the degrading enzymes. The enzymes released in this section of the trap include:

- Proteases, for the release of nitrogen from proteins.

- Phosphatases, for the release of phosphate from various organic molecules.

- Invertase, for the breakdown of sucrose (possibly present to mobilize sugars being carried down the trumpet from Zone One).

- Lipases, for the mobilization of fat reserves.

The function of Zone Four is to assimilate the soluble nutrients released as a result of digestion. This region is at the base of the trap and usually contains a small amount of fluid. If the insect falls down the pitcher to this region, then it will drown. The layer of epidermis on the leaf at this point does not bear any cuticle. This facilitates the rapid diffusion of nutrients into the plant. In addition, in this region of the trap there are fine hairs across the leaf surface. Most books on carnivorous plants argue that these hairs assist in keeping the insect in the trap (D'Amato, 1998; Pietropaolo and Pietropaolo, 1986; Slack, 2000). However, this is unlikely, as later in the season the mountain of flies that builds up in the trap will easily rise above the level of these hairs, yet the insects

still cannot escape the pitcher. In some pitchers, such as *S. psitticina*, the hairs do help hold the prey but in the upright pitchers the hairs may have a function for enlarging the surface area of the leaf for the uptake of nutrients. In many respects, Zones Three and Four can be regarded as the stomach of the pitcher plant. The sequence of events in prey capture are shown in Figure 8.5.

The *Sarracenia* pitcher plants stand out from the rest of the carnivorous plants as being the most effective at prey capture. The trumpets of *S. flava* and *S. alata* can reach over a metre in height and yet at the end of a growing season they can be literally stuffed with dead insects, crustaceans and possibly amphibians. In the summer the plants can often be heard to `buzz' with the weight of trapped flies! These plants have a voracious appetite and are ruthless killers. The image of an insect falling into a open pit does not initially sound interesting, but once viewed these plants possess a fascination all of their own. They are also easy to grow for experimental demonstrations and many species will even grow outside in colder temperate climates.

Active spring trap – bladderworts

There are three genera of carnivorous plants that possess spring traps. The most well-known has to be the Venus flytrap (*Dionaea muscipula*) and no carnivorous plant captures the imagination better than this species. Charles Darwin, in his book *Insectivorous Plants*, described *Dionaea* as `the most wonderful plant in the world' (Darwin, 1875). Unfortunately it is monotypic (only a single species in the genus) and the only close relative is *Aldrovanda vesiculosa*, the Polish waterwheel plant. Both of these species are very rare in the wild and are seriously threatened by eutrophication (nutrient pollution) of their habitat. The action of the trap mechanism of the Venus flytrap will be covered in Chapter 08. Here in this section the trap mechanism of the bladderwort (*Utricularia*) will be investigated in greater detail.

There are over 200 different species of bladderwort known, which is around 40% of all species of carnivorous plants. They have the broadest range of any carnivorous plant, being found from Australia to Alaska. The genus is very diverse and can be subdivided into aquatic (27%), terrestrial (60%) and epiphytic (13%) bladderworts (Figure 8.6).

The aquatic species *Utricularia vulgaris* was the first plant in this genus to be described. Initially it was thought that the bladders attached to the plants were a means of flotation for the plant. In many of the older bladders there are indeed air bubbles, which could act to keep the plant

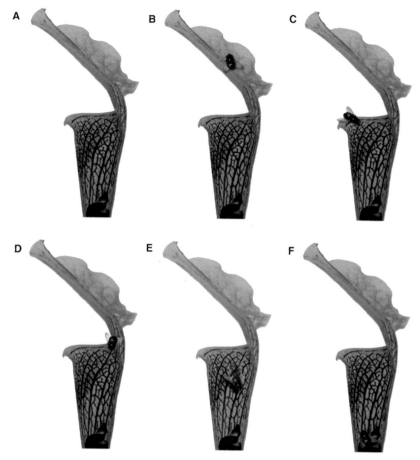

Figure 8.5   Fly capture by *Sarracenia rubra* ssp. *alabamensis*. (A) The hood of the trap is covered with secretory glands, producing a nectar secretion high in carbohydrates. (B) Flies are attracted to the pitcher through the offer of carbohydrate. (C) The fly investigates the pitcher and, finding that sugar secretion increases near to the opening of the trumpet, ventures there. (D) Gradually the fly works its way to the slippery region of the pitcher. (E) On the slippery region, the fly falls into the pitcher and is unable to recover before it is trapped. (F) Due to the trap structure, the fly cannot fly out of the trap and the slippery sides prevent climbing. The fly then starves, dies and is digested.

afloat (Figure 8.7). In the younger bladders these are absent. It is from these bladders that the plant received its name, *Utricularia*, which is derived from the Latin *utricularius*, meaning master of a raft supported by bladders. However, it is unlikely that the traps are in fact a means of flotation, since the young shoots seem perfectly able to float at the water surface despite the lack of air bubbles in the bladders. In addition, in terrestrial bladderworts, where the bladders are buried in the soil around the plant, bubbles can also be found in these bladders, although in this instance there is no possibility for the air-filled bladders to act a buoyancy agents. Even in 1875, Darwin failed to see their true nature in his book *Insectivorous Plants*, but an American botanist, Mary Treat, first proposed in 1876 that the plants might be carnivorous in nature.

Bladderworts are found in nutrient-deficient boggy areas. However, occasionally some species are known to inhabit pools of water within other plant species, such as bromeliads and pitcher plants. In their environment, the bladders are known to capture small crustaceans, copepods, rotifers, ostracods, mosquito larvae and even small tadpoles. The capture of larger prey requires several resets and triggering of the trap for the prey to be completely engulfed. The structure of the actual plant is highly unusual. They have no root system but instead have specialized leaves, bladders and stolons that branch from a central stem. In terrestrial and epiphytic bladderworts, all that is visible of the actual plant is a series of single leaves that rise out of the peat. Underground is a series of stolons, which branch upwards to form further leaves or downwards to form a series of small bladder traps in the soil. The

Figure 8.6 Photographs of aquatic and terrestrial *Utricularia* plants. (A) Detail of growing shoot of *U. vulgaris* and aquatic bladderwort species. (B) Detail of bladder trap of *U. vulgaris* filled with copepod prey. (C) Leaves of *U. livida*, a terrestrial bladderwort species. (D) Detail of bladder trap of *U. livida*.

bladder traps can vary in size from 200 μm to 10 mm. The larger traps are usually borne by the aquatic species. The plants invest a great deal of energy in the formation of the traps, with over 40% of the dry weight of some aquatic species being made up of bladders (Friday, 1992).

To understand how the bladder traps function, their structure must be investigated. The traps are generally disc-shaped and are made up of two layers of cells (Pietropaolo and Pietropaolo, 1996). In aquatic species these cells are photosynthetic. The walls to the traps are elastic. At one end of the disc there is an opening, which is covered by a hinged piece of tissue that acts as a doorway, sealing the inner chamber of the bladder from the solution outside. On the doorway there are a series of trigger hairs (usually four), and a filamentous network of tissue known as antennae or guiding hairs surrounds this doorway (Figure 8.8A and B). Within the chamber of the bladder, the inner surface is covered with quadrifid glands (Figure 8.9B). Just under the threshold of the opening there is another series of glands, known as bifid glands (Figure 8.9C). These glands are thought to be involved in the setting of the trap and the absorption of the digested contents of prey.

The essence of the trapping mechanism is the creation of a low pressure in the bladder through the pumping out of water from the inner chamber. The likelihood of prey capture is enhanced by the presence of the guiding hairs around the doorway. This has been demonstrated experimentally through the removal of hairs from around the trap. Traps treated in this way were much less efficient at capturing prey than those still possessing the hairs. In terrestrial species the guide hairs tend to be shorter and obscure the doorway. In this case, the hairs will also have the function of preventing particulate matter from being sucked into the bladders. It is not known whether the bladders have any means of attracting the prey to the trap in the first place, or whether the organisms just trigger the trap by accident. The trap is triggered by an organism coming into contact with the trigger hairs at the doorway to the trap. These hairs are stiff and, on being touched, cause the doorway to move slightly, and it is this action that opens a channel between the low pressure within the chamber of the bladder and the outside solution. The sequence of events in prey capture is shown in Figure 8.10. Water rushes into the chamber and the volume can swell by up to 60% (Figure 8.11). This whole action can occur within

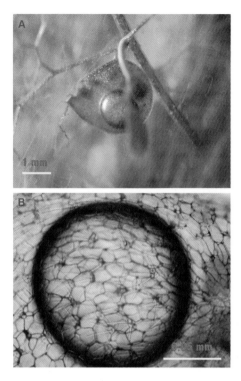

Figure 8.7 (A) Photograph of an air bubble, clearly visible in the centre of a *Utricularia vulgaris* bladder trap. Air bubbles will block the trap opening and reduce the capacity of the quadrifid glands to pump water and secrete digestive enzymes. (B) Detail of the trapped bubble.

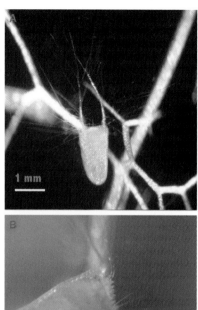

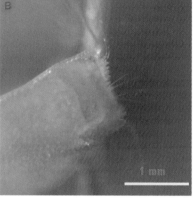

Figure 8.8 Detail of bladder guide hairs and trigger hairs, showing (A) the guide hairs (antennae) of *Utricularia minor*, and (B) the trigger hairs on a bladder of *U. vulgaris*.

2 milliseconds – possibly the fastest movement of any plant system. The movement of water into the trap pulls the doorway fully open and any organisms in the near vicinity of the mouth of the trap are sucked into the central chamber. Once the pressures equalize, the doorway springs shut and the prey is captured and held in the chamber (Figure 8.12). It is at this point that the chamber then becomes a stomach for the plant.

Once triggered, the trap rapidly resets and water is pumped out of the chamber, although it is not yet known how this is achieved, since plants lack active valves to pump water from the bladder chamber. During triggering of the trap, the tense walls expand outwards, creating the suction, but there is no reverse mechanism in the walls to expel the water again. The most likely hypothesis is that the water must pass through the cells of the chamber walls and this could involve the bifid glands. If water channels allowed water into the wall cells through osmosis, then the inner chamber could be evacuated. However, then there remains the problem of how the wall cells get rid of this additional water. Since water cannot be actively pumped from the cells, it must leave the cells via osmosis.

This could only be achieved through the creation of an ion or sugar gradient across the outer membrane of cells lining the chamber. Such a gradient would draw water from the cells via osmosis. The most likely sites for this movement are the bifid glands, as their structure acts to increase the surface area of the inner layer of cells exposed to the digestive chamber. There are still a number of complications to this model. For plants in a largely aquatic environment, how could an ion gradient be maintained between the bladder wall cells and the outside solution? One possibility is that the gradient is created between the cell layers of the bladder and helps reduce the rate of diffusion of solutes away from the bladder trap. However, this would mean that the outer layer of cells within the wall would have to be leaky to prevent a build-up of pressure in the intercellular space. The resetting procedure is very fast and can occur within 40 minutes. The trap can be sprung many times, as is evident in older bladder traps that contain over 15 organisms.

Once an organism has been trapped, digestive enzymes are secreted into the chamber of the bladder. The quadrifid

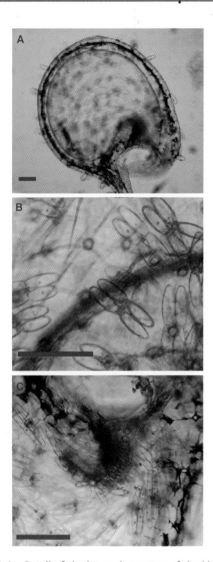

Figure 8.9 Detail of the internal structure of the bladders of *Utricularia alpina*, an epiphytic bladderwort species: (A) detail of bladder trap; (B) detail of quadrifid glands on the inner surface of the bladder chamber; (C) threshold of trap, showing bifid glands. Bars represent 50 μm.

small communities of organisms that live within the traps. It is unclear to what extent the plant and the organisms benefit from the relationship, but it is possible that the plant imprisons the invertebrates in the bladder traps and then feed off nutrients that are sucked into the bladder chamber each time the trap is triggered. The plant may then benefit from the excretory products of the organisms without having to produce digestive enzymes of its own.

Given the diversity of this genus, the bladderworts are without doubt one of the most successful of the carnivorous plants, colonizing both land and water. In cultivation the epiphytic and terrestrial species are usually very easy to grow. However, since most of the carnivorous action occurs microscopically and underground, it is often difficult to see their remarkable prey capture ability. The plants flower readily and are often grown more for their flowers than their carnivorous ability. The aquatic species tend to possess larger traps than the other groups of bladderworts and thus their trapping mechanism is easily visible, even without the aid of a microscope. However, these species are remarkably sensitive to competition by algae. This probably reflects the ability of algae to use *Utricularia* as a support for growth in the water. This leads to the plant gradually becoming overgrown with alga cells, and then the alga effectively overcomes the bladderwort in competition for the limited carbon dioxide reserves in the water.

Sticky flypaper – sundews

The use of sticky glue on plant surfaces is the most frequently used mechanism for plants to capture flies. Most species that use this method of animal capture use a modified trichome that protrudes from the leaf surface and bears a small droplet of mucilage. This mucilage is made up of a complex mixture of polysaccharides dissolved in water (Rost and Schauer, 1977). The size of prey that can be captured by the traps depends on the extent and stickiness of the mucilage. The only exception to this is the species of the genus *Roridula*, which use a sticky resin. In this section the trap mechanism of the sundews (genus *Drosera*) will be investigated in greater detail.

There are known to be around 150 different species of sundew, making this genus the second largest of the carnivorous plants. A selection of different *Drosera* species is shown in Figure 8.13. Although the genus is widespread around the world, over half of these species can only be found in Australia. The trapping mechanism of *Drosera* stands out from the other sticky flypaper-type carnivorous plants, because in sundews the tentacles that bear the mucilage can move in response to contact with

glands are thought to secrete protease and phosphatase enzymes into the bladder and these act to digest and kill their prey. Once amino acids and phosphates are released, they are rapidly taken up by the plant and mobilized to the growing shoot. However, no studies have yet established the extent to which bacterial action contributes to the digestive process.

In bladder traps of *U. vulgaris*, the captured invertebrates are always dead. However, very frequently in the bladder traps of certain species this is not so and there is now increasing evidence that some bladderwort species, including *U. livida*, *U. minor*, and *U. praelonga*, support

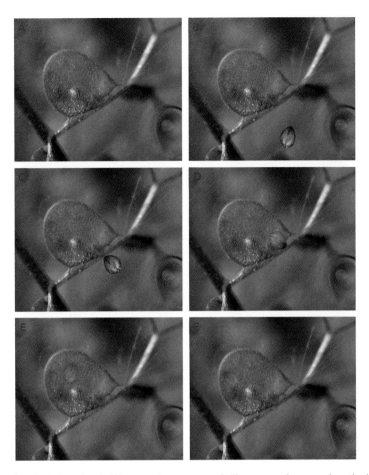

Figure 8.10 Photograph of *Utricularia vulgaris* (A) capturing a copepod. The copepod approaches the bladder trap (B), either by accident or through a chemical attractant. The prey touches the trigger hairs (C) and causes the doorway to the trap to open. The vacuum within the trap causes water to flood into the bladder chamber, carrying the copepod with it (D). The captured prey is then digested over several weeks (E, F).

prey. Furthermore, many species can also fold their leaves around prey to make a stomach for the absorption of nutrients. These abilities make the sundews a fascinating genus of carnivorous plants to study.

The shape of the leaves of *Drosera* varies between the different species but the general form differs little. Usually there is a thin petiole to the leaf that broadens at the end (Figure 8.14). At this point numerous tentacles protrude from the leaf surface; the tentacles tend to be longer at the outside edges of the leaves. The tentacles are arranged such that they fan out from the leaf surface to provide the maximum area for prey capture (Figure 8.14A). Each tentacle usually bears a small red head, which is covered by a dew drop of mucilage (Figure 8.14B–D). It is unclear why prey are attracted to *Drosera* plants, since no nectar secretion has been detected on the leaves. Suggestions have ranged from olfactory signals to the dewy appearance

of the plant. Whatever the case, insects are attracted in large numbers. Once caught, the insects struggle to free themselves. If the insects are large, the mucilage will almost certainly not be able to hold them and they will escape. However, smaller insects are not strong enough to escape the plants, and the more they struggle the more mucilage is generated by the plant. This mucilage is thick and syrupy and rapidly coats the insect as it frantically struggles and comes into contact with further tentacles. This mucilage then smothers and kills the prey by blocking the spiracles necessary for the insect's respiration (Figure 8.15). During the struggle the insect will touch the tentacle heads. In addition, the insect may excrete bodily fluids in order to effect its escape. Both of these act as signals to cause the tentacles to move.

This movement has to be one of the wonders of the plant kingdom. Few plants, as we will see later, can move in a

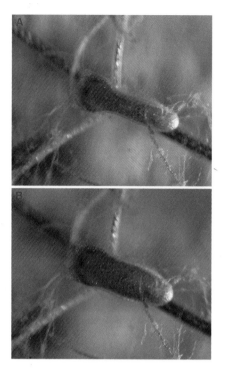

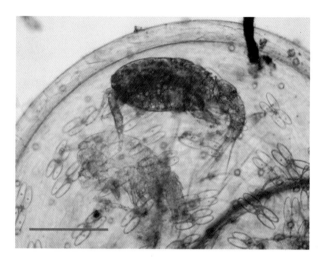

Figure 8.12   Photograph of captured crustaceans in *Utricularia reniformis* bladder trap. Close-up detail of two crustaceans within a *U. alpina* (epiphytic bladderwort species) bladder. The quadrifid glands are clearly visible on the inner surface of the bladder chamber. Bars represent 100 μm.

Figure 8.11   Plant view of in *U. vulgaris* bladder trap being triggered. Bladder trap before (A) and after (B) touching the trigger hairs and releasing the trap mechanism.

Figure 8.13   Types of sundew plants (*Drosera* spp.) from around the world: (A) rosette-forming temperate species *D. curvispata*; (B) the Australian pygmy species *D. nitidula* – pygmy *Drosera* spp. are renowned for the small size of the plants (some measuring less than 5 mm); (C) the tropical species *D. binata*; (D) the tuberous species *D. peltata*.

Figure 8.14 Detail of *D. rotundifolia* leaf and tentacles: (A) close-up of fully formed *D. rotundifolia* leaf; (B) tentacles of *Drosera* close up; (C) tentacle bearing dew drop of mucilage; (D) magnified image of *Drosera* tentacle head.

time frame to be observed without time-lapse photography. The fastest recorded movement for a *Drosera* species is for the initiation of tentacle bending to occur within 10 seconds of being stimulated. Without doubt the fastest species to observe is *D. burmanii*. Thus, an observer can easily witness the tentacle movement. The direction of movement of the tentacles varies, depending upon where on the leaf the tentacle is. Little bending is seen on tentacles at the centre of the leaf. However, tentacles on the outside of the leaf will bend towards the leaf centre. This brings struggling prey into contact with further tentacles and closer to the leaf surface. A fly captured on a *D. indica* leaf is shown in Figure 8.15.

The question of how the tentacles move has not yet been fully answered (Hill and Findlay, 1981). Certainly, the tentacles of most species of *Drosera* will only bend in the direction of the leaf centre. The stimulation of the tentacle head with either a mechanical stimulation (insect struggling) or a secretion of a nitrogenous compound (this could result from insect defecation or urea excretion) causes the tentacle to grow. This growth is created not by cell division but by cell expansion (Muravnik, 2000; Ragetli *et al*,

1972). Cells on one side of the tentacle (the side furthest away from the leaf centre) expand relative to near side (the side closest to the leaf centre) and cause the tentacle to bow. There is some evidence that the movement is caused by a loss of water from the cells on the near side of the tentacle and a gain of water by the tentacle on the far side. As has been mentioned earlier, the movement can be extremely rapid and is far too fast a process to be carried out by a compound such as an auxin. Experiments have shown that placing tentacles in the presence of a dilute acid can cause tentacles to bend in the time frame observed; thus, acidification of the apoplastic space may cause the cells on the far side to expand in response to the increased water availability. The cells on the near side of the trap must lose water as a result of a change in the ion environment within and outside the cells.

The trigger for the whole process appears to be an action potential developed in the head of the tentacle which, once developed, passes down the tentacle (Shimmen, 2001; Williams and Pickard, 1972a,1972b). This could then trigger the opening of ion channels in the near-side cells, resulting in a loss of turgor of these cells and an increase in

Figure 8.15 *Drosera indica* leaf and captured fly. (A) The species *D. indica* produces thin lanceolate leaves and bears no red pigment, as many sundews do. (B) Upon prey capture, the tentacles bend to capture and draw the prey to the leaf surface. However, the leaves do not bend to envelope the prey, as in other sundew species.

ion uptake in the cells on the far side of the tentacle, which would result in an increase in turgor through water movements. The regions of the tentacle that tend to be affected by the bending response are upper and middle sections. If the stimulation of the tentacle head is intensive enough, then neighbouring tentacles may bend without any apparent contact with the captured prey. The action potentials have never been observed to be conducted across the leaf surface and it is not understood how the signal is passed across the leaf surface. However, the other tentacles bend in an identical manner to the stimulated tentacle; thus, the signal must be relayed to the head which, in turn, causes the turgor changes that result in the movement.

Once the prey is subdued and suffocated, the plant needs to overcome one last obstacle. Since *Drosera* species inhabit wet environments, it is likely that if rain fell during the digestive process the secreted enzymes would be washed away. This would be a costly loss of nitrogen for a carnivorous plant. In response to this

problem, many *Drosera* species have the ability to bend the whole leaf surface to wrap around the prey or form a pocket (Figure 8.16). This would act to produce a stomach for the leaf, which would increase the surface area of the plant exposed to the insect prey whilst also protecting the digestion process from rain. This leaf movement process is much slower than the tentacle bending and can take 48 hours to complete. Evidence suggests that this response occurs through the release of auxin to the lower surface of the leaf, where cells expand and cause the leaf to fold over. The auxin release accumulates where the prey is on the trap and thus the trap folds around the prey. As the trap folds, more of the prey is placed in contact with the leaf surface and further folding occurs until the prey is enveloped. Some have suggested that the folding mechanism may result from the release of auxin from the tip of the leaf. However, this is unlikely, as removal of the leaf tip still permits the folding response to occur. The auxin does, however, need to be transported, as auxin transport inhibitors will prevent leaf-folding. Thus, the most likely model for the leaf-folding response is as follows. The cells on the leaf upper surface detect the presence of large quantities of nitrogenous compounds (small quantities would not trigger leaf-folding); auxins are released from these cells into the apoplastic space, then cells in the lower surface of the leaf respond to the auxin and expand, thus the leaf folds.

The digestion of captured prey is then carried out by the tentacle heads (Muravnik, 2000). These have been shown to release phosphatases and proteases for the mobilization of the insect nutrients (Amagase, 1972; Clancy and Coffey, 1977). The head also performs the bulk of the absorption of the solubilized nutrients into the leaf, which are then available for further growth of the plant. On completion of the digestive process, the carcass of the insect, which is largely made up of chitin, is usually washed away by rainwater. The leaf and the tentacles straighten by a process which is the reverse of the trapping mechanism. Given the chances of loss of nutrients from the leaf during digestion, the whole process is very fast and the prey can be fully digested within a week. On average, the leaf traps can capture three insects and then the leaf is so damaged as to be non-functional; however, by that time new traps will have developed to replace these older ones.

There is much more that can be said about the trapping mechanism of all of the carnivorous plant species that are mentioned here, but it is beyond the scope of this book to take this further. However, this group of plants has to be one of the most fascinating of all and never fails to capture the imagination of even those oblivious to the plants around them.

Figure 8.16 *Drosera capensis* leaf capturing fly: (A) leaf of the Cape sundew; (B) bluebottle fly on leaf; (C) after 30 minutes the tentacles on the leaf have responded to the insect and bend to touch the body; (D) after 1 hour the leaf starts to respond to the products of digestion – it bends at the edges of contact between the leaf surface and the insect; (E) after 2 and 4 hours, respectively, the leaf bends further to completely engulf the prey.

References

Amagase, S (1972) Digestive enzymes in Insectivorous Plants. III. Acid proteases in the genera *Nepenthes* and *Drosera peltata*. *Journal of Biochemistry*, **72**, 73–81.

Clancy, FG and Coffey, MD (1977) Acid phosphatase and protease release by the insectivorous plant *Drosera rotundifolia*. *Canadian Journal of Botany*, **55**, 480–488.

Dress, WJ, Newell, SJ, Nastase, AJ and Ford, JC (1997) Analysis of amino acids from the pitchers of *Sarracénia pupurea* (Sarraceniaceae). *American Journal of Botany*, **84**, 1701–1706.

Friday, LE (1991) The size and shape of traps of *Utricularia vulgaris* L. *Functional Ecology*, **5**, 602–607.

Friday, LE (1992) Measuring investment in carnivory: seasonal and individual variation in trap number and biomass in *Utricularia vulgaris* L. *New Phytology*, **121**, 439–445.

Hill, BS and Findlay, GP (1981) The power of movement in plants: the role of osmotic machines. *Quarterly Reviews of Biophysics*, **14**, 173–222.

Mody, NV, Henson, R, Hedin, PA, Kokpol, U and Miles, DH (1976) Isolation of the insect paralysing agent coniine from *Sarracenia flava*. *Experientia*, **32**, 829–830.

Muravnik, LE (2000) The ultrastructure of the secretory cells of glandular hairs in two *Drosera* species as affected by chemical stimulation. *Russian Journal of Plant Physiology*, **47**, 540–548.

Ragetli, HWJ, Weintraub, M and Lo, E (1972) Characteristics of *Drosera* tentacles. I. Anatomical and cytological detail. *Canadian Journal of Botany*, **50**, 159–168.

Rost, K and Schauer, K (1977) Physical and chemical properties of the mucin secreted by *Drosera capensis*. *Phytochemistry*, **16**, 1365–1368.

Shimmen, T (2001) Involvement of receptor potentials and action potentials in mechano-perception in plants. *Australian Journal of Plant Physiology*, **28**, 567–576.

Williams, SE and Pickard, BG (1972) Receptor potentials and action potentials in *Drosera* tentacles. *Planta*, **103**, 193–221.

Williams, SE and Pickard, BG (1972) Properties of action potentials in *Drosera* tentacles. *Planta*, **103**, 222–240.

Background general reading

D'Amato, P (1998) *The Savage Garden: Cultivating Carnivorous Plants*, Ten Speed Press, Berkeley, California, USA.

Darwin, C (1875) *Insectivorous Plants*, 1st edn. John Murray, London.

Juniper, BE, Robins, RJ and Joel, DM (1989) *The Carnivorous Plants*, Academic Press, London.

Pietropaolo, J and Pietropaolo, P (1996) *Carnivorous Plants of the World*, Timber Press, Portland, OR.

Slack, A (2000) *Carnivorous Plants*, MIT Press, Cambridge, Massachusetts, USA.

9

Asexual and sexual reproduction

Reproduction offers plants solutions to three major problems facing the way plants have evolved to live:

1. *Immobility*. Plants are rooted to the ground and this is essential for their life style. With this fixed root system comes immobility. During the evolution of higher plants, much of the Earth's surface was unavailable to plants, due to the lack of a means of moving to different the habitats. Reproduction offers a means of moving through new clonal plants or through the production of seeds. Therefore, reproduction in plants meant that most of the land surface of the Earth could be colonized by plants.

2. *Longevity*. Plants may, in many instances, be long-suffering and long-lasting, but if the great extinctions of the past teach us anything, it is that nothing lasts forever. No matter how old an individual plant is, one day conditions will change to make survival impossible. Plants need some means of surviving such adverse conditions, which may be local or temporary, and reproduction can offer a solution to this. Clonal plants may move away from the point of origin of the mother plant and therefore widen the spread of the plant, thereby escaping a particular stress. In addition, seeds offer even greater opportunities to disperse genetic material well away from the parent plant and also provide a structure that can remain dormant over prolonged periods. This permits a plant to survive temporary changes in conditions. Clones or seeds also produce new individuals which are free from fungal infections or damage present in the parent plant.

3. *Genetic variation*. Finally, reproduction offers opportunities for genetic variation, which is essential in a changing habitat. Asexual reproduction offers little opportunity for genetic variation but the potential from sexual reproduction is immense.

In this chapter we will look in greater detail at how plants reproduce and the merits and limitations of the different methods.

Asexual reproduction

Many plant species use asexual reproduction as a means of propagation. Asexual reproduction has several advantages and disadvantages (Figure 9.1). This form of reproduction produces individuals that are genetically identical and does not support much scope for phenotypic variation. This can lead to plants being vulnerable to changes in the habitat. In addition, offspring are generated close to the parent plant, so parent and offspring often have to compete for nutrients and light. However, asexually produced plants usually attain adulthood in advance of seedlings of the same parent and hence rapidly out-compete neighbouring plants reliant on seeds. Under many conditions, asexual reproduction is more successful than sexual reproduction for plants and a large number of species use this as their sole means of reproduction.

There are examples of most organs of plants being used for asexual reproduction. The following list summarizes the different strategies that plants use to propagate asexually.

Roots

There are many examples of plants that use their root system for asexual reproduction. For example, the quaking aspen uses its root system to form clonal plants, which has proved to be tremendously successful and allowed large stands of this tree species to form. This has arguably allowed the quaking aspen to become the largest organism on Earth! (see Chapter 19).

Figure 9.1 Asexual reproduction in plants. Plants possess a range of different means of asexual reproduction. (A) Apomixis: using this method, the flowers spontaneously produce seeds without the need for any fertilization. The actual source of the embryo can come from the egg cell or the cells surrounding it. (B) A hyacinth bulb (*Hyacinthus* hybrid); bulbs frequently produce bulbils (small offset bulbs) at the base of the bulb. (C) A gladiolus corm (*Gladiolus* hybrid); corms frequently produce small offset corms at the base of the stem. (D) Strawberry plant (*Fraxinus* hybrid), forming runners, modified stems that form new plants. (E) Black poplar (*Populus nigra*) plant forming new plants from its root system. (F) Lax-flowered orchid survives from year to year by forming underground tubers. In some seasons the plant forms multiple tubers and propagates.

Stems

Stems can be used for reproduction in a number of ways. In some plant species, stems arch over and touch the ground. These contact points take root at their tips and form new plants, e.g. blackberries. Other species, such as strawberries, produce above-ground stems known as stolons, which produce daughter plants at alternate nodes. In addition, underground stems can develop into storage organs that also serve as reproductive structures, such as bulbs (onion family), corms (anemones), rhizomes (irises) and tubers (nightshade family).

Leaves

Leaves are rarely used for asexual reproduction. There are instances where damaged leaves can form new plants, such as in the African violet and certain begonia species. However, the best example of asexual reproduction from leaves is in the genus *Kalanchoe*, where several members of the genus form plantlets at the leaf margins (e.g. *K. daigremontiana*). In horticulture many of these structures are used deliberately in order to propagate plants. This is important, since this practice produces clones which maintain parental traits, such as flower colour or fruit flavour.

Unlike in many other multicellular organisms, the cells in plants frequently remain totipotent and can be used to propagate a whole plant from a single cell or structure (see

Chapter 18). Thus, stems can be cut into sections and planted, resulting in rooted clones of the parent plant. In addition, instead of rooting cuttings, they can be grafted onto other root-stock. Many plants are grown purely for their vigorous root-stock, then the aerial tissue is removed and a small piece of tissue from a similar plant (known as a scion) is inserted into a nick in the root-stock stem and allowed to graft. This practice allows the generation of plants that have the advantages of a strong root system and produces aerial growth characteristics of the parental tissue of the scion. This is common practice for apples, roses and grapevines in Europe.

Apomixis

Apomixis is a form of asexual reproduction that lies at the boundary between sexual and asexual reproduction. There are two major recognized forms of this form of reproduction: non-recurrent and recurrent. In non-recurrent apomixis, an embryo forms from a haploid (1n) cell. In recurrent apomixis, the plant forms an egg cell that contains 2n chromosomes without fertilization. This can result from a failure in meiotic reduction or from division of other cells (Vielle-Calzada *et al.*, 1996). These cells then divide and create an embryo. In the example of recurrent apomixis, the offspring are all genetically identical to the mother plant. The process of reduction in the number of chromosomes in the egg cell is bypassed and the egg cell goes on to form an embryo without fertilization,

Figure 9.2 The flowers of plants are held on structures known as inflorescences, which can take a range of different forms, as named in the figure.

through a process known as parthenogenesis. In order to form an endosperm, sometimes the central cell divides and takes over this role, but generally fertilization is required. Thus, plants that can utilize apomixis can use seeds that have been generated in an asexual manner for reproduction. Apomixis occurs naturally in around 40 different plant families, making a total of over 400 different plant species. This phenomenon is common in citrus trees, blackberries and certain orchid species. Hybridization between different plant species frequently results in infertile offspring. It would be easy to regard such offspring as doomed in evolutionary terms, but apomixis permits reproduction. Many races of blackberries are sterile hybrids which can use apomixis to reproduce very successfully.

There is currently great interest in producing apomictic crop plants which produce clonal seed and therefore preserve the genetic identity of valuable hybrid lines. Frequently hybrid vigour is evident in crop hybrids that is lost in future generations. If the genetics behind the whole process of apomixis was understood, then it would be possible to clone indefinitely plants bearing valuable phenotypes.

A recent report (Pichot *et al.*, 2001) has identified a cypress (the Duprez cypress, *Cupressus dupreziana*) which can use paternal apomixis for asexual reproduction. In this example, the cypress produces diploid pollen grains, which form an embryo on reaching the female cones of the same species or on female cones of another cypress species (the provence cypress, *Cupressus sempervirens*). Since the Duprez cypress is very rare (around 230 individual plants in Algeria), the formation of seeds in surrogate females is a definite advantage. This is the only example known of paternal apomixis.

Sexual reproduction

Higher plants reproduce using modified leaf structures called flowers (Figure 9.3). Plants produce structures called inflorescences that bear the flowers. There is a wide range of different structures that the inflorescence can take and some of these structures are shown in Figure 9.2. Flowers possess male and female elements and the act of sexual reproduction in plants involves the simple movement of pollen from the anther to the stigmatic surface (Figures 9.4, 9.5). This process could easily be achieved if self-fertilization was the objective of flowering. However, as Charles Darwin, in his book *The Various Contrivances by which Orchids Are Fertilized* (1877) informs us, 'Nature thus tells us, in the most emphatic manner, that she abhors perpetual self-fertilization', and flowers use many means to try to increase the likelihood of cross-fertilization between different individual plants. The most basic mechanism to increase the probability of cross-fertilization is the physical separation of the anther and the stigmatic surface. This presents the first hurdle plants need to overcome; in order for a flower to be fertilized, there needs to be a means of movement of the pollen from the anther to the stigma. This obstacle has been overcome in two main ways: through wind/water-based pollination and through animal-mediated (largely insect-based) pollination. The basic structure of the two types of flowers using these different routes is shown in Figures 9.4 and 9.5. The developmental pathway for the formation of pollen and the egg cell in flowering plants is shown in Figures 9.14 and 9.15.

Figure 9.3 Floral diversity. For fertilization, a flower must achieve the movement of pollen from the anthers to the stigmatic surface. There are many ways of achieving this and ut it is this range of different methods that has given us the floral diversity we see on earth today.

Wind-based pollination

Of wind- and water-based pollination, wind is far and away the most important agent for transfer of pollen to the stigma. Mosses and algae mainly use water as a means of transferring pollen. Pines, conifers, trees in temperate climates and all species of grasses, rushes and sedges use wind pollination. Wind-pollinated species are much rarer

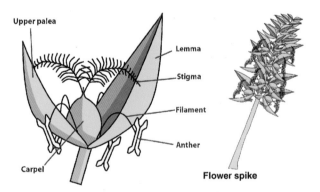

Figure 9.4 Schematic of a typical wind-pollinated flower. In order to achieve efficient movement of pollen from the anther to the stigmatic surface, the pollen is light and easily made airborne. Hence, the anthers hang from the flower. The stigmatic surface has large feathery appendages in order to maximize the chances of catching air-borne pollen. The colourful petals seen in insect-pollinated flowers are not required, so many wind-pollinated flowers are unobtrusive.

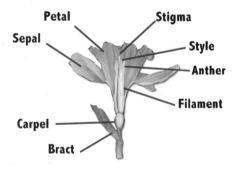

Figure 9.5 Schematic of a typical insect-pollinated flower. In order to achieve efficient movement of pollen from the anther to the stigmatic surface, the flower must attract insects. Here a daffodil flower (*Narcissus*) is shown; the flower, which provides nectar for visiting insects, is yellow and will stand out from other plants in a habitat. The nectar glands are placed such that an insect will brush both the anthers (so picking up pollen) and the stigmatic surface (so depositing pollen).

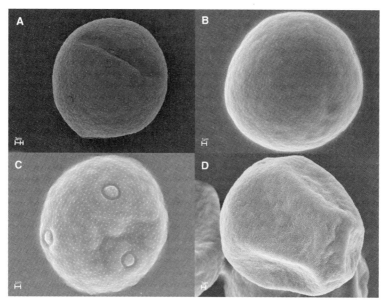

Figure 9.6 Electron micrographs of pollen from wind-pollinated flowers. The pollen is from: (A) a beech tree (*Fagus sylvatica*); (B) cocksfoot grass (*Dactylis glomerata*); (C) ribwort plantain (*Plantago lanceolata*); (D) perennial rye grass (*Lolium perenne*). The pollen for wind-pollinated plants is usually spherical and bears little surface ornamentation.

in the tropics, where there is greater species diversity and longer distances between individuals that can cross-pollinate. Wind-pollinated flowers are generally small, lack petals and have prominent stigmas and anthers (Figure 9.4). The anthers produce and shed large quantities

of pollen into the air. On average, a catkin on a birch tree produces around 5 million pollen grains (Procter *et al.*, 1996). The pollen produced by wind-pollinated species tends to be dry and smooth-coated (Figure 9.6); there are few projections from the grain surface, which is in marked

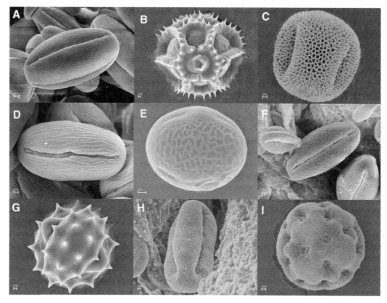

Figure 9.7 Electron micrographs of pollen from insect-pollinated flowers. The pollen is from: (A) a bramble (*Rubus fructicosus*); (B) bristly ox-tongue (*Helminthia echioides*); (C) herb robert (*Geranium robertianum*); (D) creeping cinquefoil (*Potentilla reptans*); (E) red clover (*Trifolium pratense*); (F) common poppy (*Papaver rhoeas*); (G) ragwort (*Senecio jacobaea*); (H) hogweed (*Heracleum sphondylium*); (I) great bindweed (*Calystegia sepium*). The pollen for insect-pollinated plants is very varied is shape and size. The outer surface bears surface ornamentation, which enables the pollen to stick to insects when they visit flowers.

Figure 9.8 Detail of the petal surface. The petals of insect-pollinated flowers need to attract insects and be distinctive, so that a particular plant species can easily be recognized again by an insect. The petal surface plays a role in this recognition. Here four species of plant are shown, with very different flowers. (A, B) Meadow buttercup (*Ranunculus acris*); the petals are highly reflective and shine in the sun, an effect created by the surface of the petals being very flat and smooth. (C, D) Garden pansy (*Viola* sp.); this flower has the appearance of velvet, an effect created by an array of dome-shaped cells across the petal surface, which makes the colour across each individual cell uneven (darker around the edges of the cells), causing the velvet-like appearance. (E, F) Flower from lavender (*Lavendula* sp.); this petal has a matt coloration with a furry appearance, an effect created by the presence of large numbers of trichomes rising above the petal surface. (G, H) Flower of a blue pimpernel (*Anagallis monellii*); the petal has a plain matt surface, created by an array of rough cells in the epidermis.

contrast to insect-pollinated species. In many examples, wind-pollinated species produce only a single seed per flower. The distance moved by a pollen grain in the air depends greatly on turbulence, wind strength and the height of the parent plant. Field studies with perennial rye

grass (*Lolium perenne*) revealed that few plants received pollen from a donor plant at distances of greater than 25 m from the point of origin (Griffiths, 1950). However, for a mature specimen of *Pinus elliottii*, pollen could be detected at distances of over 160 m from the donor plant (Wang *et al.*, 1960).

The production of pollen by plants is expensive in terms of carbohydrate and nutrients, particularly of nitrogen and phosphorus. At first sight, the masses of pollen required to be produced by wind-pollinated species in order to achieve fertilization would appear to be a waste of resources, but for species which dominate a habitat, such as trees in temperate regions and grasses in savannahs, the wind may actually carry pollen more efficiently than insects. Furthermore, such habitats may be unable to support the large numbers of insects required to carry out the pollination process. Thus, wind still has a major role in the pollination of many plant species. Indeed, wind pollination actually has advantages over insect pollination, as plants waste fewer resources on floral structures and nectar production and they are less dependent upon the abundance of particular insect species.

For successful pollination, airborne pollen grains are caught by feathery stigmatic surfaces of a receptive female. As mentioned earlier, the stigmas of wind-pollinated plants are prominent and usually have a large surface area. Not surprisingly, the density of pollen grains is highest near to the anther shedding pollen. Thus, it is remarkably easy for wind-pollinated plants to self-pollinate themselves. As a consequence of this, many wind-pollinated species are dioecious (Figure 9.19) and have single-sex individuals, or monoecious (Figure 9.18) and have separate-sex flowers on the same individual.

Insect-based pollination

The prospects of a flower being pollinated are raised dramatically using insect pollination, as it is more precise and focused. In marked contrast to wind-pollinated flowers, insect-pollinated flowers need to advertise their presence to attract pollinators (Figures 9.3, 9.5). Plants can use insects, birds, reptiles and mammals to fertilize flowers but the major pollinators are insects. Curiously, there are few reports of flowers adapted to birds, reptiles and mammals in Europe and Asia and in these regions flowers are mainly insect-pollinated.

The two major factors that initially attract insects to flowers are coloration and scent. It is the extent of the reward for visiting a flower that then reinforces these signals and makes a particular insect a regular visitor.

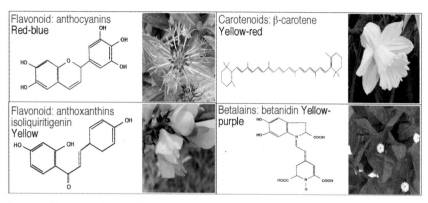

Figure 9.9 Floral pigments. To attract insects flowers possess pigments in the petal and sepal cells that give a flower an overall colour. Examples are shown of the major pigments and the colours they create.

The commonly used pigments in flowers are listed in Figure 9.9. Guiding marks are provided on flowers that steer visiting insects to the nectaries. Layers of cells containing either different pigments or varying amounts of the same pigment produce these markings (Figures 9.6,

Figure 9.10 Nectar trails. To ensure that pollination occurs as efficiently as possible, insects visiting flowers are guided to the nectaries. Many plant species produce flowers which have markings that indicate where the nectaries are and guide the insect along a specific path. Nectar trail markings are shown for four species: (A) common butterwort (*Pinguicula vulgaris*); (B) common poppy (*Papaver rhoeas*); (C) *Geranium* sp.; (D) foxglove (*Digitalis purpurea*).

9.10). It must always borne in mind that insects are particularly sensitive to UV light and therefore the image the human eye sees is not what an insect would see (Figure 9.12). The texture of the petals also influences the appearance and attraction of a flower to a particular insect. A summary of the basic forms of flower texture is given in Figure 9.8. A further important consideration of flower form is its size and its ability to support the target insect pollinator. The petals in many species are highly evolved in order to ease the passage of the insect to the nectaries and the delivery of pollen to the insect. For example, *Mimulus luteus* possesses a large yellow flower with a large opening and bright red markings (Figure 9.11). This is perfect for a bumble bee to land on and then pollinate. It is the vast range of different insect targets that has resulted in the immense variety of floral forms there are on earth today (Figure 9.3).

Flower scent can also be crucial in attracting insects to flowers, and the combination of flower appearance and smell are important cues to insects in identification of different plant species. Plants produce a range of scents including terpenes, benzoid compounds, alcohols, esters and ketones. Some plants produce scents to attract insects that are hunting for dung or carrion; these release compounds containing amines and ammonia and mimic scents of the natural process of decay (see Araceae in Chapter 15). Possibly the most remarkable plants for attracting insects are the *Ophrys* (bee orchids), where the flower mimics the appearance of a female bee and uses pheromones to complete the deception. These are discussed in detail in Chapter 15.

For insect visitors to flowers, the most important incentive is the reward. Probably the first evolutionary reward for insects from flowers was the pollen itself (Figure 9.7). Insects can be so greedy for pollen that they will even visit wind-pollinated flowers in order to obtain a meal. But

Figure 9.11   A bumblebee visiting a *Mimulus* flower. The red markings frame the opening to the flower, guiding the bee into the area of the nectaries. During feeding by the bee, the flower will be pollinated.

Figure 9.12   *Primula vulgaris* flower under UV light. The colours of flowers we see may not be the colours that insects see. Bees are sensitive to light in the ultraviolet end of the light spectrum, which means that some markings on the flowers may be more apparent under UV light. This is so with the *Primula* flower, where contrast in the flower markings is low when viewed under visible light (A), but under UV light (B) the nectar trails are much more evident.

pollen is very expensive for the plant to make. Dried pollen has been shown to contain the following components (Harborne, 1993):

- 16–30% protein.

- 0–15% starch.

- 0–15% monosaccharides and disaccharides.

- 3–10% fat.

Pollen cells have thick protein coats and are rich in nitrogen and phosphorus, which are costly elements for a plant to lose. Thus, the evolution of nectar production by flowers attracted insects more interested in carbohydrate than protein. However, pollen gathering from flowers is very important to the diet of many insects, particularly bees.

Nectar is a very cheap alternative to pollen as an attractant for flowers. The main components of nectar are sugars, such as sucrose, glucose and fructose. The actual concentration of the sugars varies, depending on the air humidity, but can range between 20% and 75% of the dry weight of nectar (Baker and Baker, 1983). The composition of the nectar influences the insects that will visit a flower (Figure 9.13). For example, bees store most of their collected sugar as honey, which is around 80% sugar. In order to concentrate the sugar in the nectar, bees have to evaporate much of the excess water from the nectar. This requires young bees to drive air through the hive by beating their wings. It is not worth bees collecting nectar of lower than 20% sugar concentration, since it

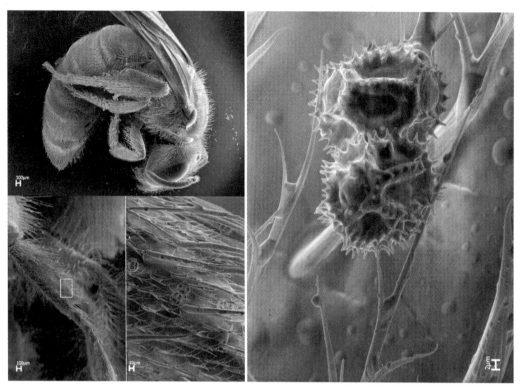

Figure 9.13 Solitary bee (*Andrena* sp.), captured after visiting a smooth hawksbeard flower (*Crepis capillaris*). Once a bee species has located a plant species that offers a good reward in nectar, the insect will be very faithful to these flowers. There are many hundreds of pollen grains on this insect, and every one is from the hawksbeard flower. This demonstrates that some insects can be very faithful to a particular plant species, which is just what the plant needs.

requires too much effort to concentrate the nectar to make honey. Thus, bees usually target flowers which produce high concentrations of sugar in their nectar (usually 40% sugar). In order to collect nectar, bees need to work very

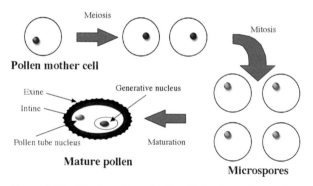

Figure 9.14 The formation of pollen. Pollen forms from pollen mother cells, each of which undergoes meiosis to form a tetrad of microspores, which mature to form the pollen grains. The mature pollen grain contains either two or three nuclei; one nucleus will form the pollen tube nucleus and the other one or two will form generative nuclei.

hard. To make 0.5 kg of honey, a bee will visit between 2 and 8 million flowers! This represents 17 000 foraging trips, visiting on average 500 flowers per trip (Ribbands, 1949). Seen in this context, the true success of insect pollination of flowers can be appreciated. Further details on insect interactions with plants and non-rewarding flowers are given in Chapter 15.

Nectar should not be seen only as a sugary solution. It also contains amino acids at concentrations of around 1 mg/l (Baker and Baker, 1983a). The amino acid composition also has major implications for the visiting insects. Carrion and dung flies prefer nectar with high concentrations of amino acids (12 mg/l), whereas butterflies and bees are satisfied by a much lower concentration (around 1 mg/l). In a small number of instances, oils are produced by flowers as a reward, e.g. certain orchid species.

Pollination

Pollination occurs when compatible pollen is deposited on the stigmatic surface of a flower and fertilization of the

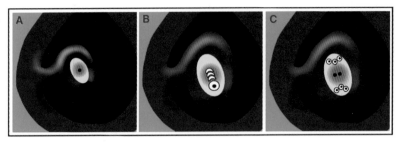

Figure 9.15 The formation of the egg cell. In a similar way to pollen formation, a single cell (A) divides to form four cells. Three of these cells wither and die, leaving room for a megagametophyte (egg cell) to form (B). The nucleus continues to divide to form eight nuclei, which arrange themselves in the pattern shown in (C).

female gametophyte occurs. However, as has been portrayed above, the stigmatic surface of any particular flower is likely to pick up pollen from many different plant species, particularly in the instance of wind-pollinated species. To some extent the range of different pollen types being deposited on the stigmatic surface of insect-pollinated species is offset by a phenomenon known as flower constancy. Insects learn that a particular flower is rewarding and tend to be faithful and visit a suitable species repeatedly, so a flower is likely to receive mainly pollen from the same species. In addition, it is becoming clear that strong adhesion of pollen to the stigmatic surface may be crucial for successful reproduction.

The stigmatic surface of flowers is made up of papilla cells (cells bearing finger-like projections which are receptive to pollen). The stigmatic surfaces can be either dry (e.g. for the Brassicaceae and Graminae) or wet (Orchidaceae and Solanaceae). Dry stigmatic surfaces pose a considerable barrier to pollination, whereas wet stigmas bind pollen in an indiscriminate manner. For dry stigmas, adhesion to the stigmatic surface is crucial for pollination. It has been demonstrated in *Arabidopsis* that there is stronger adhesion of its own pollen to an *Arabidopsis* stigma than other *Brassica* species (Zinkl and Preuss, 2000). The factors influencing the specificity of binding of pollen to the stigmatic surface are not yet fully understood, but it is likely that there is great control over what pollen will bind to the surface. Once on the stigmatic surface, compatible pollen becomes hydrated within a matter of minutes and a pollen tube will usually form within 20 minutes (Figure 9.16). Pollen hydration is again very specific and non-compatible pollen, even in the

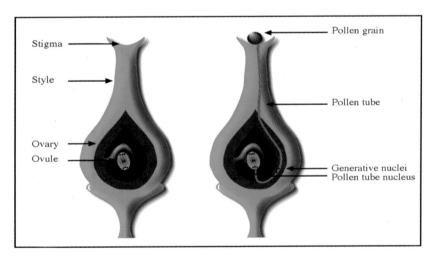

Figure 9.16 Fertilization of the egg cell. When pollen is deposited on the stigmatic surface, it germinates and a pollen tube forms. This penetrates the stigmatic surface and grows through the style. The pollen tube is led to the ovaries, where the pollen tube terminates at an egg cell. The tip of the pollen tube dissolves and penetrates the egg cell; the pollen tube nucleus then dissolves and the two generative nuclei enter the egg sac. One nucleus fuses with the egg cell nucleus and the other fuses with the polar nuclei. The former goes on to divide and form the embryo of a seed, and the latter goes on to form the endosperm of the seed.

close vicinity of compatible pollen, will remain dry. For wet stigmatic surfaces, there is growing evidence that the process of recognition occurs at the level of secretions made by the stigma and interactions between the pollen tube and the style.

The first level of recognition between the flower and the deposited pollen begins on the stigmatic surface. Usually only pollen from the same species will germinate and develop a pollen tube on a flower. It is becoming increasingly clear that this recognition is based on secretions from and adhesion of the pollen tube to the stigmatic surface. The stigmatic surface has been shown to secrete a range of sugars, oils, pectins and proteins, depending on the species (Fahn, 1974). Since the pollen grain contains very limited energy reserves, the pollen tube will use the maternal secretions to fuel growth and development of the pollen tube. In lilies it has been shown that complex polysaccharides known as pectins interact between the surface of the pollen tube and the flower stigma, which then allows the pollen tube to penetrate the stigmatic surface and grow towards the ovaries. As the pollen tube penetrates the conducting tissue in the style, it continuously obtains nutrients from the tissue.

The mechanism for directing the pollen tube to the ovules has yet to be determined. Two main hypotheses have been put forward to explain how the pollen tube locates the egg cell. It has been proposed, first, that the ovule releases chemical signals that direct the pollen tube to grow to it; and second, that the structure of the cell walls of the cells in the style is such that it forms tracks which direct the pollen tube growth to individual ovules (Welk *et al.*, 1965). The latter is the favoured hypothesis, since it accounts for why multiple fertilizations are a rare event. The pollen tube usually penetrates the ovules through the micropyle region (Figure 9.16). At this point the pollen tube nucleus dissolves and one

of the generative nuclei fuses with the polar nuclei (this will form the triploid endosperm of the seed) and the other fuses with the egg cell (this will form the embryo).

Mechanisms for improving cross-fertilization

The basic problem with the design of around 90% of flowers is that they are hermaphrodite and contain both male and female parts. The anthers and the stigmatic surface are perilously close together. The whole process of sexual reproduction offers little in terms of rearrangement of the genetic material of a plant without cross-fertilization between different individuals. It is certainly true to say that many plant species make little attempt to ensure cross-fertilization. The close proximity of anthers and the stigma makes self-fertilization very likely in flowers. Simple changes in the architecture of the flower can vastly improve the chances of cross-fertilization. Physical separation of the anthers and the stigma is an adaptation which makes self-pollination less likely, but could be considered only a minor measure. The best example of the rearrangement of the flower shape to minimize self-fertilization is in the case of the pin-eyed and thrum-eyed primroses (Figure 9.22). In this example, the combination of flower structure and other mechanisms prevents self-fertilization. There are three major further steps that plants take to minimize self-fertilization:

1. Some plants are what is known as *monoecious*, i.e. they possess separate-sex flowers on the same individual plant (Figure 9.17). Possibly the best example of this is in corn (*Zea mays*), which bears the anthers at the top of the plant (tassels) and the female flowers lower down

Figure 9.17 Example of a monoecious flower. Maize plants (B) produce a male flower at the top of the plant (C) and a female flower near the middle of the plant (A). This mechanism raises the likelihood of cross-pollination.

Figure 9.18 Examples of dioecious flowers. Dioecious plants possess separate male and female individuals. Three examples are shown: (A, B) white willow (*Salix alba*); (C, D) holly (*Ilex* sp.); (E, F) red campion (*Silene dioica*).

the stem. Some plants have taken this a step further and the male and female parts of the flower mature at different times. Usually the female is receptive first, followed by the shedding of pollen, and thus self-fertilization is impossible. A good example of this is in the Araceae and this is shown in Figure 15.13.

2. Some plants are *dioecious*, i.e. they possess male and female individuals (Figure 9.18). This effectively eliminates any possibilities of self-fertilization. However, this then presents difficulties in ensuring that different-sex individuals are within a suitable distance for pollination. There are also partially dioecious plants with intermediate characteristics, known as gynodioecious (Figure 9.19) and androdioecious plants. Examples and definitions of the different types of flower produced by higher plants are listed in Table 9.1.

3. Around 30% of higher plants exhibit *incompatibility mechanisms* that act to prevent self-fertilization.

Self-incompatibility mechanisms

The two forms of incompatibility mechanisms recognized in plants are sporophytic and gametophytic incompatibility. The essential features of the two mechanisms are summarized in Figure 9.20. Both mechanisms are governed by an S-allele which is made up of a single or very closely linked set of genes. A list of species exhibiting these forms of incompatibility is given in Table 9.2. These mechanisms merit further comment, since they are among the few cell recognition systems known in plants.

Figure 9.19 Gymnodioecious plants. There are some plants that are not quite dioecious but appear to be evolving in that direction. The ribwort plantain (*Plantago lanceolata*) comes in two forms, a hermaphrodite form (A) and a female-only form (B). This condition is called gymnodioecious.

Table 9.1 Varying forms of flowering in higher plants.

Type of flower	Description	Proportion of known plant species (%)	Examples
Hermaphrodite (Figure 9.5)	Bisexual flowers	90	Daffodil, daisy, cornflower, etc.
Monoecious (Figure 9.17)	Separate-sex flowers on same individual	5	Birch, corn, arum lily, cucumber
Dioecious (Figure 9.18)	Separate-sex flowers on different individuals	<5	Holly, white bryony, white and red campion
Gynodioecious (Figure 9.19)	Female and hermaphrodite individuals	<1	Bladder campion, thyme, ribwort plantain and wild strawberry
Androdioecious	Male and hermaphrodite individuals	Very rare occurrence	Durango root (Datisca glomerata), Fraxinus lanuginosa

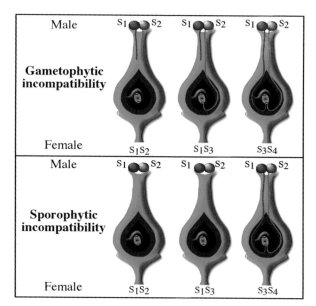

Figure 9.20 Sporophytic and gametophytic incompatibility, showing the main features of these two incompatibility mechanisms, which are governed by the S allele. *Gametophytic incompatibility*: if pollen from a S_1S_2 is deposited on an S_1S_2 female flower, the pollen will germinate but will terminate and die in the style. The female flower recognizes the common alleles. If the same pollen arrived on the stigma of a S_1S_3 female, the pollen containing the S_2 allele would be permitted to fertilize the egg cells. *Sporophytic incompatibility*: if pollen from a S_1S_2 female flowere is deposited on an S_1S_2 female flower, the pollen will not germinate but will terminate and die on the stigma. The female flower recognizes the common alleles. If the same pollen arrived on the stigma of a S_1S_3 female, the same would happen again, as the female can recognize that the male parent plant had a common S allele, even though the pollen grain itself may have a completely different S allele within its nucleus. For fertilization to occur, the male parent must have two different S alleles from the female flower.

Table 9.2 List of plant species exhibiting differing compatibility mechanisms.

Family	Species	Incompatibility system
Brassicaceae	Cauliflower (*Brassica oleracea*)	Sporophytic self-incompatibility
	Thale cress (*Arabidopsis thaliana*)	Nil
Gramineae	Rice (*Oryza sativa*)	Nil
	Rye (*Secale cereale*)	Gametophytic self-incompatibility
Leguminosae	Pea (*Pisum sativum*)	Nil
	Trifolium repens	Gametophytic self-incompatibility
Malvaceae	Apple (*Malus*)	Gametophytic self-incompatibility
Papaveraceae	*Papaver rhoeas*	Gametophytic self-incompatibility
Primulaceae	*Primula* species	Sporophytic self-incompatibility
Solanaceae	*Nicotiana alata*	Gametophytic self-incompatibility
	Nicotiana tabacum	Nil
	Potato (*Solanum tuberosum*)	Gametophytic self-incompatibility
	Petunia hybrida	Nil
	Petunia inflata	Gametophytic self-incompatibility

The phenomenon of self-incompatibility has arisen several times in the evolution of higher plants. However, for each family only a single self-incompatibility mechanism is represented.

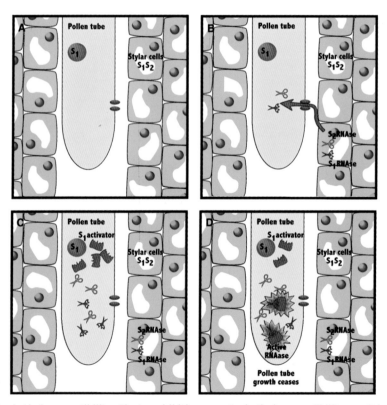

Figure 9.21 The role of SRNase in gametophytic incompatibility. Plants exhibiting gametophytic incompatibility use S locus-encoded RNase (SRNase) in order to control the process. If the pollen and the female parent plant contain the same S allele, then SRNase becomes active in the pollen tube as it grows down the style. This inhibits growth and kills the pollen tube, thus preventing fertilization. How the S allele is recognized in this mechanism is not understood.

Figure 9.22 *Primula* flowers and heteromorphic incompatibility. *Primula* flowers exhibit distyly. There are two flower forms that are controlled by the S allele. In the dominant (thrum-eyed) form (A, B), the anthers are at the top of the flower and the stigma is at the bottom. In the recessive (pin-eyed) form (C, D), the anthers are at the bottom of the flower and the stigma is at the top. *Primula* plants exhibit sporophytic incompatibility.

Gametophytic incompatibility

The phenomenon of gametophytic self-incompatibility is summarized in Figure 9.20. In most instances, plants showing gametophytic incompatibility are binucleate (members of the Gramineae being the exception). In this form of incompatibility, pollen which possesses an identical S-allele to the female fails to fertilize the flower. Thus, if pollen was shed by a S_1S_2 individual onto the stigmatic surface of a S_2S_3 flower, the pollen containing the S_2 allele would fail to fertilize the ovules. However, pollen containing the S_1 allele would fertilize the ovules. On further analysis it has been shown that the S_2 pollen germinates but fails to penetrate the full length of the style and reach the ovaries. It is clear that some interaction occurs during passage of the pollen tube through the style. The model for the nature of this interaction is illustrated in Figure 9.21. As the pollen tube extends through the style, the cells of the conducting tissue produce SRNase (a specific RNA degrading enzyme which is part of the S-allele). The SRNase is transported out of the style cells and is taken up by the pollen tube. If an incompatible reaction occurs, the SRNase is active and prevents further growth of the pollen tube, so that fertilization cannot occur. However, if a compatible reaction occurs, the SRNase is inhibited by the

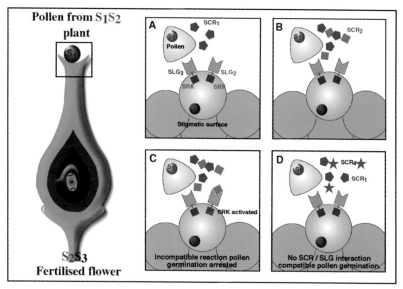

Figure 9.23 Sporophytic incompatibility. In plants exhibiting sporophytic incompatibility, the female plant can recognize whether the male parent of the pollen has a common S allele with itself. If a female (S_2S_3) encounters pollen from a S_1 male, no reaction occurs and pollination can occur (A). However, SCR protein is deposited on the developing pollen grain in the anther, so if the pollen came from a S_1S_2 plant then SCR_2 will also be present on the grain surface (B). If there is a common SCR allele between the pollen and the fertilized flower, an interaction occurs (C). In the figure, SCR_2 is a common allele between both plants. The SCR protein interacts with a protein called S-locus glycoprotein (SLG) on the epidermal layer on the stigmatic surface. The SLG protein is linked to another protein, S-receptor kinase (SRK), which is inside the cells on the stigmatic surface. If a common S allele is present in the SCR proteins, it will bind to SLG and this will activate the associated SRK protein. This then inhibits germination of the pollen grains. (D) Fertilization of a S_2S_3 flower by S_1S_4 pollen; since there is no common SCR allele, the pollen is able to fertilize the flower.

presence of an inhibitor. The nature of the inhibitor has yet to be identified, but it is clear that the gene product is closely linked to the SRNase. There is a very rare phenomenon in some gametophytic self-incompatible plants which also possess heteromorphic flowers. In flax and *Primula* species there are two flower forms which serve to enhance the probability of cross-fertilization; the pin- and thrum-eyed primulas are shown in Figure 9.22.

Sporophytic incompatibility

The phenomenon of sporophytic incompatibility is summarized in Figure 9.20. Plants possessing sporophytic incompatibility are always trinucleate. This form of self-incompatibility is more complex than gametophytic self-incompatibly, but there is a greater understanding of how self-recognition occurs. In this form of incompatibility the pollen arrives on the stigmatic surface and a protein known as S locus cysteine-rich protein (SCR) is released. If the SCR is from an individual possessing an identical S allele, the SCR binds to a receptor, S locus glycosylated protein (SLG), bound to the cell wall of the cells of the stigma (Figure 9.23). This interaction with SLG activates a protein linked to SLG in the cytosol of the stigma cells, known as S locus-related kinase (SRK), which initiates a chain of events that prevent pollen germination. However, if the pollen is from an individual with different S alleles, then recognition does not occur and fertilization is permitted. One crucial feature of sporophytic incompatibility is that SCR protein is produced by the pollen and the tapetum cell layer in the anther (Figure 9.24). Thus, a pollen grain will possess SCR protein from both S alleles present in the parent plant and recognition is not based solely on the presence of the S allele in the pollen grain nuclei.

Seed development

Higher plants undergo double fertilization. One generative nucleus from the pollen fuses with the egg cell and the other with the polar nuclei to form a triploid cell; then the zygote develops from the egg cell fusion and the endosperm from the triploid cell. The pathway of development of the mature seed is shown in Chapter 10 (Figure 10.3). The zygote, after many cell divisions, develops into the embryo of the seed. Closely linked to this is the food-rich endosperm. Usually the cotyledons form an interface between the embryo and the endosperm (Figure 9.25). However, in some seeds the cotyledons absorb the food reserves of the endosperm and the embryo thus fills the whole seed coat. The integument layer around the egg cell is compressed and forms the testa (the coat of the seed).

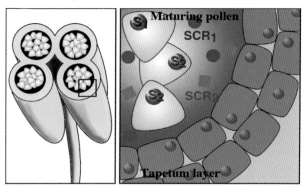

Figure 9.24   Deposition of SCR protein in sporophytic incompatibility. The SCR protein is deposited on the surface of the pollen while the pollen grains are developing in the anthers. Therefore, all of the different SCR proteins encoded by the S alleles within the male parent will be deposited, hence the female can detect whether the source of pollen possessed a common S allele with itself.

The micropyle is still visible in the seed and usually serves as a weak point in the testa to allow the emergence of the radicle on germination. The wall of the ovary matures during seed development to form the fruit. The fruit is often adapted to aid in the process of seed dispersal.

Plant reproduction overcomes the three major problems plants face as a result of their life style: immobility, longevity and genetic variation. Obviously, on the basis of planet biomass, plants have been spectacularly successful at overcoming these hurdles. Through reproduction, plants can move from one location to another, they can resist severe stresses and they can evolve. Despite having a fixed position, it would not be true to say that plants cannot move; they can move as a result of growth and in response to certain stimuli. These are the topics of the next two chapters.

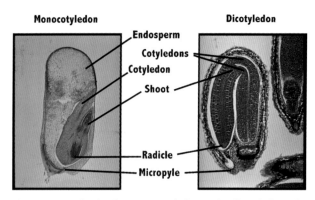

Figure 9.25   Seeds of a monocotyledon and a dicotyledon. The major structural features of a monocotyledon (*Zea mays* seed, left) and a dicotyledon seed (*Capsella bursa-pastoris* seed, right) are shown. The *Capsella* seed cotyledons have absorbed the endosperm of the seed.

References

Baker HG and Baker I (1983) A brief historical review of the chemistry of floral nectar, in *The Biology of Nectaries*, (eds BL Bentley and TS Elias), Columbia University Press, New York, pp. 126–152.

Darwin C (1877) *The Various Contrivances by which Orchids Are Fertilised*, 2nd edn, John Murray, London.

Fahn A (1974) *Plant Anatomy*, 2nd edn, Pergamon, New York.

Griffiths DJ (1950) The liability of seed crop of perennial ryegrass (*Lolium perenne*) to contamination by wind-borne pollen. *Journal of Agricultural Science*, **40**, 19–36.

Harborne JB (1993) *Introduction to Ecological Biochemistry*, 4th edn, Academic Press, London.

Lord EM, Walling LL and Jauh GY (1996) Cell adhesion in plants and its role in pollination, in *Membranes: Specialized Functions in Plants*, (eds M Smallwood, JP Knox and D.J. Bowles), BIOS Scientific, Oxford, UK. pp. 21–37.

Pichot C, Maataoui M, Raddi S and Raddi P (2001) Surrogate mother for endangered *Cupressus. Nature*, **412**, 39.

Procter M, Yeo P and Lack A (1996) *The Natural History of Pollination*, Harper Collins, London.

Ribbands CR (1949) The foraging method of individual honeybees. *Journal of Animal Ecology*, **18**, 47–66.

Vielle-Calzada JP, Crane CF and Stelly DM (1996) The asexual revolution. *Science*, **274**, 1322–1323.

Wang CW, Perry TO and Johnsson AG (1960) Pollen dispersion of slash pine (*Pinus elliottii*) with special reference to seed orchard management. *Silvae Genetica*, **9**, 78–86.

Welk M, Sr. Millington WF and Rosen WG (1965) Chemotropic activity and the pathway of the pollen tube in lily. *American Journal of Botany*, **52** (8), 774–782.

Zinkl GM and Preuss D (2000) Dissecting *Arabidopsis* pollen–stigma interactions reveals novel mechanisms that confer mating specificity. *Annals of Botany*, **85**, 15–21.

10

Plant growth

The size of most animals is determined genetically and cells divide a certain number of times and then cease. This type of growth is termed *determinate*. The number of divisions will vary little, whether the organism is starved or stressed. This is not so with plants, since their growth is influenced tremendously by their environment. Plants continue to grow throughout their lives.

The growing regions of plants are known as *meristems*. These cells divide and produce new cells that continue to grow and mature. As a result of this style of growth, it is difficult to say how old a plant is. As the meristem divides new tissues are produced which will have a limited lifespan of perhaps one season. For example, deciduous trees grow new leaves each year from meristems, but as autumn approaches these leaves die and fall. Thus, some tissues are replaced with new ones each year but the growing meristem lives on. The size of many plants is completely dependent upon how they interact with their environment and as a consequence plant growth can be described as *indeterminate*. An extreme example of indeterminate growth is that of the bristlecone pine (*Pinus longaeva*). Some specimens of this tree are thought to be almost 5000 years old. The oldest tree ever recorded is 4789 years old and is named the Methuselah tree, on account of its age. Many generations of cells in the leaves, stem, branches and roots will have died but they have been constantly replaced by meristems throughout the tree. Hence, at least part of the plant must be just under 5000 years old, but the majority of the cells will be less than a few years old. This may be the secret of eternal life for trees: by constantly replacing cells, a tree can retain its youth.

This reveals another unusual property of plants and this is that a meristem can return the plant to a juvenile form and in effect start all over again. This property is exploited in horticulture, through plant cuttings. Cuttings take on the form of a young plant; the leaves are reduced in size, a new root system forms and often they lack the maturity required for flowering which the parent possessed. An example of this is the Cox apple. Discovered in 1825 by Richard Cox, branches from the original tree have been grafted ever since onto other apple tree rootstocks. Hence, every Cox apple tree is at least 180 years old! Grafting is essential in order to maintain the Cox apple phenotype, since sexual reproduction will result in new and unpredictable characteristic flavours in the apples.

The size of a plant is not fixed. A plant such as a *Petunia* grown in nutrient-deficient clay soil will be small, but the same plant left in nutrient-rich soil in a greenhouse will be huge by comparison. This indeterminate property of plants allows them to respond to, and take advantage of, their environment. However, not all plant organs are indeterminate. Leaves and flowers have a very defined size and so indeterminacy is not universal in plant cells. In addition, certain plants produce determinate growth every year. For example, *Rhododendron* buds develop branches that grow to a certain size and then cease. New buds for next year's growth then begin to develop. Thus, one season's growth may be determined but overall, considering many seasons' growth, the structure of the plant is indeterminate.

In this chapter we will be looking at how plants grow from a single cell (the zygote, which is generated after fertilization) to an embryo and ultimately to a whole plant. This process requires growth and differentiation to form all of the different tissues present in a plant. The process of growth necessitates movement of the plant, but in this chapter the topic of formation of the whole plant will be discussed, and movement will be left to the next chapter.

Types of growth

There are several types of growth to consider in plants. An organism can be considered to have grown in terms of length or area, but such growth does not necessitate an increase in dry matter. Expansive growth can be a result of an increase in volume of a cell, as a result of taking in water.

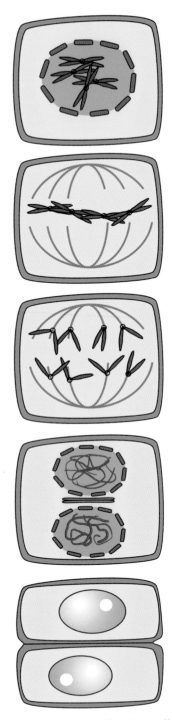

This is a very fast and easy form of growth, since it requires no synthesis of new proteins, nucleic acids or carbohydrates. This is precisely how plants put on much of their growth. As plant cells develop and mature, they begin to expand as a result of the increasing volume of their vacuoles and a relaxation in the structure of the cell walls (see Chapter 11). For many plants, the turgor developed through the expansion of the vacuole is the major force that supports the whole plant. The other form of growth requires the synthesis of new cells and materials, and this involves cell division.

Cell division and the cell cycle

Plant cells divide just as in any other eukaryotic organism. Chromosomes duplicate, condense and aggregate around the centre of the cell, then they are pulled apart at the centromere and migrate to opposite poles of the cell. A cell plate forms in the middle of the cell and then the cell divides to produce two identical cells (this is depicted in Figure 10.1). There are four distinct phases to the cycle of cell division and these are shown in Figure 10.2. The G_1, S and G_2 phases make up interphase (Doerner, 1994). The cells in a meristem spend 90% of their time in interphase and the rest in mitosis (M phase). In G_1 phase the cell prepares for replication. Most of the cellular processes are functioning and the cell is gathering reserves for growth. Nucleotides are made in large quantities in this phase to fuel the DNA duplication in the subsequent S phase. G_1 is a very variable phase and cells can enter this phase for a matter of minutes or days. In S phase the entire DNA in

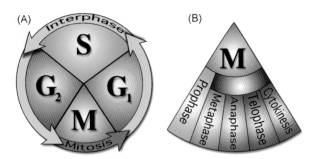

Figure 10.2 Cell cycle in plants. (A) The cell cycle in plants divides into two phases, interphase and mitosis. Interphase takes up around 85% of the time required for cell division in plants. During interphase a cell grows and then there is a gap (G_1 phase). The cell then replicates its chromosomes (S phase) and then there is another gap (G_2 phase). The cell then enters mitosis, during which the replicated chromosomes are separated and two cells are formed. The sequence of events occurring in mitosis is summarized in (B).

Figure 10.1 Division of plant cells. Plant cells divide in an identical manner to animal cells. From interphase, the chromosomes aggregate in the centre of the cell (prophase) and then lie along the metaphase plate (metaphase). The mitotic spindles then shorten, separating the sister chromatids (anaphase). A plate then forms between the new cells (telophase) and in plants a new cell wall is deposited between the new cells.

the nucleus is replicated and new histones are synthesized to maintain chromosome architecture. After this phase the cell then has a second period, the G_2 phase, in which it prepares for the last events before mitosis. The chromosomes begin to condense and the microtubules begin to form. This phase is generally short and lasts on average around 3–5 hours. G_2 phase ends with the beginning of prophase. The cell then enters M phase, in which mitosis occurs. Here the chromosomes are split and two genetically identical daughter nuclei are formed. Mitosis is split into five distinct events and these are described in Figure 10.2.

Polyploidy in plants

It is worth mentioning at this point that around one-third of flowering plants are polyploid, i.e. they contain more than two sets of homologous chromosomes (Bretagnolle and Thompson, 1995). In most examples the cells are tetraploid and contain four sets of homologous chromosomes. Polyploid plants tend to be larger than diploid ones, but not always. Polyploidy can also allow the plant to extend its habitat range by being able to inhabit more severe climates, such as drought-stressed regions. This phenomenon is not clearly understood but may be a result of the presence of multiple copies of advantageous genes (Adams *et al.*, 2003). Many cultivated plants are polyploid. Different organs in the same plant can possess different ploidy levels. An excellent example of this is in the seed of all flowering plants. Plants are unique in that they undergo double fertilization during sexual reproduction. The fusion of one generative nucleus with the ovule gives rise to a zygote which will form the embryo of the seed. Then a second generative nucleus fuses with two polar nuclei to form a triploid cell, which divides to form the endosperm or food reserves of the seed.

Seed formation and germination

Development of a new plant begins with embryogenesis. There are instances where embryos result from cells other than the zygote, and this is known as somatic embryogenesis (see Chapter 18 on Tissue Culture). The cell division of the zygote rapidly produces a multicellular miniature plant in the form of an embryo (Fosket, 2002). Initial division forms a terminal and a basal cell. The basal cell divides further to form the suspensor tissue, which holds the developing embryo in position in what was the embryo sac of the ovule. After the formation of seven cells, the division of this cell ceases. The terminal cell divides to form the embryo structure (see Figures 10.3 and 10.4 for the sequence of events leading to the formation of the embryo and the mature seed).

The triploid cell resulting from the fusion of one of the sperm cell nuclei and the polar nuclei divides in the structure that was the embryo sac of the ovule, to form the endosperm (Bewley and Black, 1985). In some species this tissue will be maintained, e.g. peas and castor beans. During cell division a structure known as a cotyledon (or seed leaf) forms as an interface between the endosperm and the embryo; in monocotyledons one such structure is formed and in dicotyledons two are formed. However, in other species, such as members of the Brassicaceae, the endosperm is reabsorbed by the cotyledons and in the seed these contain the food reserves. The processes involved in germination are discussed in detail in Chapter 14. The contents of the seed imbibe water and the internal tissues swell, which places great pressure on the testa of the seed and ultimately allows the radicle to emerge. Germination is said to have occurred at the time when the radicle emerges from the testa of the seed.

In some dicotyledon seedlings the first structure to emerge from the soil is the hypocotyl (region below the cotyledons). The hypocotyl elongates, forming a stem, and the cotyledons are dragged through the soil to the surface. These cotyledons form the first leaves of the seedling. This form of germination is known as epigeal germination. In other dicotyledons and a majority of monocotyledons, the hypocotyl does not elongate and instead the epicotyl (region above the cotyledons) extends and emerges from the ground, bringing with it the shoot apical meristem, which forms the first leaves of the seedling. This form of germination is known as hypogeal germination. The next organs to form from the shoot apical meristem are normal leaves. For most seeds, if the radicle meristem is damaged then new lateral roots readily form. However, there are few seeds that have more than one shoot apical meristem; thus, if that is damaged then the seedling is destroyed. One example of an exception to this is in runner beans (*Phaseolus coccineus*). If the plumule is damaged, then the dormancy of two further axillary meristems on the cotyledons is relaxed and these then emerge from the soil.

Now we need to look at how the plant structure is generated by the meristems of the plant.

The dividing meristem

A meristem in a plant is a population of cells which divide to maintain the population but also generate cells that continue to divide but have a determined fate (Hopkins,

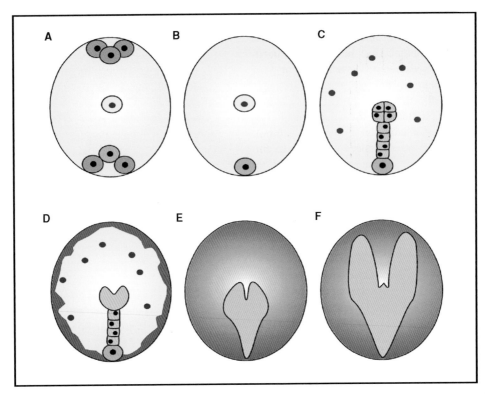

Figure 10.3 Seed development in plants. After fertilization of the egg cell (A), two cells are involved in the overall formation of a seed (B). The fertilized egg cell divides to produce an immature embryo, linked to the endosperm by a series of cells known as the suspensor (C). The cells at the top of the suspensor divide to produce first a heart-shaped embryo (D, E) and then a mature embryo (F). The endosperm is derived from the central cell in (A, B). The cell migrates to the periphery of the space where the seed will form and then divides to slowly surround the embryo. The cotyledons form an interface between the embryo and the endosperm. In some plants the endosperm is absorbed by the cotyledons and is not present in the mature seed.

1999). There are two forms of apical meristems of plants; the shoot apical meristem and the root apical meristem.

Shoot apical meristem

The shoot apical meristem is made up of a small dome of cells (around 200 µm in diameter) surrounded by curled-up leaves at various stages of development (Figure 10.5). The size of the meristem remains constant and is carefully regulated by the plant. The shoot apical meristems can be long-lived, e.g. there are numerous examples of trees which are many centuries old and yet the shoots on the plant are still growing. As mentioned earlier, the bristle-cone pine, at 4789 years old, must possess the longest-lasting shoot apical meristem (Fritts, 1969). Thus, meristems can go on dividing for long periods of time. In contrast, there are also examples of very short-lived meristems, such as those in annual plants. *Arabidopsis thaliana* is an annual plant and can have a life cycle as

short as 30 days. Hence, the meristem only has the capacity to form a limited number of structures before ceasing to grow.

The cells in the meristem possess thin walls and lack central vacuoles. The meristem contains three distinct layers of cells L1, L2 and L3 (Hopkins, 1999). The physical arrangement of these layers is shown in Figure 10.6. The cells of L1 and L2 form a region known as the tunica, and these cells divide anticlinally (perpendicular to the shoot); this produces a greater surface area to the meristem (Figure 10.7). The cells in L3 form a region known as the corpus, and these cells divide both anticlinally and periclinally and as a consequence increase the volume of the meristem and push it upwards. Layers L1 and L2 form the epidermis of the plant, whereas L3 divides to form the internal structures to the stem and leaves.

The meristem is made up of a series of zones. In the very centre are the apical initials or stem cells, which form the central zone. These are cells that do not differentiate and in some plants can, theoretically, divide indefinitely.

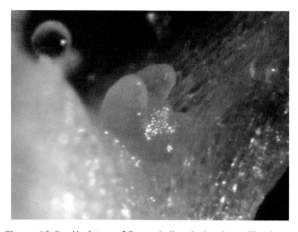

Figure 10.5 Meristem of *Drosophyllum lusitanicum*. The dome-shaped meristem typical of plants is shown. The lumps forming around the dome are the primordia of leaves, which will expand as they mature. The apical initials of the plant are contained in the central dome; it is these cells that decide the structures that will form as a result of cell division.

Figure 10.4 Mature seeds. Once formed, the seeds are mature. (A, B) Seeds coming out of the ripened ovary from two tree species. (C) Seeds of a *Protea* breaking off the base of the ripened ovaries. These seeds are linked to a protective coat to facilitate their dispersal across a habitat.

This group of cells does not get bigger and on division one daughter cell remains part of the apical initials and the other begins to differentiate and follows a particular developmental sequence. However, this second cell does continue to divide but the number of divisions of which it is capable is limited. Around this central zone is a doughnut-shaped region of tissue known as the peripheral zone. In this zone the dividing cells begin to form structures such as leaf primordia and axillary buds. Underneath the central zone is the rib zone. This part of the meristem divides to form the central pith of the stem. Cells produced by the shoot apical meristem can continue dividing for some time after formation. The final stage of

growth of the shoot is the elongation of the stem in the internode region. The internode region is the length of stem between leaf attachment points. In some species, such as *Primula veris*, there is no elongation of the internode and thus the plant adopts a rosette habit and is very short. However, in other species the internodes are long, such as in *Nicotiana tabacum*, making a tall plant. In some plants it is only the youngest internode that elongates, whereas in other species many internodes can be elongating simultaneously.

Leaf primordia

As the shoot apical meristem divides, it produces a series of protrusions on the surface, the leaf primordia (Cleland, 2001). The initiation of leaf primordia follows a very definite pattern or phyllotaxy, which depends on the species of plant. The different arrangements of leaves on the stem that plants use are shown in Figure 10.8. Thus, some plants initiate only a single primordium at a time, whilst others initiate several. Once initiated, cells in the leaf primordia continue to divide through what are called marginal meristems to form the whole leaf. This spatial relationship is determined genetically in the meristem, but little is known about how the plant regulates this process. The cell divisions are largely anticlinal, expanding the blade of the leaf, and are determinate (capable of a limited number of divisions). L1 forms the epidermis of the leaf; L2 forms the photosynthetic mesophyll layer; and L3 forms the vascular tissue of the leaf. The leaves are initiated

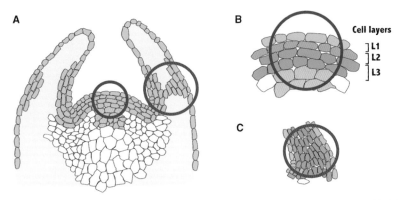

Figure 10.6   Cross-section (A) showing the detailed cell structure within a plant meristem. The apical initials are shown in pink in the centre. These cells are capable of constantly dividing but do not differentiate unless the meristem becomes determinate (i.e. makes a specific structure, such as a flower). The layer of cells across the dome forms the outer structures of the stem (B) and the leaves (C). This layer can be separated into three; L1 will divide and form the epidermis of the plant; L2 will form part of the mesophyll of the leaf and the layer immediately beneath the epidermis; and L3 will also form mesophyll tissue, as well as vascular tissue. Mutations within these layers can lead to interesting forms in some plants. If the L2 layer loses its ability to make chlorophyll, a variegated leaf results.

in a very ordered way and usually forms a recognizable pattern (Figure 10.9).

During leaf formation, small groups of cells maintain their meristematic characteristics at the junction between the leaf and the stem. These cells divide to form an axillary bud, which can grow to form branches to the plant's aerial structure. The leaf and the axillary bud form a node on the stem, and the regions of stem in between nodes are known as internodes. The shoot apical meristem for a particular plant species produces a constant repeated series of structures made up of one or more leaves, axillary buds and an internode. This repeated unit is known as a phytomer (Figure 10.10).

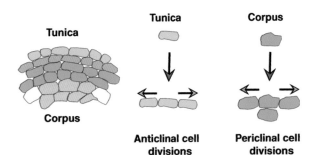

Figure 10.7   Division of cells within the meristem. The meristem can be divided into two regions, the tunica, which is made up of layers L1–L3, and the corpus, which divides to form the central pith to the plant. Cells in the tunica only divide anticlinally, producing more surface to the dome. The corpus cells divide anticlinally and periclinally, thus producing more volume to the meristem and raising the height of the plant.

Root apical meristem

The root apical meristem is less complex than that of the shoot (see Figure 10.10). The development of the root can be split into four developmental zones (Steeves and Sussex, 1989):

1. *The root cap.* The tip of the root is protected by the root cap, which is made up of around four layers of cells whose function is to allow the root tip to force its way through the soil and prevent soil particles damaging the root apical initials. As the root grows, the cells of the root cap are sloughed away by soil particles and are constantly replaced by division of the root cap stem cells of the root apical meristem. The root cap also secretes mucigel, a mucilaginous polysaccharide, which lubricates the root tip and root axis as it extends through the soil. The role of the root cap in the sensing of gravity by the plant is discussed in Chapter 11.

2. *The meristematic zone.* The apical initials in the root are known as the quiescent centre. These cells divide both forward (to produce more root cap cells) and backward (to produce further root axis). This group of cells lies just below the root cap. The cells extend for just over 200 μm into the root. The meristem only generates further root axis; lateral root formation is not controlled by the root apical meristem (Figure 10.11).

3. *The elongation zone.* After the cells of the root axis are formed they undergo further divisions and then begin to

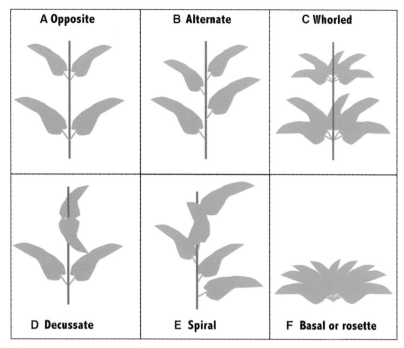

Figure 10.8 Leaf initiation by the meristem. Periodically, the tunica cells divide more rapidly in a synchronous manner, some to form leaf primordia. Leaves are formed in a very ordered way by the meristem and some of the described forms of leaf arrangements are shown.

differentiate. During this process they undergo cell expansion. This expansion is the major force that drives the root through the ground to explore new volumes of soil.

4. *The maturation zone.* Although during elongation cells can begin to differentiate, they are not fully mature until expansion is complete and the cells enter the maturation zone. At this stage the pattern of differentiated root cortex and stele, and vascular tissue become obvious. In this zone root hairs begin to form.

The origin of root hairs

Root hairs are one of the fastest growing sections of the plant and only pollen tubes can approach their rate of extension. They are derived not from a meristem but from a root epidermal cell. Whether a cell produces a root hair is dependent upon its position on the surface. If there are two cells forming a junction below the epidermal cell then it can form a root hair; the signal that allows this to occur is as yet unclear. A root hair can be over 2 mm long and thus is many times longer that the cell that develops it. Root hairs vastly increase the root surface area and the volume of soil exploited by the root system. They are vital for efficient water and nutrient uptake and are also important in the

Fabaceae for the formation of root nodules with symbiotic rhizobia. Root hairs are short-lived structures, often lasting only a matter of days. However, they are constantly replaced by growth of new hairs as the root advances.

The origin of lateral roots

Lateral roots form from the pericycle (a ring of meristematic cells just below the endodermis) in mature regions of the root. The growing lateral root grows through the tissue of the root cortex to emerge through a tear in the root epidermis. The lateral root may use simple force to exit the root axis or it may use enzymic digestion of the cell wall region in the root. The pericycle lies adjacent to the root vascular tissue and thus allows intimate contact between the lateral root and the main vascular bundles of the primary root system. The lateral root allows the root system to exploit larger volumes of soil. The role of auxins in stimulating lateral root formation in the presence of nitrates or phosphates has been discussed in Chapter 14.

Flower development

The production of leaves by the shoot apical meristem is called vegetative growth. Flowers can originate singly

Anticlockwise   **Clockwise**

Figure 10.9   Arrangement of leaves around the stem. The arrangement of leaves around the stem is known as the phyllotaxy, and examples of such arrangements have already been shown in Figure 10.8. However, the initiation of leaves in plants is more complex than the earlier figure would indicate. In plants with very short internodes, the pattern of developing leaves becomes clearer, as shown here in *Sedum* sp. The leaves are arranged in whorls. In this figure the individual leaves have been labelled in a schematic diagram to make them clear and the whorls formed in the anticlockwise and clockwise direction have been indicated in different colours. It was observed by early botanists that the repeated spiral pattern always forms a number of spirals that are part of the Fibonacci series. Here the *Sedum* forms three anticlockwise spirals and five clockwise spirals. This ratio appears to be necessary in order to generate the space required for new leaves to develop. The order within this phenomenon generates the extreme beauty of plants to the human eye.

directly from the apical meristem. However, for many plant species they are produced in clusters on inflorescences. For most plants there is some environmental signal which can switch the identity of the meristem from being vegetative to an inflorescence meristem. Mitotic activity rises in the meristem and the overall shape changes to become larger.

The inflorescence meristem can be determinate or indeterminate, depending on the species. Plants that produce cymose (paniculate, simple or scorpioid cymes) have a determinate inflorescence meristem. Other types of inflorescence meristem are indeterminate, e.g. spikes, racemes and umbels. The different forms of inflorescence produced by plants have been discussed in Chapter 9 (see legend to Figure 9.2). The role of the inflorescence is to place the

flowers in as advantageous a position as possible and this is dependent on the reproduction strategy of the plant. In most cases the inflorescence possesses longer internodes than the rest of the plant to raise the flowers over the rest of the plant. It will form a certain number of structures and then cease growing.

Periodically the meristem will form a floral meristem (Coen and Meyerowitz, 1991). Once this is initiated, the floral meristem is determinate, and will produce a series of organs in a set pattern and then cease activity. Often the conversion of the inflorescence meristem into a floral meristem is also associated with a change in phyllotaxy. Where more than one organ is initiated per node they are formed in whorls. Furthermore, the internode length is negligible and this gives rise to the tight rosette observed in

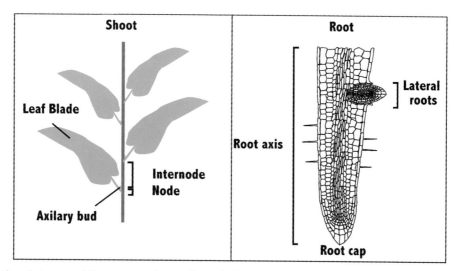

Figure 10.10 The phytomers of the shoot and root. The apical meristems of the shoot and root produce repeated structures (phytomers) in a regular pattern. In leaves the repeated pattern is a node, consisting of a leaf or leaves, axillary bud(s) and an internode. The internode length dictates the overall structure of the whole plant. In the root the phytomer is less well defined; the repeated structure here is root axis and lateral roots.

most flowers. The development of flowers is shown in Figures 10.12–10.17.

Vascular cambium and secondary growth

The vascular cambium forms a lateral meristem which extends along the stem and root of a plant. Cell division in the tissue forms secondary xylem on the inside and secondary phloem on the outside. The meristem does not form new organs but instead forms new stem and root tissue. This allows the stem to undergo secondary growth and swell outwards. A second lateral meristem is present in the stems and roots and that is the cork cambium. This layer forms within the secondary phloem and the root cortex. The layer of cells divides to produce the outer protective tissues of the stem and root, e.g. the bark of trees.

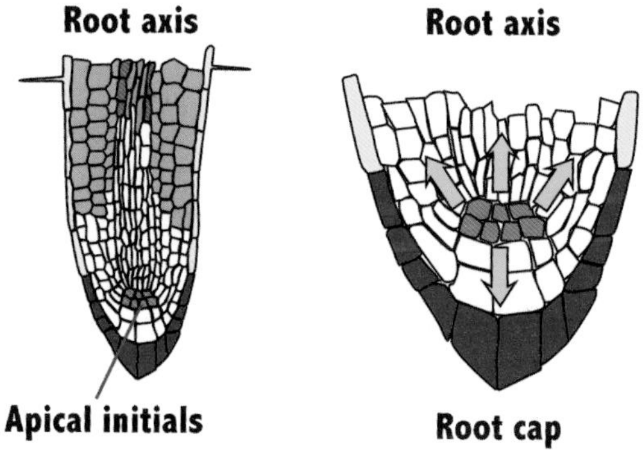

Intercalary meristem

These meristems are generally located at the bases of certain plant organs. The most familiar example is found in members of the grass family (Poaceae). The leaves of grasses have an intercalary meristem at their base. As a consequence, if the leaves are damaged, e.g. through grazing, the leaf will continue to extend and grow from the base. This is why grasses continue to grow throughout the year, despite being cut repeatedly with a lawnmower.

Figure 10.11 The root apical meristem, which lies at the tip of the root, beneath the root cap. This structure divides to produce more root. The cells divide to form more root axis, thus lengthening the root. In addition, the meristem produces new cells to replace the root cap. The root cap is a very strong structure that must resist soil particles. The structure is gradually worn away as the lengthening root axis forces the root through the soil. The root apical meristem acts to replace lost root cap cells.

Cell death

When a tissue becomes old it undergoes a process called senescence. This is where the tissue itself initiates a metabolic process whereby it dies. The process is active and controlled by the plant. There are many reasons why senescence can occur. The tissue may be old, or environmental

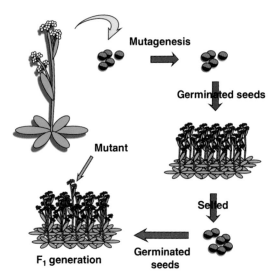

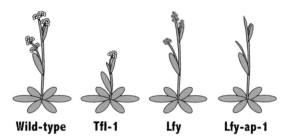

Wild-type   Tfl-1   Lfy   Lfy-ap-1

Figure 10.14 Mutants of the inflorescence structure. There have been numerous mutants of the inflorescence structure reported and this figure shows a few of these. *Terminal flower 1* (*Tfl-1*) produces a very short meristem, which is not maintained for long and forms a terminal flower very early on in the inflorescence development. *Leafy* (*Lfy*) forms a leafy structure instead of an inflorescence. *Leafy apetala 1* (*Lfy-ap-1*) fails to initiate the floral meristems. These mutants give information concerning which genes are required to form an inflorescence.

Figure 10.12 The use of *Arabidopsis thaliana* to study plant development. This species has been used to study many aspects of plant biology and possesses many features that make it useful for this purpose. Mutation analysis has been one of the earliest and most productive of the areas of research, simply because a developmental mutant is very easy to spot. One way of using *Arabidopsis* is to mutagenize seeds so as to introduce a single mutation. In the heterozygous state the mutation is unlikely to be observable, but if plants are allowed to self-pollinate, the subsequent generation (F_1) will have some individuals that are homozygous for the mutation. If the mutation is in a developmental gene, a phenotype is likely to be readily observed. Flowering mutants were some of the first to be identified in this way.

factors dictate that the structure is no longer required, e.g. leaves on deciduous trees in autumn, or when a flower is pollinated it dies rapidly to make way for seed capsule development. Many plant species are monocarpic and once flowers have been pollinated and seed pods have begun to form they die. This is so for annual plants, which complete their life cycle in a single year. However, a common horticultural practice is to remove the flowers as soon as they die and this prevents the formation of the seed pod. The plant will continue to grow and produce new flowers in order to reproduce, which is precisely what is required from annual

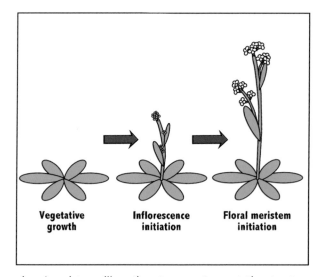

Vegetative growth   **Inflorescence initiation**   **Floral meristem initiation**

Figure 10.13 Mutations in flowering. A meristem will continue to generate vegetative structures until it is triggered to change its fate. Flowering is one such conversion that can occur during plant growth. Mutation of the ability of *Arabidopsis* plants to flower was an easy observation to make and hence flower developmental mutants were the first to be observed, so that a great deal has been learned about the control of flowering in plants. Studies in *Arabidopsis* have suggested that there are three stages to flowering: first, conversion of the vegetative meristem into an inflorescence meristem; second, maintenance of the inflorescence meristem and the initiation of the floral meristems; finally, production of the floral meristems and the elements of a flower.

Figure 10.15 Floral diagram of a *Primula*. The flower can be described pictorially in such a floral diagram. The flower is made up of four whorls of structures; the sepals, petals, stamens and carpels.

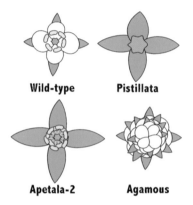

Figure 10.16 *Arabidopsis* mutants of the floral meristem. Three of the main mutants of the floral meristem are shown. The *pistillata* mutant fails to form either petals or stamens. The *apetala-2* mutant forms carpels and stamens instead of sepals and petals. The *agamous* mutant forms petals and sepals repeatedly, instead of stamens and carpels, and is therefore sterile. In each instance two whorls of structures are lost in the mutant and are replaced by two others.

Figure 10.17 Mechanism of control of organ identity in the floral meristem. (1) As the meristem divides and forms the sepals, the gene *apetala-2* is expressed. This controls *agamous* and prevents it from operating. (2) As the next whorl is formed, *apetala-2* continues to be expressed and the gene *pistillata* is switched on, which directs the second whorl to form petals. In the next whorl, *apetala-2* expression is switched off and *agamous* is expressed at the same time as *pistillata*. This directs this whorl to form into the stamens of the flower. (3) In the final whorl, only *agamous* is expressed and the carpels form.

garden plants. Perennials, on the other hand, do not respond in the same way; if the flowers are removed, the plant is most likely to respond by increasing the reserves stored for subsequent years of growth, rather than producing more flowers.

Unlike animals, plants can grow throughout their life. This ability underpins the phenomenon of totipotency discussed in Chapter 18. The growth of a plant also allows it to move; however, in the next chapter we shall be focusing on how plants move in response to stimuli.

Plant growth regulators and cell growth

As has been mentioned earlier, cell expansion plays a large role in the growth of plant tissues. One of the major factors that influence this cell expansion is the plant growth regulator, auxin. This will be discussed in the next chapter on the topic of plant movement.

References

Adams KL, Cronn R, Percifield R and Wendel JF (2003) Genes duplicated by polyploidy show unequal contributions to the transcriptome and organ-specific reciprocal silencing. *Proceedings of the National Academy of Science*, **100**, 4649–4654.

Bewley JD and Black M (1992) *Physiology and Biochemistry of Seeds in Relation to Germination.* Springer Verlag, New York.

Bretagnolle F and Thompson JD (1995) Tansley Review No. 78. Gametes with the stomatic chromosome number: mechanisms of their formation and role in the evolution of autopolypoid plants. *New Phytologist,* **129**, 1–22.

Cleland RE (2001) Unlocking the mysteries of leaf primordia formation. *Proceedings of the National Academy of Science of the United States of America,* **98**, 10981–10982.

Coen ES and Meyerowitz M (1991) The war of the whorls: genetic interactions controlling flower development. *Nature,* **353**, 31–37.

Doerner PW (1994) Cell cycle regulation in plants. *Plant Physiolology,* **106**, 823–827.

Fosket DE (2002) Growth and development. In *Plant Physiology,* 3rd edn (eds Taiz L and Ziegler E), Sinauer Associates, Sunderland, MA, pp.339–374.

Fritts HC (1969) Bristlecone pine in the White Mountains of California: growth and ring-width characteristics. *Papers of the Laboratory of Tree-Ring Research,* No. 4. University of Arizona Press, Tucson, AR.

Hopkins WG (1999) *Introduction to Plant Physiology,* 2nd edn. Wiley, New York, 287–307.

Steeves TA and Sussex IM (1989) *Patterns in Plant Development,* 2nd edn. Cambridge University Press, New York.

11

Plant movement

There are no examples of mature plants that can move from one location to another. The problems associated with removing the root system, translocating to a new environment and then re-establishing itself are just too great to overcome. In order to get to other locations, plants use seeds or asexual propagules, such as tubers, runners, rhizomes or baby clonal plants, but the parent plant does not physically move in these examples. It may be thought that plants do not actually move at all unless they grow, but this is incorrect. Plants move a great deal, but they move in a different time frame from animals. Plants rarely require fast movement and there are few examples of this. Only in unusual examples, such as the Venus flytrap and the *Mimosa* plant, is rapid movement used. Most plants appear to lack the proteins used in the energy-demanding processes for movement found in animals. Neither is it necessary for most plants to move rapidly. Plants move in response to many different stimuli, such as light, gravity, touch and chemicals. These stimuli do not demand an immediate response and hence plants do not need to move rapidly. In order to respond to these stimuli, plants have evolved completely different ways to move. How plants achieve these movements is the topic of this chapter.

Tropism and nastic movements

There are two main forms of movement in plants:

1. A *tropic movement* is the directional movement of a plant, or part of a plant, in response to a stimulus; examples include a plant's response to light or gravity. If the plant grows towards the stimulus, this response is described as a positive tropism and if it grows away it is described as a negative tropism. There is a wide range of tropisms in plants, including phototropism, gravitropism, heliotropism, hydrotropism, thigmotropism and chemotropism.

2. A *nastic movement* in a plant occurs when a plant responds to an external stimulus but the movement is independent of the direction of the stimulus. Examples of such movements include the opening of flowers in the daytime, and the movement of tentacles and leaf blades in sundew plants. All of these movements are triggered in response to a stimulus but the direction of the stimulus and the response are irrelevant.

Tropic movements

Auxins and plant movement

It is difficult to discuss auxins without referring to the history behind their discovery, and this is intimately linked to phototropism.

It has been known for centuries that the shoots of most plants grow towards a light source. However, it was Charles Darwin who started to understand the mechanism by which this occurred (Darwin, 1880). He observed that illuminating an oat coleoptile from the side caused the tip to bend towards the light source (Figure 11.1). He also identified the tip of the shoot as being the crucial plant tissue that sensed the illumination, and that the signal moved from the tip of the plant to the root (basipetal movement; see Figure 11.4). The work of Boysen-Jensen 1913 took this research further by interrupting this movement, using mica sheets, to demonstrate that the signal could be prevented from moving down the shoot by introducing a block. This established that the signal was water-soluble and moved down the stem from the shoot tip (Figure 11.2). The signal compound was finally identified by Went 1926, who experimented with dissolving the signal compound in agar and studying its nature. The active compound was given the name auxin (derived from the Greek word *auxein*, which means to grow) (Figure 11.3). Auxins formed the first group of plant hormones identified,

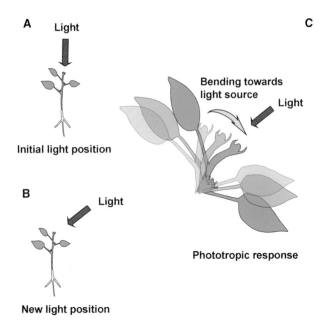

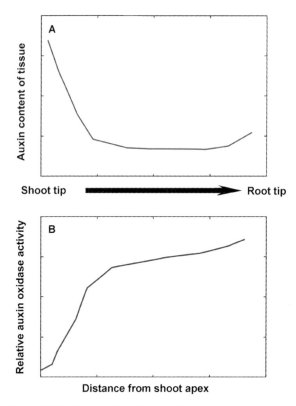

Figure 11.1   Phototropic response of a shoot. The shoot of a plant is generally positively phototropic, i.e. it will bend towards the light. A plant illuminated from the top will grow upwards (A), but if the light is moved to the side (B), the shoot tip will bend (C), so that the shoot grows towards the light source. This bending is not new growth but the expansion of tissues on the less-illuminated side. The mechanism behind this observation is explained in Figure 11.6.

Figure 11.2   Distribution of auxins in a plant. Auxin concentrations are at their highest in the shoot tip (A), where they are synthesized, then the concentration gradually declines. A second, smaller peak is found at the root tip. The presence of this gradient is partially explained by the presence of auxin oxidase, which has high activities in tissues away from the shoot tip (B).

because they satisfied the definition of a hormone, i.e. a regulatory chemical substance that signals the coordination of cellular functions in other tissues. Indole-3-acetic acid is the most frequently encountered natural auxin (Figure 11.3A, B). Since its discovery, many other compounds have been identified which mimic auxin activity in plants and are much more potent. These unnatural auxins have been called plant growth regulators (Figure 11.3C, D). There is some debate as to whether plants possess hormones in the sense that animals possess hormones. This is because plant hormones frequently promote a response at one concentration and inhibit it at another (Figure 11.5). As a consequence, it is difficult to predict how a cell type will respond to a plant hormone. This is different from what is observed with animal cells.

One of the major roles of auxin is controlling cell expansion. Auxins stimulate H^+/ATPase pumps in the plasma membrane, creating a fall in the pH of the apoplastic space (see Figure 11.7). This drop in pH activates expansions which sever specific cross-links in the cell wall. This reduces the pressure the cell exerts on the cell wall and causes the cell to take up water through osmosis. The cell then expands until its osmotic pressure is raised.

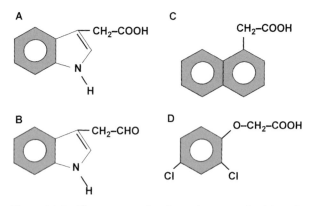

Figure 11.3   The structure of auxins and compounds with auxin activity. (A) Indole acetic acid (IAA); (B) indole acetaldehyde; these two are naturally occurring auxins. (C) 1-Naphthaleneacetic acid (NAA); (D) 2,4-dichlorophenoxyacetic acid (2,4D); these two are synthetic compounds with auxin activity. NAA is frequently encountered as a component of rooting powder for horticultural cuttings. 2,4D is used as a selective herbicide for killing dicotyledons and not monocotyledons (as in lawn weed killer).

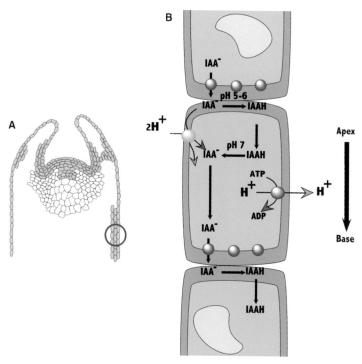

Figure 11.4 Basipetal movement of auxins. Auxins are transported from the shoot tip (A) downwards (known as basipetal movement). There are transporters for IAA⁻ ions in the plasma membrane on the lower side of cells, just below the stem epidermis (B). Once these ions are pumped from the cells, they lose their charge and become IAA, which can freely diffuse across the plasma membrane into the neighbouring cells. Once in the cell, the lower pH of the cytosol allows IAA to become ionized again and form IAA⁻. This procedure is repeated down the stem. Technically, IAA can freely diffuse back into the cell that pumped it out in the first place but, because of the presence of IAA⁻ pumps on the lower sections of these cells, the net flow is down the plant.

As a consequence, auxins control the expansion of cells in the shoot tip and developing leaves.

It is this process which governs most of the movements observed in plants. In plants, auxins can be viewed as the equivalent of the nervous system in animals and cellular hydraulics can be viewed as the muscle of the plant.

Phototropism

It is only recently that the mechanism behind the process of phototropism has been largely unravelled. It has been known for many years that auxin produced by the shoot tip is distributed evenly down the sides the growing shoot. Auxin displays basipetal movement in a layer of cells just underneath the epidermis of the stem (Figures 11.4, 11.6). This works through the shoot tip, itself synthesizing auxin, which then enters the apoplastic space underneath the epidermis. Due to the low pH of the apoplast, the auxin tends not to be in the form of an ion, and as a result, freely diffuses into neighbouring cells. The relatively high pH in the cytosol of the cell favours the auxin to form an anion

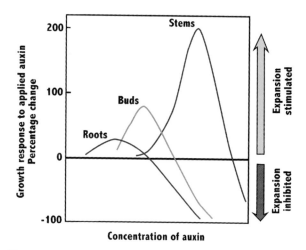

Figure 11.5 Sensitivity of plant tissues to auxins. The different tissues in a plant are differentially sensitive to auxin concentrations; the shoot is least sensitive and the root is the most sensitive. Up to a certain point, an increase in auxin concentration causes cells in these tissues to expand, but above a threshold the auxin concentration begins to inhibit expansion. This is particularly relevant in the root system for the gravitropic response.

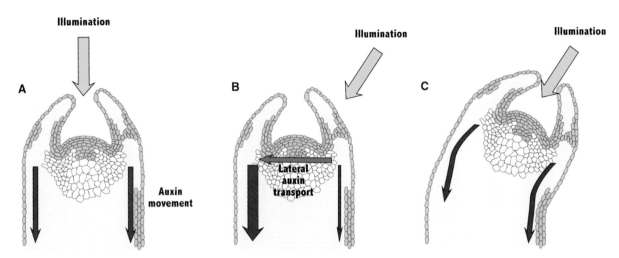

Figure 11.6 Mechanism of the phototropic response. With illumination from the top of a plant (A), auxins are transported down the stem in a balanced manner. However, if the illumination strikes the shoot tip at an angle (B), one side of the shoot will receive more light than the other. This causes an imbalance in the stimulation of lateral auxin transporters. Auxin is preferentially transported to the side of the shoot with least light. This causes the cells on this side to expand in response to the elevated auxin concentration and the shoot begins to bend towards the light source (C). This will occur until the light balance on the shoot is equivalent on all sides.

and thus it can no longer freely diffuse across the cell membrane. On the lower face of the cells underneath the epidermis, there is a transporter which pumps the anionic form of the auxin out of the cells. This whole process continues down the stem into the roots and sets up a gradient of auxin concentration down the plant. This gradient is important for many plant processes, one of which is the phototropic response.

However, if the shoot is illuminated from one side, then auxin tends to build up in concentration on the dark side of the shoot. This causes expansion of the cells on the dark side of the stem and the shoot tip then begins to move in the direction of the light; this is known as the Cholodny–Went theory, first put forward by Went and Thimann 1937 (Figures 11.6, 11.7). What has puzzled botanists for decades is why the concentration of auxin falls on the

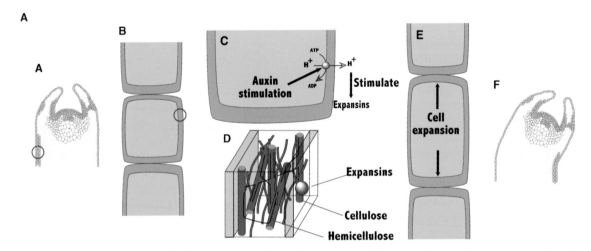

Figure 11.7 Control of cell expansion by auxins. Just below the epidermis of the shoot, the cells are very sensitive to the concentration of auxins (A, B). Auxins stimulate a H^+/ATPase pump in the plasma membrane of these cells (C). This movement of H^+ ions lowers the pH in the apoplastic space and this fall stimulates the activity of a group of enzymes called expansins (D). These enzymes act to alter the bonding between cellulose and hemicellulose in the cell wall in an ordered manner, which leads to relaxation of the cell wall structure at the sides of these cells. The osmotic pressure of these cells then causes the cells to stretch longitudinally (E). This extension, if uneven, causes the shoot to bend (F).

illuminated side of the shoot tip and rises on the dark side of the tip. Recent research has now provided a clearer answer (Liscum and Stowe-Evans, 2000). Two flavin-based photoreceptors, phototrophin 1 and 2, are now known to be the primary sensory mechanisms for detecting the amounts of UVA/blue light falling on the shoot tip (Briggs and Christie, 2002). It is still unclear how the signal from the photoreceptor is transmitted, but auxin efflux transporters on the lateral side of endodermal cells are activated and pump auxins laterally rather than basipetally. As a consequence, auxins are moved towards the dark side of the shoot and the phototropic response occurs until the amounts of light on both sides of the shoot are equal. Since the auxins on the dark side are moved down the shoot, a large length of the stem responds to the bending.

Other plant organs exhibit a negative phototropic response (Figure 11.8), such as root tips. In *Arabidopsis* it has been demonstrated that both red and blue light photoreceptors cause the root tip to grow away from the light source (Ruppel *et al.*, 2001). This response is weaker than the gravitropic response (see later) and is frequently masked by it. Many climbing plants exhibit a negative phototropic response and grow towards shade, in an effort to find possible supports for growth. The underlying mechanisms behind these movements have not been thoroughly studied.

Heliotropism

Many plants track the sun during its movement in the sky and the movement of the plant is known as heliotropism. These movements are rapid and reversible. The response is frequently observed in flowers such as lesser celandine (*Ranunculus ficaria*) and sunflowers (*Helianthus* spp.; Figure 11.9). The plants appear to obtain some advantage from getting optimal illumination of the flower during the day and hence track the sun. This is much less well understood than the other tropic movements of plants. However, what is clear is that it is not the flower itself that

Figure 11.8 Negative phototropism in ivy-leafed toadflax (*Linaria cymbalaria*), which lives in nooks in stone walls, shown here in the mediaeval walls of St Emilion, France. When the flower initially emerges, it grows towards the light. Once it is pollinated and the seed pod begins to form, the old inflorescence becomes negatively phototropic and hunts out crevices in the wall, where it places the seeds to allow them to establish themselves in new areas on the wall. This strategy can be very effective and the only great difficulty must be in establishing a colony on a wall in the first place.

Figure 11.9   Heliotropism in sunflowers (*Helianthus annuus*). The flowers of sunflowers follow the sun in order to always have the flower facing the sun during the day.

senses the light. If the flower is removed, then the stem will continue to track the sun. If the stem is covered, it is clear that only the upper portion of the stem is required for sensing, even though the bending of the stem occurs along its length. In addition, the bending is a direct result of changes in cell length of the cells along the stem, with the larger cells being on the dark side. Finally, red light does not elicit a tracking response, but blue light does. Thus, the whole mechanism is likely to be very similar to that described for the phototropic response.

A second form of heliotropism is exhibited by leaves, in which they maintain the upper surface of the leaf perpendicular to a light source (known as diaheliotropism). This ability is most common in members of the families Fabaceae and Malvaceae. This response maximizes the amount of light incident on the leaf surface and thus allows high photosynthetic rates. Organs connected to the leaves or petioles, called pulvini (singular, pulvinus) control these movements. The mechanism for function for these organs is not understood.

Another form of heliotropism has also been described, known as paraheliotropism, and is frequently observed in plants from the family Fabaceae. In this form of heliotropism, instead of tracking the sun, a plant places its leaves so they are at an angle to minimize the incident light on them. This strategy is usually seen as an adaptation of plants to drought stress. In one study, two bean plants, *Phaseolus acutifolius* and *P. vulgaris*, were studied (Yu and Berg, 1994). *P. acutifolius* (a bean adapted to arid regions) exhibited paraheliotropism when subjected to high light intensities and drought conditions, whereas *P. vulgaris* showed little ability to carry out a paraheliotropic response.

Gravitropism

Gravitropism is the response of a plant to grow either towards or away from the force of gravity. Shoots exhibit a negative gravitropic response and grow away from the earth, to find light; while roots exhibit a positive gravitropic response and grow towards the earth, to provide anchorage and to find water and mineral supplies. Thus, with shoots and roots there is an opposite response. Many other plant structures also exhibit a gravitropic response by altering their angle, depending on the angle of the plant. For example, leaves of some plants tend to maintain a horizontal position with respect to the rest of the plant. This is often referred to as a gravitational set point angle.

For the shoot, there is good evidence that the shoot tip and tissue just behind is necessary for perception and response to gravity. It has been demonstrated that auxins redistribute across the shoot tip to the lower side and this stimulates expansion of the cells on that side. As a result, the horizontal shoot tip is restored to the vertical position. This is very similar to what is observed in the phototropic response.

In roots, the sensing of gravity occurs in the columella cells within the root cap. If the root cap is removed, the root can no longer sense gravity until it has regenerated a new cap. Auxin arrives in the root through the stele tissue. At the root cap the auxin is transported laterally to the upper portions of the root cap. Auxin then enters the cortical parenchyma cells and moves basipetally up the root, but only in the elongation zone. If the concentration of auxin received by the cortical parenchyma cells is radially equivalent, then the root cell expansion will be equal on all sides. However, if there has been a lateral redistribution of auxin, the root will begin to bend. If a root is tipped sideways, auxin will move (by stimulation of lateral auxin transporters) so that there is a higher concentration on the lower side (this is identical to the shoot). Unlike in the shoot, however, in the roots the concentrations of auxins involved inhibit cell expansion rather than stimulate it, and hence the root expands more on the upper side than the lower side, and the root bends downwards.

Why is the cell extension inhibited in roots and stimulated in the shoot? This question has not yet been fully answered, but there are some indications as to the mechanism of control. It has been known for a long time that auxins generally stimulate a response at low amounts but inhibit it at higher concentrations, and cell extension is one of these examples. At low concentrations auxins stimulate the ATPase/H^+ pump; however, at higher concentrations, the pump is inhibited. It is thought that auxin interacts with the protein binding to the ATPase/H^+ pump (ABP_{57}) and stimulates the pumping of H^+ across the plasma membrane (Figure 11.7). At higher concentrations auxins bind to a second site on ABP_{57}; this relieves the initial stimulatory action of auxins at lower concentrations and therefore reduces the activity of the H^+ pump. In the roots of plants, auxins generate a response at much lower concentrations than would occur in the shoots; thus, a small change in the auxin concentration can inhibit root extension rather than promoting it.

Sensing gravity

Plants sense gravity in particular cells in the shoot and the root, using starch granules within amyloplasts (known as statoliths). The gravity-sensitive cells in the shoot are known as statocytes and in the root the columella cells of the root cap are the gravity sensors (Blancaflor and Masson, 2003). In the shoot, the statocytes form an endodermal cell layer in the hypocotyl, which encircles the vascular tissue. Thus, even with the shoot tip removed, the shoot can sense gravity, In the root, the statocytes form a layer in the columella cells of the root cap (Figure 11.10). Removal of the root cap or destruction of the columella cells using a laser can stop the root from exhibiting a gravitropic response (Blancaflor et al., 1998).

Having a greater density than the rest of the cell, the statoliths move when the root or shoot is moved with respect to the vertical axis. It is not yet certain how the plant senses the movement of the statoliths, but the statoliths only move slightly and do not move freely around the cell. It is thought that it is the unequal pressure caused by these statoliths on the cytoskeleton that signals that the plant has been moved from the vertical plane. This unequal pressure is then thought to bring about lateral transport of auxin, so that a higher concentration of auxin arrives at the lower section of the root or shoot. The shoot then bends upwards and the root bends downwards.

In grasses, the shoot senses and responds to gravity as described earlier. However, the stem structure of grasses is particularly vulnerable to being blown over and, in order to be effective for producing an inflorescence and for seed dispersal, the stem needs to be in a vertical position. To achieve this, grasses periodically have pulvini at the nodes of the stem. The mechanisms by which gravity is sensed and how this is relayed to the cells in the pulvinus are not understood, but the net result is the reversible extension of cells on the lower portion of the pulvinus, causing the stem to be returned to an upright position.

Hydrotropism

Hydrotropism is the response of plant to moisture. This is invariably associated with the positive hydrotropic response of roots in their search for moisture in the soil. It was first put forward that the roots of plants were hydrotropic by Loomis and Ewan (1936). This early work contained many flaws and, although it is essential that roots can detect and respond to the water content of the soil, proving that hydrotropism exists has been remarkably difficult. There have been several studies in Arabidopsis plants to try to discover the mechanism for hydrotropism. It is clear that there are three important factors that dictate how roots orientate themselves in the soil: first, the roots respond to gravity and grow downwards; second, the roots respond to touch (see thigmotropism); third, roots respond to the moisture content of the soil. These three factors need to be balanced in consideration of root orientation. The hydrotropic response in roots has proved to be very weak and is easily masked by the gravitropic response. It has only been through the production of Arabidopsis plants that are deficient in their ability to respond to gravity that the hydrotropic response has become clear. It has been suggested that in Arabidopsis hydrotropism interacts with gravitropism through the degradation of the starch granules in the columella cells of the root cap (Takahashi et al., 2003). This weakens the gravitropic response and then allows the root to bend in response to the moisture content gradient in the soil. The mechanism by which the bending occurs is not understood, but is thought to involve auxin redistribution across the root tip.

Thigmotropism

There is growing evidence to support the view that plants can be extremely sensitive to touch. This statement is not limited to exotic species such as the Venus flytrap (Figure 19.2) or the stigma of the devil's claw plant (Figure 11.12). To put this into context, human skin can sense a thread weighing 2 µg when it is drawn across it. However, tentacles of Drosera species have been shown to be sensitive to a thread weighing less than 1 µg and the

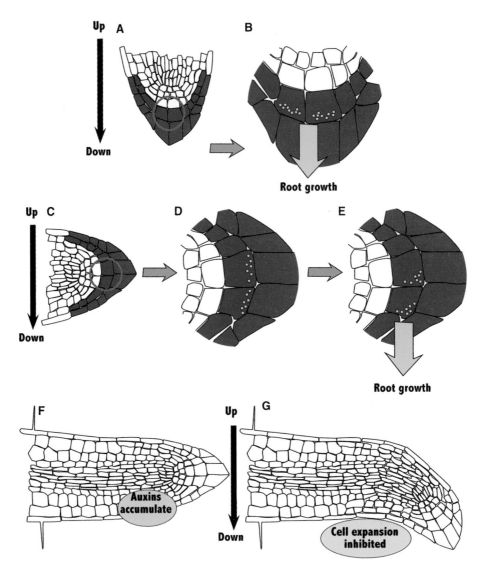

Figure 11.10 The gravitropic response in roots. Auxins also play a role in the sensing of gravity in both shoots and roots. In the roots there is a layer of columella cells (A) in the root cap, which sense gravity through the presence of starch granules (or stratoliths) in these cells. In the vertical position, these starch grains gravitate to the lower areas of the cells (B: the figure exaggerates this for clarity). However, when the root is placed at 90° to the vertical (C), these starch granules move in response to the change in the gravitational field (D). This movement of the granules initiates a redistribution in the auxin concentration in the root (E). A high concentration builds up on the lower side of the root. The presence of high concentrations of auxin in the cells below the shoot tip (F) causes an inhibition of cell expansion (root cells are much more sensitive to auxin concentrations than shoots and at high enough concentrations cell expansion is inhibited rather than stimulated) (G). Thus, the upper section of the root expands more than the lower section and causes the root to bend downwards to restore its vertical position in the gravitational field. The gravitropic response in roots overrides most other tropic responses by roots.

climbing plant species *Sicyos* can sense and respond to a thread of just 0.25 μg (Figure 11.11). Thus, some plants possess even more sensitive outer surfaces than humans!

The notion that plants possessed the sense of touch was first discussed by Charles Darwin (1880) in his book *Movement in Plants*. It was observed that many climbing species of plant could sense whether they were touching something or not in order to use it as a means of support. Plants such as *Bryonia dioica*, *Convulvulus* species, *Clematis* species and some plants in the family Fabaceae do not form a sufficiently strong stem to support their

Figure 11.11 Plants can be very sensitive to touch. Two particularly sensitive species are: *Sicyos* sp. (A), which can sense and respond to a 0.25 µg thread drawn across its leaf surface; and *Drosera capensis* (B), which can sense and respond to a 1 µg thread drawn across the leaf surface.

Figure 11.12 The flowers of several plant species possess touch-sensitive stigmatic surfaces. (A) The V-shaped stigma of *Ibecella lutea*. A bee visiting this flower will rub any pollen from its back onto the stigma, thereby pollinating the flower. However, the stigma, once touched (B), closes up as shown (C) within 5–10 seconds, hence it is highly likely that as the bee leaves the flower the stigma will have closed and the chances of self-pollination are much reduced.

Figure 11.13 Movement in climbing plants. There are numerous examples of plants that, instead of producing a rigid support structure to allow the plant to grow upwards, instead use other plants or structures for support. Here are three different examples of such climbing plants: (A) *Smilax asper*, which uses tendrils to grip other plants; (B) *Ipomea* sp., which uses its stem to encircle neighbouring plants; and (C) *Clematis* sp., which uses the leaf petioles to wind around the stems and branches of other plants.

growth and without external supports they flop and creep along the ground. However, in most habitats there is an abundance of physical supports that these plants can use, such as stone surfaces and other plants. These plants possess structures to enable them to utilize these supports and grow upwards. For most of these species, the plants use exaggerated circumnutation movements (see below). This periodic oscillatory movement of the shoot allows it to

come into contact with neighbouring objects (Figure 11.14). Once a point of contact is established, the plant changes its behaviour and bends at that point, such that the stem or plant structure encircles the point. Through this mechanism, the climbing plant can use the object touched as a means of vertical support.

It is not always the stem that forms the point of contact with support structures. In plants such as peas (*Pisum*

Figure 11.14 Runner bean (*Phaseolus vulgaris*) movement. The shoot of a young runner bean seedling is shown against a circle divided into segments, allowing the movement of the shoot of the bean to be followed over 2 hours, during which the shoot has turned through almost a complete circle. This rotational movement allows the plant to sense any structures nearby that it can encircle and use for support as it grows.

sativum) and white bryony (*Bryonia dioica*), a modified stem structure known as a tendril forms. These structures make an oscillatory motion until a support structure is touched. Then the tendril bends and begins to coil around the support. In white bryony an impressive long coil-like spring is formed (Figure 11.13). The tendrils grip onto neighbouring plants and force them to support the growing climber. In *Clematis* species, instead of a tendril the petioles of the leaves twist around neighbouring structures and form the equivalent of knots. Both tendrils and petioles, once they have entwined their target, harden and cease to be thigmotropic.

The mechanism by which plant movement occurs is still unclear. Darwin (1880), in his experiments, showed that pea tendrils would respond within 30 minutes if touched on one side; this response could be cancelled if touched on the other side shortly afterwards. He also demonstrated that the tendril would not respond if it touched something flat or if a droplet of water was placed on it. However, if a small grain of sand was added to the water droplet, the tendril would respond. Tendrils need a pressure point when they contact a potential support and this elicits bending behaviour. Once touched, the cell membrane of the epidermal cells is distorted. This distortion causes an action potential to be conducted along the epidermal cells. It is at present unclear how this action potential then elicits a thigmotropic response. However, the net effect is differential growth across the tendril, with the point touched becoming compressed and the opposite surface expanding. If a tendril fails to find a support, it can exhibit a number of responses, depending on the species of plant. It can become rigid, begin to coil regardless of whether it touches a support or it can wither away.

In *Arabidopsis*, four genes have been identified that are upregulated when the plant is touched (*TCH1–4*). Three of these encode proteins linked to calcium signalling but *TCH4* encodes a xyloglucan, endotransglycosylase, which modifies the structure of the xyloglucan component of the cell wall. If these proteins are present in climbing plants, it is possible that mechanoreceptors in the cell membranes of touch-sensitive cells trigger a release of calcium in the cell and signal an increase in expression of these genes. In white bryony it has been shown that treatment of the tendrils with specific jasmonates causes them to coil in an identical manner to touched tendrils (Kaiser *et al.* 1994). This effect was shown to be independent of ethylene and auxins.

As has been mentioned earlier, roots also exhibit a negative thigmotropic response. When roots come across obstacles such as stones in the soil, the gravitropic response needs to be inhibited and the root needs to bend to move around the obstacle. This allows the root to follow the path of least resistance. Darwin performed experiments with roots and showed that there is an interaction between gravitropism and thigmotropism. A root growing vertically exhibited a strong thigmotropic response when touched. However, a root that was growing horizontally tended to exhibit a stronger gravitropic response. Thus, a root will always try to grow downwards until an obstacle is hit and then the gravitropic response will be inhibited and a thigmotropic response becomes dominant. However, as a root grows horizontally around an obstacle it will continuously seek to restore vertical growth.

Chemotropism

Chemotropism is the response of a plant to a particular chemical. Shoots do not exhibit a chemotropic response but it has been suggested that roots do. It would seem logical that roots would respond to gradients of different chemical components of the soil and that they would grow towards mineral ions particularly needed by the plant. However, there is little evidence that roots do bend towards specific mineral ions. When a root encounters a higher concentration of a mineral ion that is valuable to the plant, the root system responds by initiating lateral roots in that volume of soil to take up as much of the ion as possible (see Chapter 5).

A chemotropic response is exhibited in the tentacles of certain *Drosera* species (see Figure 11.7). If phosphates or nitrates are applied to the tentacles of *Drosera capensis*, the tentacles will bend towards the centre of the leaf. Strictly speaking this is a chemonastic response, since the tentacles always bend in the same way, regardless of the direction of the chemical signal. However, for certain species with long thin leaves, such as *D. indica* or *D. binata*, the tentacles appear to have greater mobility and exhibit a positive chemotropic response to phosphate or nitrate ions applied to the tentacles, with neighbouring tentacles bending towards the point of application of the nutrients. In addition, if the ions are applied beyond a threshold concentration, most *Drosera* species will show a positive chemotrophic response by bending their leaves to enfold the nutrients.

Nastic movements

Nastic movements occur irrespective of the direction of the stimulus causing them. Such motions include sleep movements in leaves and flowers and the rapid motions in the Venus flytrap (see Chapter 19) and the *Mimosa* plant.

Nyctinasty (sleep movements)

Many species of plant exhibit nyctinasty, whereby the leaves or flowers close up at night time. In some species, such as clovers (*Trifolium*), the leaflets rise at night and press together, and in other species, such as *Oxalis*, the leaflets fold downwards. Some flowers also open and close during the day (Figures 11.15 and 11.16). All of these motions are dictated by a circadian rhythm influencing a pulvinus organ at the base of the leaves. These rhythms occur daily and are controlled by an endogenous pacemaker that keeps time independent of the temperature. The rhythms are influenced by light and, although they will usually persist for several days, if a plant is kept in constant darkness they begin to disappear.

The pulvinus at the base of the leaflets controls the movements. In species such as *Albizia* there have been a number of studies on the mechanism of leaf closure. On both sides of the leaf there are two types of motor cells, the ventral and the dorsal. When the pulvini are exposed to blue light, K^+ ions are pumped out of the dorsal motor cells along with Cl^- ions, and water leaves the motor cells by osmosis. This leads to a collapse in cell turgor in the dorsal motor tissue. At the same time, the ventral motor cells pump K^+ and Cl^- into the cells and water enters via osmosis, which causes the ventral motor tissue to become turgid. This pushes the leaflets apart, opening the leaves during the day. When the pulvini are exposed to red light followed by darkness the reverse happens and the leaves fold up at night. The mechanism is likely to be very similar in all plants that exhibit nyctinastic movements. The topic of how plants tell the time will be covered in detail in Chapter 13.

Seismonasty

Seismonasty is the movement of a plant in response to mechanical stimulation. Mechanical stimulation can be in the form of touching (Figure 11.17), shaking, wounding or burning the plant. Some examples of plants exhibiting seismonasty are the Venus flytrap, *Mimosa*, *Drosera* species and *Utricularia*. In this chapter we will look in more detail at *Mimosa pudica*.

Mimosa pudica is one of several hundred closely related Fabaceae species that react rapidly to touch. These species can actually move in a time frame that can be observed. When touched, the leaflets of the pinnate leaf fold upwards together and then the petiole bends downwards. The folding motion takes less than a second to occur and usually, once one pair of leaflets have closed, the next pair quickly follows and so on, like a set of stacked dominoes falling over. The leaves also exhibit nyctinasty. The function of the seismonastic response is not understood. It has been hypothesized that it could be a defence against herbivory, drying winds or heavy rainfall damage.

At the base of each leaflet and at the base of the petiole there is a pulvinus. It has long been supposed that the movement in the leaves of *Mimosa* was the result of a collapse in turgor of cells in the ventral region of the pulvinus in the leaflets and the dorsal region of the petiole. This collapse is precipitated by a massive rapid efflux of ions and sugars into the apoplastic space, followed by water by osmosis. This would be similar to the mechanism for sleep movements in *Albizia*. However, such mechanisms are usually slow and for *Mimosa* leaves to move so quickly they would need to have some major modifications. An entirely new hypothesis has been put forward,

Figure 11.15 Nyctinasty movements in *Osteospermum* flowers, which move in response to the light. The flowers open in the daylight (B, C) but close at night (A). Simple experiments can probe this response and some of the environmental parameters can be varied, in order to identify the signal for flower opening. During the day, light intensity and temperature rise and the humidity falls, and all of these could play a role in the opening response. Simply varying the humidity has no effect on the flowers, but a rise in temperature and light do cause the flowers to open. If the plants are kept at a constant temperature in the light they fail to open. Therefore, the most likely reason why *Osteospermum* flowers open is because the light heats the flowers and causes the florets within the flower to expand. This process is reversed as the day ends.

based on evidence suggesting that the rapid movements in *Mimosa* may be caused by a protein in the cells forming a gel (Kagawa and Saito, 2000; Pollack, 2001). In the turgid state the protein is stretched, with water bound to it. However, upon stimulation the protein rapidly coils and creates disorder in the water molecules, causing the cells to become flaccid very rapidly. An action potential is propagated from cell to cell, causing neighbouring leaves to respond in a similar manner. The protein then takes around 10 minutes to achieve the stretched state again. This hypothesis would make the motor movement in *Mimosa* very similar to that in animal muscles. There is a long way to go before we understand the motor mechanism of one of the world's fastest moving plants. A discussion of how the Venus flytrap achieves its rapid movement is discussed in Chapter 19.

Thermonasty

This is usually seen as the reversible opening and closing of certain flowers in response to temperature. Such plants are easily identified, as the flowers tend to stay closed unless the sun is shining, but if they are cut and brought indoors they will open even in the absence of light, e.g. tulips and crocuses. The opening motion is a result of permanent growth at the base of the petals and as a consequence the petals get longer and longer.

Epinasty

This is the downward bending of leaves and branches as a result of unequal concentrations of auxins supplied to the

Figure 11.16 Sleep movements in flowers. There are a large number of plant species whose flowers open and close their flowers at various times during the day or night, independent of weather conditions (these are known as aequinoctales), so this phenomenon is not restricted only to the day–night transition. The timing of the response of flowers to a point over a 24 hour period was deemed to be so accurate that Linnaeus, in his volume *Philosophia botanica* (1751), made reference to a 'horologium florae' (floral clock). Such a clock, Linnaeus predicted, would be able to determine the time accurately to within half an hour through the use of a large number of aequinoctales species. Such a clock was probably impractical, as many species that would be used would be considered weeds. All the plants reported to be aequinoctales are insect-pollinated and therefore the flowers will be open when a visitation by their primary pollinators is most likely. The clearest example of this is with several night-flowering tobacco species; these flowers are pollinated by moths, which will only be around at night. Plants flowering during the day and closing at night clearly do not attract pollinators at night. Keeping flowers open at night would also expose the flowers to cooler temperatures or heavy dews that may be damaging. As a flower represents a considerable investment by the plant, it is important that its reproductive potential is not wasted. Clearly, not all plants use this strategy and this may just reflect the variability of hardiness of a flower for particular plant species, or that some plants are just less fussy and have never evolved a mechanism for protecting the flower in this manner. The species shown are: (A) Barbary nut (*Iris sisyrinchium*); (B) daisy (*Bellis perennis*); (C) meadow buttercup (*Ranunculus acris*); (D) Californian poppy (*Eschscholzia californica*).

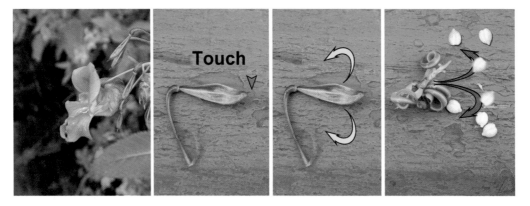

Figure 11.17 Explosive movements in seeds. The dispersal of seeds at maturity is an important method for plants moving to other environments. Many seed pods possess explosive ability; here the Indian balsam (*Impatiens glandulifera*) is shown. When the seed pod is mature, a fault line is created at the tip and along its sides. This region is exceedingly weak and is put under great tension. A slight touch can trigger this fault to open, and the tension within the seed pod is released explosively. The sides of the pod rapidly curl backwards, propelling the seeds violently outwards. This force can throw the seeds several metres away from the parent plant.

dorsal and ventral surfaces of the petiole. This can occur as a response to ethylene gas, and is a permanent growth response.

Circumnutation

Virtually every part of a plant makes circular or elliptical movements around the vertical axis and this motion is known as circumnutation. In climbing plants this motion is exaggerated in order to maximize the opportunity of the shoot coming into contact with possible support structures (see thigmotropism). These motions continue in the light and dark (although their frequency reduces at night time). The motions vary in the angle of bending and the period required for a full circle to be described. Usually the movements are small and a full rotation takes between 2 and 4 hours. The circumnutations can occur in either a clockwise or an anticlockwise direction, and plants have even been known to swap direction. Other than in climbing plants, the function of these motions is unknown. There have been two main hypotheses to explain them:

1. The motions are a result of overcompensation of the gravitropic response, whereby the statoliths take a finite time to move within the statocytes and as a consequence the shoot bends slightly too far and generates the circumnutation motion.

2. The motions are the result of an endogenous rhythm, as a result of growth regulator imbalances across the shoot tip set by the plant as it grows.

In 1984 an opportunity to test whether gravity did influence circumnutation motions presented itself (Brown and Chapman, 1984). Some sunflower plants were transported into space, where the gravitational field was negligible. The plants continued their circumnutation motions, suggesting that gravity does not play a role in these movements. As a consequence, it seems likely that, although light can affect the rate of circumnutation motions, they are a result of uneven cell divisions and cell expansions within the growing shoot tip and are not produced in response to a stimulus, unlike tropic and nastic movements.

Plants can move both in terms of growth and in response to stimuli. Only in a handful of examples is plant movement thought to represent a defence of the plant. For protection, plants rely on defensive structures and a cocktail of poisonous chemicals and this is the subject of the next chapter.

References

Blancaflor EB and Masson PH (2003) Plant gravitropism. Unravelling the ups and downs of a complex process. *Plant Physiology*, **133**, 1677–1690.

Blancaflor EB, Fasano JM and Gilroy S (1998) Mapping the functional roles of cap cells in the response of *Arabidopsis* primary roots to gravity. *Plant Physiology*, **116**, 213–222.

Boysen-Jensen P (1913) Uber die Leitung des phototropischen Reizes in der Avenakoleoptile. *Berichte der Deutschen Botanischen, Gesellschaft,* **31**, 559–566.

Briggs WR and Christie JM (2002) Phototropin 1 and phototropin 2: two versatile plant blue-light receptors. *Trends in Plant Science*, **7**, 204–210.

Brown AH and Chapman DK (1984) Circumnutation observed without significant gravitational force in spaceflight. *Science*, **225**, 230.

Darwin C (1880) *The Power of Movement in Plants*, John Murray, London.

Kagawa H and Saito E (2000) A model of the main pulvinus movement of *Mimosa pudica*. *JSME International Journal Series C – Mechanical Systems, Machine Elements and Manufacturing*, **43** (4), 923–928.

Liscum E and Stowe-Evans EL (2000) Phototropism: a 'simple' physiological response mediated by multiple interacting photosensory-response pathways. *Photochemistry and Photobiology*, **72**, 273–282.

Loomis WE and Ewan LM (1936) Hydrotropic responses of roots in soil. *Botanical Gazette*, **97**, 728–743.

Pollack GH (2001) *Cells, Gels and the Engine of Life: A New Unifying Approach to Cell Function*, Ebner, Seattle, WA.

Ruppel NJ, Hangarter RP and Kiss JZ (2001) Red-light-induced positive phototropism in *Arabidopsis* roots. *Planta*, **212**, 424–430.

Takahashi N, Yamazaki Y, Kobayashi A, Higashitani A and Takahashi H (2003) Hydrotropism interacts with gravitropism by degrading amyloplasts in seedling roots of *Arabidopsis* and radish. *Plant Physiology*, **132**, 805–810.

Kaiser I, Engelberth J, Groth B and Weiler EW (1994) Touch- and methyl jasmonate-induced lignification in tendrils of *Bryonia dioica* Jacq. *Botanica Acta*, **107**, 24–29.

Went FW and Thimann KV (1937) *Phytohormones*, Macmillan, New York.

Went FW (1926) ''On growth-accelerating substances in the coleoptile of *Avena sativa*''. *Proceedings of the Koninklijke Nederlandse, Akademie van Wetenschappen* **30**, 10–19.

Yu F and Berg VS (1994) Control of paraheliotropism in two *Phaseolus* species. *Plant Physiology*, **106**, 1567–1573.

12

Plants and stress

The notion that plants can be stressed conjures images of armies of plant psychologists going to fields to dispense therapy to them. In fact, this was precisely what US President Bill Clinton thought when he tried to prevent funding to support research into how plants survive stresses. However, he was corrected quickly by US scientists, who explained that plant stresses are conditions that cause plants to function suboptimally. An individual plant cannot move from position to another and therefore must cope with whatever conditions it is exposed to or die. Most plants will be stressed in some way or other, whether it be from too few or too many nutrients in the soil, too high or too low a temperature, etc. In this chapter, discussion is focused on the major plant stresses, which are:

- Drought.

- Salinity.

- Flooding.

- Low temperature.

- High temperature.

As plants have evolved, some have become very well adapted to specific niches so that, for example, they can survive very dry environments. Many such species have become so well adapted that they can no longer survive in environments without the stress. Examples of this are many of the carnivorous plants, which have adapted to low-nutrient boggy soils by turning to carnivory. The root systems of these plants are so effective at scavenging the available nutrients from the soil that, if placed in an environment with high nutrient levels, the plants rapidly die as a result of nutrient poisoning or dehydration. The same is true for alpine plant species, which are adapted to dry summers and cold winters. In poorly drained soils many alpine plants die rapidly, due to waterlogging of the root system, but on well-drained composts they are fine. So plants can become dependent on the stresses they are adapted to resist.

Adaptations to drought stress

Many stresses to which plants are subjected manifest themselves through water availability. Drought stress is clearly the lack of water as a result of little rainfall or poor irrigation. There is a series of responses plants made to a lack of water which ultimately lead to the death of the plant (Figure 12.1). One of the first responses is for a plant to reduce aerial growth and promote root growth. The reduction in aerial growth may take several forms. The plant may reduce the production of new leaves, or allow some leaves to abscise, to reduce the area of the leaf canopy. This, of course, makes sense, since it is the roots which harvest water from the soil and the leaves are the major losers of water as a result of transpiration losses, which are necessary to support the gas exchanges of photosynthesis. However, there are limits to this, as it is essential that a plant does not produce too extensive a root system that can no longer be supported by photosynthetic metabolism in the leaves. Once this point is reached, the plant needs to alter its metabolism to cope with a reduced water content. This leads to the accumulation of compatible solutes, such as sucrose, sorbitol, proline or glycine betaine, which lead to osmotic adjustment within the plant cells (Figure 12.2). These compounds are non-toxic in relatively high concentrations and do not interfere with metabolic processes. As the osmotic potential of the cell falls during drought stress, these compatible solutes accumulate and help to maintain water flow into cells and as a consequence maintain cell turgor (Figure 12.3). Some plants can accumulate solutes to combat solute loss, e.g. sugar beet (*Beta vulgaris*) accumulates betaine as a compatible solute,

Physiology and Behaviour of Plants   Peter Scott
© 2008 John Wiley & Sons, Ltd

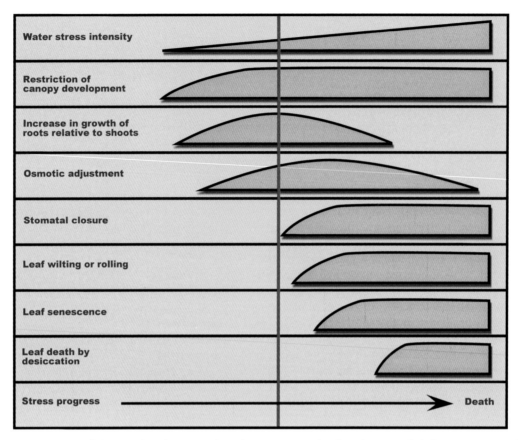

Figure 12.1   The response of plants to drought stress. When plants are exposed to drought stress, their first response is to limit aerial growth and promote root growth. This makes sense, since it is the leaves that lose water and the roots that pick up water from the soil. The plants then begin to accumulate compatible solutes (see Figure 12.2). These compounds accumulate without affecting the metabolic activity of the plants, but they help to lower the water potential of the plant cells, so that water will move by osmosis into the plant. If the drought period continues beyond this, then the plant will have to roll its leaves (in some cases) or wilt. At this point the drought stress will do serious damage to the plant, as it can no longer photosynthesize. Tissue death occurs after this point. The figure is based on Nobel (1988).

whereas other plants, such as cowpeas (*Vigna unguiculata*), are unable to do so. The advantages of accumulation of compatible solutes are short-lived and these only serve as a temporary respite.

If water is not provided, the plant needs to respond by closing its stomata. Under drought stress the stomatal aperture size is controlled by abscisic acid (ABA; Figure 12.4). ABA is synthesized in the mesophyll cells in leaves and is stored in the chloroplasts. As water availability drops, the water potential in leaf cells falls and ABA is synthesized and released from the chloroplasts (Hartung *et al.*, 1998). Elevated amounts of ABA in the guard cells trigger stomatal closure. This has the net effect of conserving water in the leaves, but the plant is left unable to photosynthesize. From this point onwards the plant tissues die, first the leaves and then the stem and roots.

Crassulacean acid metabolism

Given that water stress is probably the most serious stress that plants encounter, there has been considerable evolutionary pressure to evolve means to reduce its impact on plant growth and survival. One such mechanism is Crassulacean acid metabolism (CAM; Figures 12.5–12.8). This is mainly an adaptation to the C3 photosynthetic pathways to reduce water losses as a result of transpiration. For photosynthesis to occur, there need to be plenty of the substrates for the chemical reactions to occur. One of the essential substrates for photosynthesis is CO_2. This gas cannot diffuse across membranes or leaf cuticles in sufficient quantities to fuel photosynthesis, and as a consequence plants need to leave their stomatal pores open during the day to allow the exchange of gases into and out of the leaf (Figure 12.9). Opening these pores results in

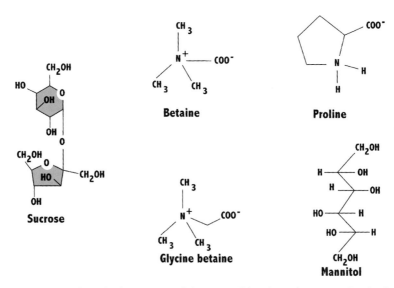

Figure 12.2 Example of compatible solutes in plants. Some of the compatible solutes that accumulate in plants as a result of drought stress are shown.

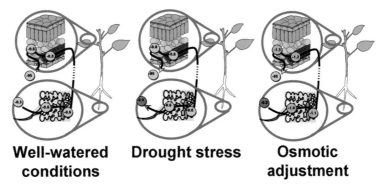

Well-watered conditions **Drought stress** **Osmotic adjustment**

Figure 12.3 Drought stress and plants. A series of diagrams are shown to represent the accumulation of compatible solutes in plants. If the water potential of the soil falls, it could fall such that the water potential in the soil is lower than that in the plant. At this point, water will leave the plant via osmosis rather then the other way around. Plants respond to this by accumulating compatible solutes in their tissues, which lowers the water potential, such that the plant is again able to take up water from the soil.

large quantities of water being lost from the leaf, and it is often estimated that for every 1 g of CO_2 fixed by a C3 plant during photosynthesis, up to 1 litre of water is lost. In many habitats this huge water loss is bearable except under extreme conditions, but in environments subjected to

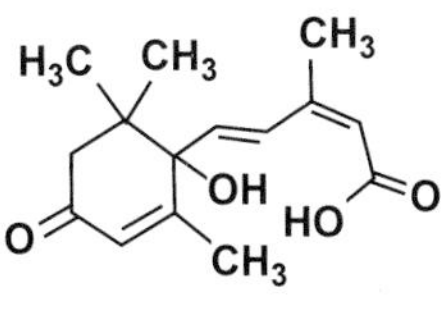

Figure 12.4 The structure of abscisic acid, a plant growth regulator associated with drought stress in plants. It causes the closure of stomata in order to conserve water.

frequent drought, plants would not be able to survive without significant metabolic and morphological adaptations. CAM is a metabolic adaptation which is often observed in plants growing in drought-stressed areas of the world.

CAM plants are easily distinguished from C3 plants because the first product of CO_2 fixation is not the three-carbon compound 3-phosphoglyceric acid but the four-carbon compound oxaloacetic acid (this is rapidly converted to malic acid), and in addition the leaf anatomy is different from that of C3 plants, as it displays succulence (Figure 12.7). The titrateable acidity in leaf extracts of CAM plants varies throughout a 24 hour period, with high acidity at night and lower acidity during the day

Night time

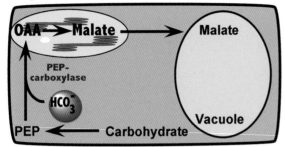

Day time

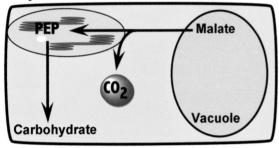

Figure 12.5 A simplified CAM cycle. CAM plants open their stomata at night, when temperatures are lower than during daytime, and fix carbon dioxide onto a PEP molecule to make malic acid, using the enzyme PEPcarboxylase. This is stored throughout the night in the vacuole. When the day begins, the stomata close and the plant mobilizes the malic acid reserves, converting them back to PEP and releasing the carbon dioxide at the same time. The C3 photosynthetic pathway then fixes the carbon dioxide. The PEP molecule is metabolized to starch or a soluble carbohydrate to be used again the next night. The closing of the stomata during the day makes CAM plants 90 % more efficient with water use than C3 plants.

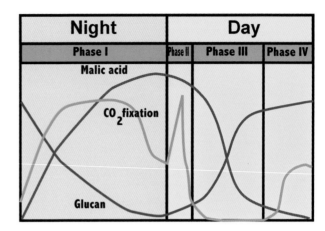

Figure 12.6 Graphical representation of the CAM cycle, which is split into four phases. In Phase I, a glucan is used to make PEP to allow carbon dioxide fixation and malic acid synthesis. When the day begins, the malic acid is not metabolized instantly and there is usually a period where C3 photosynthesis occurs, known as Phase II. In Phase III, malic acid is metabolized as a source of carbon dioxide for photosynthesis. The PEP generated is converted back to the glucan. As the day progresses, malic acid reserves can run out and the plant again opens its stomata and uses C3 photosynthesis; this is known as Phase IV. The extent to which Phases II and IV occur varies considerably between different plant species.

whole mechanism to be of benefit to the plant, at night CO_2 is added on to a phosphoenol pyruvate (PEP) molecule, using an enzyme called PEP carboxylase to give the four-carbon compound oxaloacetic acid. This in turn is rapidly converted to malic acid and is stored in the vacuole of the plant. This leads to the build-up of acidity

(Figures 12.5, 12.6). This feature of varying acidity in the leaves of desert plants was observed by Heyne in 1815, but its significance was not realized until much later (Kunitake and Saltman, 1958). It was discovered that CAM plants accumulate an organic acid during the night (usually malic acid) and then metabolize it during the day. This form of metabolism was named Crassulacean acid metabolism as it was first noticed in plants of the family Crassulaceae, which frequently exhibit it.

The function of the variation of acidity on leaf extracts was discovered to be the result of a remarkable adaptation to the C3 photosynthetic pathway. CAM plants close their stomata during the day, when air temperatures are high and humidity is low, to conserve water. They then open their stomata at night when temperatures are low and air humidity is high. The only problem with this strategy is that CO_2 is needed during the day, when photosynthesis is active, and not at night. To deal with this and allow the

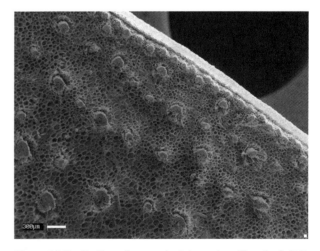

Figure 12.7 Section of a succulent *Agave tequiliana* leaf. CAM plants use their leaves to store large quantities of water and this makes them succulent.

in the leaf at night. Then, as dawn begins, the plant starts to close the stomata and convert the malic acid back into pyruvate, with the release of CO_2. This CO_2 is then used in the C3 photosynthetic pathway described in Chapter 12.

The carbon in the PEP is also cycled. At night the PEP is synthesized through the metabolism of soluble carbohydrate or starch reserves (the actual source varies from species to species). Then, as the malic acid is metabolized to release CO_2, the three-carbon organic acid (pyruvate) is converted back to the same carbohydrate from which it was generated initially. Therefore there are two cycles with a reverse phase occurring in CAM plants: that of malic acid, and that of a carbohydrate source which generates the PEP. The CAM cycle is divided into phases. In Phase I, malic acid accumulates at night. In Phase III, malic acid is metabolized to release CO_2 during the day. The other two phases are transition phases, in which malic acid mobilization is not switched on but malic acid accumulation has ceased. Phase II is a point at the start of the day where temperatures are still relatively cool and the stomata of the plant are open. At this point, ordinary C3 photosynthesis occurs. Phase IV occurs at the end of the day, when the pool of malic acid in the leaves becomes exhausted. In order to continue photosynthesis at this point, the plant needs to open its stomata. Again, this should be a point at which the high midday temperatures have been avoided. Phases II and IV are usually short and in some species they may be missing altogether.

To accommodate the accumulation of malic acid during the day, CAM plants possess large vacuoles, which take up around 90 % of the volume of the cells (in a C3 plant the vacuole normally occupies 70–80 % of a cell's volume). This makes the leaf cells of CAM plants very large and succulent (Figure 12.7). This succulence affects the whole appearance of the plant, and CAM plants have thick fleshy leaves and are usually slow-growing compared to C3 plants. They have major structural and metabolic adaptations that allow them to grow effectively in drought-stressed habitats. Even in well-watered environments, CAM plants grow slowly.

Figure 12.8 Examples of CAM plants; five different species are shown.

However, the CAM pathway allows major savings of water by the plant. A CAM plant typically loses around 100 g of water for every 1 g of CO_2 fixed. This is 90 % more efficient that a typical C3 plant. Hence, CAM plants are much better adapted to drought-stressed habitats than C3 plants. As a result of the costs of being a CAM plant, some plant species have evolved to be able to switch CAM on and off, depending on the environmental conditions. Plants such as *Kalanchoe daigremontiana* are obligate CAM plants and they use CAM under all environmental conditions. However, plants such as *Kalanchoe blossfeldiana* and *Mesembryanthemum crystallinum* are facultative CAM plants, and can switch on the metabolic pathway when presented with specific external stimuli. *K. blossfeldiana* activates its CAM pathway when subjected to short days (a photoperiodic response). The plant is native to southern Africa and the winter time is the period of low rainfall, therefore there is a great advantage to being CAM in the winter. *M. crystallinum* lives in coastal habitats across the world. When the plants are sprayed with salt water, as frequently happens in coastal areas, the plant activates the CAM pathway. As will be seen later, the presence of high salt in the soil causes severe drought stress.

CAM appears to have arisen several times in evolution and is present in many plant families. It has been estimated that around 8 % of plant species exhibit CAM at some point in their life cycle. It is almost exclusive to the angiosperms, with only a few known exceptions, e.g. *Welwitschia mirabilis* (a gymnosperm), some polypodiaceous ferns and *Isoetes* (a lycopod). In the angiosperms, CAM is represented in all of the members of the family Cactaceae and most of the members of the Asclepiadaceae, Bromeliaceae, Orchidaceae and some Euphorbiaceae (Figure 12.8). Some plants exhibit CAM but live in very damp habitats, such as epiphytic orchids and bromeliads. However, these plants do live in very drought-stressed places, since epiphytes live in positions such as on the branches of trees. Rainwater does not collect in such places and the plant is entirely dependent on rain and air moisture to obtain water, therefore CAM is an advantage for such plants.

There is a single known example of CAM being used for a purpose other than for restricting water loss from the plant, i.e. in the freshwater aquatic plant *Isoetes*. Since water availability is not restricted for such a plant, it is thought that CAM gives the plant an advantage in photosynthesis. In aquatic habitats, the availability of CO_2 is frequently limited due to competition between aquatic plants. As a consequence, the levels of CO_2 dissolved in the water drop markedly during the day. *Isoetes* uses CAM to store CO_2 at night, when it is abundant, and then utilizes the malic acid to generate CO_2 during the day, giving it a competitive advantage in such a habitat.

There are only a few crop plants that use the CAM pathway, e.g. pineapple, plants of the genus *Agave* (used in the tequila industry in Mexico and for fibre production) and *Opuntia* species (prickly pears).

C4 photosynthesis

The C4 photosynthetic pathway is an adaptation to oxidative stresses rising during photosynthesis and drought stress. C4 plants exhibit a modification to photosynthetic metabolism very similar to that of CAM plants (Figure 12.10). The first product of photosynthesis is a four-carbon acid, oxaloacetic acid, which is rapidly converted into malic acid, hence the name C4 as opposed to C3. The difference between CAM and C4 plants is that the CO_2 fixation occurs during the day in C4 plants, so their stomata stay open during the day. The C4 photosynthetic pathway was first discovered by Hatch and Slack 1966.

C4 plants are adapted to reduce the flow of carbon lost to the atmosphere by the plant through a pathway known as photorespiration (see Chapter 12). Photorespiration is the reaction of ribulose 1,5-bisphosphate with oxygen, catalysed by the enzyme RUBISCO. Large quantities of fixed carbon can be lost in this reaction, and photorespiration is the pathway plants have evolved to minimize these losses. This reaction is favoured in certain environmental conditions, such as low CO_2 and high temperatures. These conditions are particularly common in tropical regions in the world and it is in these environments that C4 plants have evolved. These plants have evolved mechanisms by which the interaction of RUBISCO with oxygen is minimized, by maintaining very high CO_2 concentrations in photosynthetic cells. To achieve this, C4 plants have a CO_2 chemical pump mechanism. There are three major forms of this chemical pump known, but the simplest one will be discussed here (Figure 12.11).

A majority of the photosynthesis in the leaf occurs in the bundle sheath cells (a layer of cells which is frequently not photosynthetically active in C3 plants), and very little photosynthetic activity occurs in the mesophyll cells of a C4 plant. Under a microscope this is clearly visible, since the bundle sheath cells are dark green as a result of the large number of chloroplasts in them, and they appear as wreaths around the vascular tissue (Figure 12.9). This has led to the name 'Kranz anatomy' for this feature (after the German *Kranz*, which means wreath). The mesophyll cells contain high activities of the enzyme PEP carboxylase, which fixes CO_2 and makes malic acid. The malic acid builds up to high

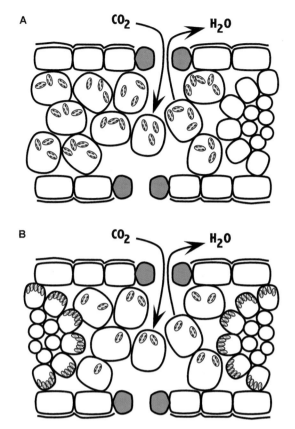

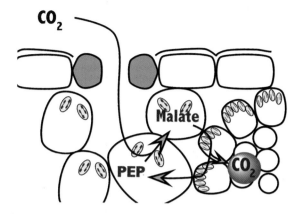

Figure 12.10 The C4 cycle. Carbon dioxide entering the mesophyll cells is fixed onto PEP to make malate. The malate diffuses into the bundle sheath cells and is converted back to PEP, and carbon dioxide is released. The PEP then diffuses back to the mesophyll cells.

Figure 12.9 Schematic representations of (A) a C3 and (B) a C4 plant. There are several modifications in leaf form in C4 plants compared with C3 plants. In C3 plants, the mesophyll cells are the primary sites for photosynthesis and the bundle sheath cells (if present) are non-photosynthetic (A). In C4 plants, the bundle sheath cells are the primary sites for photosynthesis (B). The chloroplasts in these cells are all clustered in a ring on the outer edge of these cells. There are also structural changes in the chloroplasts (shown in Figure 12.12).

concentrations and moves by diffusion into the bundle sheath cells. Here, the malic acid is decarboxylated by NAD malic enzyme and CO_2 is released (Figure 12.10). The pyruvate formed diffuses back to the mesophyll cells and is recycled to form more PEP. The CO_2 released in the bundle sheath cells is then fixed by RUBISCO; however, because of the chemical pump mechanism, the CO_2 concentration in the bundle sheath cells is maintained at a high level and photorespiration cannot be detected in C4 plants. C4 plants also possess a second mechanism to reduce photorespiration occurring. High levels of oxygen favour photorespiration, and one of the major sources of oxygen is photosynthesis itself. Oxygen is released by photosystem II as it reacts with water. The bundle sheath cells lack photosystem II and can only carry out light harvesting

using photosystem I. This presents great difficulties, since photosystem II generates not only oxygen but also NADPH, which is a substrate for the Calvin cycle. In order for the bundle sheath cells to make carbohydrate, 3PGA is exported from the chloroplast in bundle sheath cells to the mesophyll cells, where it is reduced to triose phosphate and exported back to the bundle sheath chloroplasts (Figure 12.12). The bundle sheath cells can then carry out the remainder of the Calvin cycle function for the regeneration of ribulose 1,5-bisphosphate and the synthesis of carbohydrate. Alternatively, some C4 plants can use the NADPH generated from the decarboxylation of malic acid.

The exchange between the cells occurs through plasmodesmatal links between the different cell types. There is also a gas-impermeable barrier around the bundle sheath cells, to prevent oxygen getting into them and the released CO_2 from escaping from the bundle sheath cells. The C4 pathway is very effective at concentrating CO_2 and the C4 plants are generally very efficient at photosynthesis. It is therefore no surprise to find that several of the world's major crops use the C4 photosynthetic pathway, e.g. maize, sugarcane and sorghum. Only around 0.8% of plants are known to possess the C4 pathway and it is thought to have evolved recently. It is present in several families of plants and is thought to have occurred through convergent evolution. The C4 pathway has been observed in 16 different families of flowering plants, in both monocotyledons and dicotyledons. There are some examples of genera of plants which exhibit partial C4 ability. In the genus *Flaveria*, some species are C4, some C3 and others are intermediate C4 plants. The actual mechanism used in these plants is shown in Figure

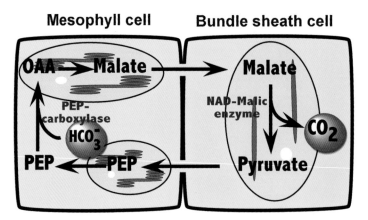

Figure 12.11 Schematic representations of the C4 cycle. Carbon dioxide entering the mesophyll cells in a C4 plant is bound onto PEP by PEPcarboxylase to make malate. This diffuses into the bundle sheath cells, where it is decarboxylated, and the carbon dioxide is released in these cells and then is used for C3 photosynthesis. The remains are reconverted back into PEP and diffuse back to the mesophyll cells. This whole process acts as a carbon dioxide pump mechanism, concentration carbon dioxide in the bundle sheath cells.

12.9. In basic terms, the pathway for photorespiration is split between the mesophyll cells and the bundle sheath cells. When photorespiration occurs, the glycine diffuses into the bundle sheath cells and is converted to serine. During this step, CO_2 is released but, because it is generated in the bundle sheath cells, the CO_2 is not lost and is refixed by RUBISCO in those cells. Therefore in intermediate C3–C4 plants, the photorespiratory pathway acts as a CO_2-concentrating mechanism.

The C4 pathway is thought to be primarily for reducing photorespiration; however, there are other benefits to the pathway. Plants using C4 also have a higher water use efficiency than C3 plants. C4 plants, on average, fix 2–5 g of CO_2 for every 1 kg of water lost. This is a direct result of

the high photosynthetic rates supported by the C4 pathway. The plants can afford to reduce their stomatal aperture, thus reducing water loss from the leaves as a result of transpiration, and still maintain high photosynthetic rates. As a consequence, C4 plants are a very successful group of plants, highly adapted to coping with stresses.

Resurrection plants

There are many adaptations that plants possess to respond to drought stress, such as CAM, but few are as remarkable as the poikilohydric or 'resurrection plants'. Like carnivorous plants, this is a heterogeneous group, made

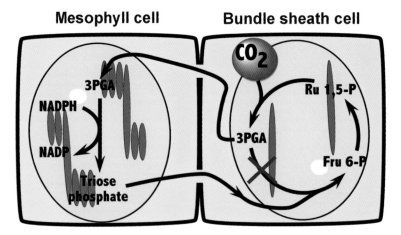

Figure 12.12 The Calvin cycle in the bundle sheath cells. As these cells lack photosystem II (in order to minimize oxygen generation), they cannot catalyse the Calvin cycle fully. As a result of the absence of photosystem II, chloroplasts in the bundle sheath lack granal stacks in the thylakoid membrane. In order to achieve the full cycle, 3PGA diffuses out of the bundle sheath cells and is reduced in the mesophyll cells. Then the triose phosphate diffuses back into the bundle sheath cells to complete the Calvin cycle.

Figure 12.13   Structures in plants that can survive tissue dehydration. Most plant tissues are severely damaged if their water content falls below 55 %. However, there are three major structures in most plants that can survive severe tissue dehydration, namely buds, pollen grains and seeds.

up of monocotyledons and dicotyledons, and if we restrict examples to the flowering plants, there are only around 45 different species that belong to this group. However, many lichens, mosses and ferns are poikilohydric. A poikilohydric plant can lose virtually all of its water and then remain dormant for several years until the rains return; then leaves take up water, unfurl and begin photosynthesis, apparently as if nothing had happened (Scott, 2000). One unusual feature of this ability is that a majority of plant species already possess it to a limited extent in seeds, buds and pollen (Figure 12.13). During maturation, all of these tissues lose a large percentage of their water content to enter a dormant state. For a limited period they can then remain dormant. As explained earlier, seeds can lie in this state for many centuries, with the authenticated record holder being the sacred lotus (*Nelumbo nucifera*) which gave a rate of 75 % germination after 1300 years storage (Shen-Miller *et al.*, 2002). Most plants cannot deploy this ability to other tissues, such as leaves, flower buds and roots. However, the resurrection plants are able to do this. The longest-surviving resurrection plant without water is 7 years in my own laboratory, but the conditions were not ideal and probably plants can survive for much longer periods in the dried state in their natural habitat. Resurrection plants have been discovered in southern Africa, Australia, and deserts in North America. There are two species in the Gneseraceae, *Ramonda myconi* (Pyrenean primrose) and *Haberlea rhodopensis*, which are found in Europe (Figure 12.14). Some species related to these, such as *Haberlea fernandii*, are partially poikilohydric and many of the aerial tissues die during dehydration.

The best-studied resurrection plant is *Craterostigma plantagineum* from southern Africa. There are 10 species in the genus *Craterostigma*, six of which are resurrection plants. The sequence of events which occur in *C. plantagineum* during drying are shown in Figure 12.15. As the plant loses water, the leaves literally shrink, with the oldest leaves dying and forming a cage around the dried living

leaves. The outer leaves are pigmented and protect the surviving shoot from prolonged exposure to sunlight. In the dried state, the plant is dormant and metabolism proceeds very slowly; as a consequence, photosynthesis cannot cope with the rate at which light falls on the plant and this could be very damaging. Therefore, by shading the surviving leaves, damage is minimized.

One common feature in all resurrection plants is the accumulation of the carbohydrate sucrose during the drying period (Figure 12.16). Some authors have suggested that trehalose also accumulates, as it does in some poikilohydric animals, but there is little evidence to support this. Sucrose appears to need to accumulate to around 100 µmol/g dry weight in the leaves, shoot and roots to confer protection. Sucrose stabilizes the membranes and proteins in the dried state, such that they do not denature. To some extent it is believed that the OH bonds on the sucrose molecule can act in a similar manner to the OH bonds in the water molecule. This stabilization prevents the denaturation of enzymes in dried tissues and means that enzyme activity can be assayed in the dried plant tissues. Proteins called dehydrins and late embryogenesis-abundant proteins also accumulate in the mature tissues of resurrection plants, but their role has not yet been fully identified.

In their natural environment, resurrection plants dry during times when there is no water. If a small amount of rain falls the plants partially rehydrate, but they can quickly dry again if the water availability is limited. However, when sufficient rain falls the leaves quickly take up the water, and in *Craterostigma* the plant is ready to photosynthesize again within 24 hours. Other species can take longer to recover. In addition, *Craterostigma* usually flowers within 2 weeks of recovering from the dehydrated state. Thus, resurrection plants can occupy a habitat which has periodic rainfall but lengthy periods of drought. This means the plant can respond to water rapidly and outcompete other species in the habitat growing from

Figure 12.14 Poikilohydric plants, such as the Pyrenean prim-rose (*Ramonda myconi*), often called a resurrection plant, can survive dehydration. The tissues of this species can survive water loss down to 5 % relative water content, allowing it to survive in environments where the conditions can cycle from being well-watered to arid. These specimens were found in the Ordessa National Park in Spain (A). The plants live on vertical walls of rock (B). Periodically the habitats dry out in the summer and the plants shrivel (C). When the rains return in the autumn, the plants rapidly regain their water and grow.

seeds. They also can flower rapidly and reproduce to release seeds before the onset of further drought.

Adaptations to salt stress

Salt stress might be expected to be a problem only in coastal regions where seawater is blown in mists across vegetation. However, salt stress for plants is a much wider problem. Water running over mountains and hillsides gradually accumulates small quantities of mineral ions from the soil. This build-up is usually not high enough to threaten plants, but if the water stands for any length of time, e.g. in a lake or in underground reserves of water, the evaporative losses cause the salt concentrations in the water to rise. Frequently underground reserves of water are used for irrigation of crops and, through evaporation and transpiration, the salt concentrations in the soil rise and crop productivity is seriously threatened. Soil salt levels may threaten crops but there are many wild species, known as halophytes, which thrive on saline soils. There are two major issues for plants subjected to salt stress. First, the major salt in saline soils is sodium chloride and the high availability of Na^+ ions near the roots tends to lead to a rise in tissue Na^+ ions. High concentrations of Na^+ ions in cells can affect the function of certain enzymes. In addition, a rise in Na^+ ions causes a drop in the osmotic potential of the water in the soil, therefore it is much harder for the plant to obtain water from the soil, hence the effect of salt on plants is very similar to that of drought stress, as detailed in Figure 12.3.

The Na^+ ions enter the roots through K^+ transporter and non-selective cation channels. Some plants can regulate the import of Na^+ ions by controlling these channels but other plants, such as rice, allow Na^+ ions into the apoplast and they then directly enter the transpiration stream and reach the growing shoot and aerial tissues. Plants such as *Arabidopsis* possess a Na^+/H^+ antiport channel, which moves Na^+ ions back out into the soil. This reduces the build-up of Na^+ ions in the cytoplasm. Na^+ ions are also pumped into the vacuole, which helps to lower the cell's osmotic potential. Plants also respond to elevated concentrations of Na^+ by expressing genes which will maintain the ratio of K^+ ions to Na^+ ions. The normal concentration of K^+ in plant cells is around 150 mM, and 5 mM for Na^+. As Na^+ ions enter a cell, this balance is upset and the function of many metabolic processes and protein synthesis is inhibited. The plant alters gene expression to try to maintain this ratio as close as possible to the normal balance by stimulating pumps to exclude Na^+ ions. Enzymes extracted from halophytic plants are just as sensitive to Na^+ ions as non-halophytes; however, halophytes are much more effective at preventing levels of Na^+ building up in the cytoplasm to such a level where cellular metabolism is inhibited.

Figure 12.15   The cycle of drying and rehydrating in the poikilohydric plant *Craterostigma plantagineum*. The drying process can take several days so the relative water content of the plant is shown as it dries. The leaves literally shrink to around 10 % of their original size. On rewatering the plants, the rehydration process is very rapid; the leaves can be photosynthetically functional within 12 hours after adding water. These plants have been known to survive for up to 7 years in the dehydrated state.

There are many mechanisms by which Na$^+$ ions can be excluded from the plant. Active pumping mechanisms can be employed in the roots to exclude Na$^+$ ions; however, such mechanisms are likely to be energetically expensive. For Na$^+$ ions to get to the leaves or growing shoot tip, they need to enter the transpiration stream. One mechanism of exclusion is that Na$^+$ ions are excluded from crossing the Casparian strip in the roots and therefore are less likely to be transported to the aerial tissues. Other plants take up the Na$^+$ ions in the xylem and then in the leaves the Na$^+$ is sequestered into salt glands, in which the salt crystallizes. An example of such a plant is what is frequently called the ice plant (*Mesembryanthemum crystallinum*). On its leaves there are numerous crystalline glands, containing salt, which give the plant a frosted appearance. This maintains low Na$^+$ concentrations in the meristem, expanding leaves and photosynthesizing leaves.

The means by which plants take up water have been discussed in Chapter 4. However, it is clear that, in order to take up water from a saline soil, a plant needs to have a lower osmotic potential than the soil. Many plants use compatible solutes to achieve this. The accumulation of metabolically inert compounds, such as proline, glycine, betaine and sorbitol (just as in drought-stressed plants), helps to maintain a low osmotic potential in plants such as mangroves. Many of the adaptations discussed earlier for drought stress can also be used to protect a plant from salt stress. The non-halophytic plants may use these

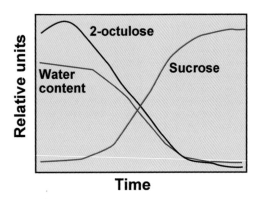

Figure 12.16 Sugar metabolism in dehydrating *Craterostigma plantagineum*. One of the major methods this plant uses to protect itself in the dried state is through the accumulation of sucrose. The plant uses a carbohydrate (2-octulose) store in the leaves in order to rapidly accumulate large amounts of sucrose. This produces a crystalline structure in the cells and enables the plant to survive long periods of severe drought.

mechanisms to some extent but in most cases they do too little too slowly and are damaged or die as a result of salt stress.

Adaptations to flooding stress and anoxia

Flooding stress is at the other extreme of drought stress, and to many plants too much water can be just as dangerous as too little. The major problem for plants during flooding is the availability of oxygen to support aerobic respiration. The roots are mainly non-photosynthetic and are completely dependent upon respiration to provide ATP and NADH to fuel essential processes. In the absence of oxygen (anoxia) or limited oxygen (hypoxia), the Krebs cycle ceases due to the lack of a terminal electron acceptor for the oxidation of NADH (Figure 12.17). In order to continue metabolizing sugars under anoxic conditions, the organism must oxidize NADH by some other means. The response of plant cells to anoxia is to ferment, using metabolic intermediates as electron acceptors. This process leads to the accumulation of lactic acid, alanine and ethanol (Davies, 1980). Whilst lactic acid is the main compound accumulated as a result of fermentation in mammals, plants produce mainly ethanol.

When the Krebs cycle activity in plants is inhibited, pyruvate builds up in the cytosol and NADH concentrations rise. Initially lactate dehydrogenase converts the pyruvate to lactic acid with the concomitant oxidation of NADH to NAD. This leads to a fall in cytosol pH and gradually during this process lactate dehydrogenase is inhibited and pyruvate decarboxylase is activated (pH optimum 5.0; Davies, 1980; Roberts *et al.*, 1985). The latter produces ethanal, which is then rapidly converted into ethanol. The ethanol can diffuse across the plasma membrane and thus be lost into the solutions surrounding the roots. Some ethanol does accumulate in the roots but it is thought that this is not particularly toxic. However, as

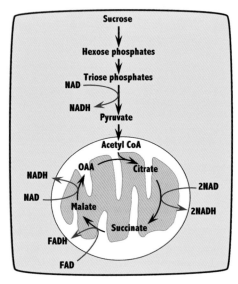

Aerobic respiration

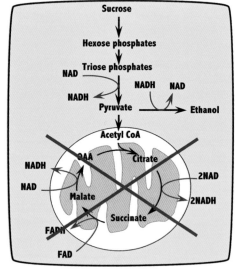

Anaerobic respiration

Figure 12.17 The functioning of the Krebs cycle in anoxia. The Krebs cycle generates most of the energy needs of a plant, in terms of ATP, in non-photosynthetic tissues. However, if the tissue is deprived of oxygen, the Krebs cycle can no longer function. Plants accumulate lactic acid and ethanol as an alternative under these circumstances.

Table 12.1 The ATP:ADP ratio in roots of a range of plant roots in air and under anoxic conditions.

Tissue	ATP:ADP ratio	
	Air	*Anoxia*
Soybean (*Glycine max*)	1.5	0.3
Maize (*Zea mays*)	4.0	0.9
Pea (*Pisum sativum*)	1.6	0.7
Rice (*Oryza sativa*)	1.2	1.2

Most plant roots exhibit a dramatic fall in the ATP:ADP ratio when subjected to anoxic conditions. Only rice, which is flood-tolerant, can maintain this ratio.

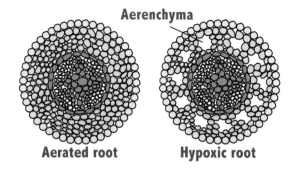

Aerenchyma

Aerated root   **Hypoxic root**

Figure 12.18 The presence of aerenchyma tissues in roots. Some plants live in environments where flooding is frequent. In many cases, they respond by producing aerenchyma tissue in their roots. Cells within the roots die to produce channels down the root system, as shown, through which air can diffuse. This can allow the root system to respire even when flooded.

ethanol concentrations rise, so do the concentrations of ethanal and it is thought that this is the major toxic agent during plant anoxia. In addition, under aerobic conditions a plant cell produces up to 38 ATP molecules per glucose molecule respired, whereas during anaerobic respiration this is cut to just two ATP molecules! As a consequence, most plants under anoxia cannot maintain high levels of ATP (see Table 12.1). This severe drop starves anoxic tissues of energy for cellular processes, therefore many metabolic functions are severely disrupted and this can lead to cell death.

When flooding occurs, the temperature can have a critical effect on whether or not a plant survives. If the root system is flooded when temperatures are low, then little damage is done, since metabolism is so slow under these conditions. However, if flooding occurs in the summer, when temperatures are high, then severe anoxic damage to a plant can be done very quickly. Plants such as potatoes or peas do not survive for more then 3 days under flooded conditions.

One of the best adaptations of plants to anoxia is the avoidance of it altogether. Certain plants such as rice and maize can, through selective cell death, produce aerenchyma channels through the roots during root development (Figure 12.18). These channels permit oxygen to reach the roots even during flooding, so anoxic conditions are not experienced by the root system.

Another adaptation is to maintain the ATP at a high level in anoxic cells. The actual mechanism by which this is achieved is still unclear, but those species that can maintain high energy levels during anoxia are among the few known species whose seeds can germinate and grow under complete anoxia, e.g. rice and *Echinochloa*. However, in order to form roots and leaves, oxygen is required and therefore the plants must reach the surface of the water if they are to survive.

Adaptations to cold stress

Plants are generally poikilothermic, i.e. unable to regulate their temperature, and so their temperature will vary depending on their environment. In a small number of instances there is evidence that some plants can regulate their temperature. For example, the skunk cabbage (*Symplocarpus foetidus*) from North America flowers in the springtime and is able to keep tissue temperatures above 15 °C, even when the temperature falls to −15 °C. The heat generated by the plant can lead to snow melting around the plant. However, this is an exceptional example and the mechanism by which it occurs is not yet understood. Plant species are usually highly adapted to particular habitats and cold temperatures easily kill plants from tropical regions of the world and vice versa, whereas plants from temperate regions of the world can, in many instances, tolerate even tissue freezing (Figure 12.19). Chilling has three major effects on plants: it inhibits photosynthesis and promotes photoinhibition; it inhibits translocation in the phloem; and it influences membrane fluidity and can result in damage to biological membranes (Figure 12.19). The lowest temperature to which plants have been exposed experimentally is just above −273 °C. Seeds have been treated in liquid helium for up to 10 hours and have still germinated (Hopkins, 1999). In my own laboratory, dried *Craterostigma plantagineum* plants have been placed in liquid nitrogen (−196 °C) for 5 hours and have been revived within 24 hours by watering. Other tissues, such as a germinating pea, have been exposed to −183 °C for 1 minute and then continued to grow (Street and Opik, 1984). The coldest temperature ever recorded on Earth is around −89 °C, so it is clear there are plant structures which can survive the coldest temperatures on the planet.

Figure 12.19 Frost damage of plants. Freezing temperatures can be very damaging to plants. Freezing affects metabolism, membrane function and the osmotic relations of the plant: (A) leaves of a meadow buttercup (*Ranunculus acris*); (B) a bristly oxtongue (*Helminthia echioides*). The cold temperatures have caused the plants to wilt, due to the lack of available water for transpiration.

The deciduous trees in temperate environments have adapted to colder winter-time temperatures by dropping their leaves (Figure 12.20). Well before freezing temperatures are experienced, the contents of the leaves are reabsorbed into the branches. Some plants are particularly successful at this reabsorption and only brown husks are left, e.g. beech and horsechestnut trees. Other species, such as the plane tree, frequently drop green leaves where all of the contents have obviously not been reabsorbed. For most European deciduous tree species, as the chlorophyll is reabsorbed the green coloration of the leaves fades and anthocyanin breakdown products in them become apparent, giving them a yellow appearance. In North America, many tree species convert chlorophyll gradually into anthoxanthins, which slowly turn the leaves a spectacular bright red colour. The whole process is influenced by a combination of cold night-time temperatures and warm day-time temperatures and, given perfect conditions, glorious colourful displays are to be seen across New England. However, these are short-lived because, as the leaf components are reabsorbed, a corky abscission layer forms between the parent plant and the leaves. When reabsorption is almost complete, extracellular enzymes digest the cell walls and this junction weakens. Then a small gust of wind loosens the leaves and they fall. Scale-like buds are left for next year's growth. The water content in tree buds is greatly reduced and they are therefore protected from the cold winter temperatures. Leaf fall may seem like a great waste, but it is not worth the expense of maintaining a full canopy of leaves in winter, when temperatures and light intensity are low and water availability is severely restricted (as a result of freezing). In addition, high winds and snow can easily damage leaves.

Conifers are much better adapted to colder temperatures. The small raised leaves offer little wind resistance and snow easily slips off them. In addition, the small waxy needle-like leaves conserve valuable water during the winter. These leaves can photosynthesize during cold periods but they are not as efficient at photosynthesis as deciduous trees during the summer time. As a result they tend to be found mainly in the temperate regions of the world.

Other species that inhabit cold environments possess special physiological adaptations. The European edelweiss plant possesses hairs to reduce water loss and to protect the plant from great swings in temperature between the day and the night. Other species, such as the cabbage groundsel, protect the growing shoot tip by closing the leaves at night to prevent cold air penetrating the more sensitive areas of the plant.

Temperatures below 5 °C damage plants such as potatoes or maize. However, the damage that occurs varies, depending on the plant species. Sometimes necrotic spots are evident on leaves, the plant wilts or growth is reduced. It is thought that chilling damages central processes in chilling-sensitive plants, which has very broad implications for the whole plant. The most likely explanation is that chilling damages biological membranes. The ratio of unsaturated to saturated fatty acids in membranes in plants bears a direct correlation to the chilling sensitivity of a plant (Lyons, 1973). Membranes possess a property known as a transition temperature, i.e. the temperature at which the membrane changes from a fluid to a gel. This alteration has major implications for the biological functions of the membrane, many of which cease at temperatures lower than the transition temperature. A membrane containing a higher proportion of unsaturated fatty acids has a lower transition temperature than one with a lower proportion. Plants which are chilling-sensitive tend to have membranes with lower proportions of unsaturated

Figure 12.20 Leaf fall in trees. As freezing temperatures can be very damaging to plants, many trees respond by metabolizing the contents of the leaves and allowing them to fall over the winter. (A, B) A range of birch (*Betula*) species in Denali National Park, Alaska, USA. (C) Ornamental *Acer* and (D) *Prunus* species. The yellow colourations in the leaves are the carotenoids that accumulate over the summertime. The red colours in leaves are caused by the accumulation of anthoxanthins as chlorophyll is broken down.

fatty acids than chilling-tolerant plants. Plants can acclimate to survive lower temperatures. If *Arabidopsis* plants are subjected to cold for one night, acclimation commences; however, the whole process will take over a week to become fully activated (Thomashow, 1999). After return to normal temperatures, acclimation is rapidly lost again. Acclimation involves the progressive changing of the fatty acid composition of membranes in cells in order to lower the transition temperature of the membranes. A nice example of what happens at the transition temperature is seen with mung beans (*Vicia radiata*). In this plant the membranes have a transition temperature of 14 °C, and at this temperature and below, the rate of growth of seedlings drops drastically (Raison and Orr, 1986).

Plants are also exposed to freezing stress. A species of larch (*Laryx dahuria*) can survive in frozen areas of Siberia, where temperatures as low as −70 °C are experienced regularly. Therefore, plants possess the ability to survive very cold temperatures indeed (Figure 12.18). However, many plants are damaged by temperatures below 0 °C. It is the formation of ice crystals in the plant that causes most of this injury. It has already been mentioned that desiccated tissues, such as seeds, can survive very cold temperatures and this is precisely because of the low tissue water content, preventing ice formation. Plant tissues can survive very rapid freezing in liquid nitrogen, since only small ice crystals form and this causes minimal cellular damage. This type of freezing does not occur in nature and slow gradual freezing results in much larger crystals forming, and these can cause a great deal of damage to the cells. With very slow freezing, a drop of 10 °C/min, ice begins to form where the dissolved solids in the solution are at their minimum, and this is usually in the apoplastic space (Hopkins, 1999). Ice crystals that form grow and tend to draw water out of the neighbouring cells. This leads to tissue wilting but, provided the ice crystals are prevented from forming in the protoplast of the cells, freezing damage does not occur. At freezing rates between these two extremes (slow freezing at 10 °C/min and virtually instantaneous freezing in liquid nitrogen), freezing damage can occur.

There has been some progress towards understanding the changes in gene expression by exposing *Arabidopsis* plants to cold (Kreps *et al.*, 2002). Approximately 30 % of the transcribed genes in *Arabidopsis* increase in expression during cold exposure. Many of the genes sequenced are identical to genes expressed during drought stress and salt stress, suggesting that a plant's response to these stresses may be very similar. A couple of genes, such as COR15a, have been targeted and expressed in plants continually (Artus *et al.*, 1996). This has resulted in a moderate increase in the ability of the plants to tolerate cold stress.

However, it is becoming clear that tolerance to cold is controlled by groups of genes, which all need to be altered to confer a high degree of protection.

Adaptations to heat stress

Plants can tolerate brief exposure to high temperatures, but a temperature above around 50 °C is severely damaging for plants not acclimatized to these temperatures (Hopkins, 1999). The air temperature rarely rises above 45 °C, so such heating effects may be expected to be rare; however, high light intensities regularly increase the temperature of a leaf by a further 5 °C or more. Plants maintain cool leaf temperatures through transpiration and evaporation from the leaves. If water availability is limited, then evaporative cooling is reduced, so plants are much more susceptible to heat damage during drought conditions. Certain plants, such as desert plants, can tolerate very high temperatures. Plants which use CAM as a means of photosynthesis, however, cannot use evaporative cooling; instead, the leaves convert light hitting the leaves (which normally

would lead to heating) to longer wavelengths of light and radiate these. By doing this, species such as the agave and cacti can survive temperatures of up to 60 °C (Nobel, 1988).

Other species move their leaves to reduce the incidence of light on the leaf and thereby reduce the heating effect. Species such as beans can raise their leaves so that they are in line with incident light (see Chapter 11). Grasses show a similar response by rolling their leaves. Large gland cells on the upper surface of the leaf lose their turgor, and this leads to a reduction in surface area of the leaf surface and the leaves roll. How grasses interpret the high light signal and initiate the response is not known. Other species are adapted to hot environments through the possession of reflective hairs or reflective wax surfaces, e.g. sage (*Salvia officinalis*) and rosemary (*Rosemarinus officinalis*) (Figure 12.21). One plant is known to produce two forms of leaf in different seasons; the white brittlebush (*Encelia farinosa*) produces hairy leaves in the summer but hairless leaves in the winter.

High temperatures are thought mainly to affect enzyme function through denaturation. However, raised temperatures also affect the fluidity of the biological membranes.

Figure 12.21 Coping with high temperatures and high light levels. High temperatures and light levels are just as damaging as freezing can be to some plants. Some plants have adapted to hot environments by forming hairs on the undersides of their leaves. The leaves curl in bright light to expose their undersides to the light. The silvery hairs on the leaves act to reflect the sunlight and hence reduce the temperatures experienced by the leaves. Here are shown the Mediterranean species sage (*Salvia officinalis*) and rosemary (*Rosmarinus officinalis*). Detail of the hairs is shown on the electron micrographs of the undersides of the leaves.

In contrast to chilling acclimation, membranes require a higher proportion of saturated fatty acids in order for membrane permeability not to be affected by high temperatures (Raison *et al.*, 1980).

Brief periods of exposure to high temperature, heat shock, in plants cause a drop in the synthesis of proteins. However, there is one group of proteins, known as heat shock proteins (HSP), which are not expressed under normal conditions but are expressed at high levels during heat shock. These proteins have been identified to be molecular chaperones associated with the import of proteins into the various compartments of the plant cells, e.g. mitochondria, chloroplasts, etc. Another protein, ubiquitin, is expressed and this protein is known to function as a protein which marks other proteins for proteolysis. These proteins aid in the restoration of normal cellular functions after exposure to high temperatures by turning over damaged proteins and aiding in movement of newly synthesized proteins into the correct cellular compartment.

Plants possess a remarkable ability to survive a range of different environmental stresses and this is an essential consequence of their sedentary life style. As has been mentioned in Chapter 12, this has also led to plants being able to synthesize a wide range of organic compounds to defend themselves through secondary metabolism. The profound ability to make a huge range of chemicals takes on a whole new dimension when considering possible uses of these chemicals by humankind. In the next chapter we will be considering the use of plants for medicinal purposes.

References

Artus NN, Uemura M, Steponkus PL, Gilmour SJ *et al.* (1996) Constitutive expression of the cold-regulated *Arabidopsis thaliana COR15a* gene affects both chloroplast and protoplast freezing tolerance. *Proceedings of the National Academy of Science of the United States of America,* 93, 13404–13409.

Davies DD (1980) Anaerobic metabolism and the production of organic acids, in *The Biochemistry of Plants – A Comprehensive Treatise, vol 2, Metabolism and Respiration* (eds PK Stumpf and EE Conn), Academic Press, New York, pp.581–611.

Hartung W, Schiller P and Karl-Josef D (1998) Physiology of poikilohydric plants. Cell biology and physiology. *Progress in Botany*, 59, 299–327.

Hatch MD and Slack CR (1966) Photosynthesis by sugarcane leaves. A new carboxylation reaction and the pathway of sugar formation. *Biochemical Journal*, 101, 103–111.

Heyne B (1815) On the deoxidation of the leaves of *Coltyledon calycina*. *Transactions of the Linnean Society of London*, 11, 213–215.

Hopkins WG (1999) The physiology of plants under stress, In *Introduction to Plant Physiology*, Wiley, Chichester, UK, 460–461.

Kreps JA, Wu Y, Chang HS, Zhu T *et al.* (2002) Transcriptome changes for *Arabidopsis* in response to salt, osmotic, and cold stress. *Plant Physiology*, 130, 2129–2141.

Kunitake G and Saltman P (1958) Dark fixation of CO_2 by succulent leaves: conservation of the dark fixed CO_2 under diurnal conditions. *Plant Physiology*, 33, 400–403.

Lyons JM (1973) Chilling injury in plants. *Annual Review of Plant Physiology*, 24, 445–466.

Nobel PS (1988) *Environmental Biology of Agaves and Cacti*, Cambridge University Press, New York.

Raison JK, Berry JA, Armond PA and Pike CS (1980) Membrane properties in relation to the adaptation of plants to temperature stress, in *Adaptation of Plants to Water and High Temperature* (eds NC Turner and PJ Kramer), Wiley, New York.

Raison JK and Orr GR (1986) Phase transitions in liposomes formed from polar lipids of mitochondria from chilling-sensitive plants. *Plant Physiology*, 81, 807–811.

Roberts JKM, Andrade FH and Anderson IC (1985) Further evidence that cytoplasmic acidosis is a determinant of flooding intolerance in plants. *Plant Physiology*, 77, 492–494.

Scott P (2000) Secrets of eternal leaf. *Annals of Botany*, 85, 159–166.

Shen-Miller J, Mudgett MB, Schopf JW, Clarke S and Berger R (1995) Exceptional seed longevity and robust growth: ancient sacred lotus from China. *American Journal of Botany*, 82, 1367–1380.

Thomashow MF (1999) Plant cold acclimation: freezing tolerance genes and regulatory mechanisms. *Annual Review of Plant Physiology*, 50, 571–599.

13

Plant senses and perceiving the world

In order to succeed in a constantly changing world, all organisms need to be able to sense their environment (Figure 13.1). Mammals use their senses of sight, touch, taste, smell and hearing to gather information about the world around them, and they then respond appropriately. The same is true for plants; they need to use similar senses to be aware of their environment. However, plants rarely need to respond rapidly to signals and, as a consequence, everything occurs in a slower time frame than with animals. In this chapter we shall explore how plants sense their surroundings.

Sensing light (sight)

Plants can see but their sensory mechanism for light lacks the resolution that is observed in a mammal. The autotrophic nature of plants makes them very dependent upon light and there are only a few plants species that cannot photosynthesize. Therefore, it is crucial that plants can sense light and respond to it. Plants need to locate light sources and grow towards them. Then they need ensure that their leaves are orientated in the correct way to maximize light incidence on the photosynthetic organs (Figure 13.2). The process of photosynthesis and phototropism has been handled in earlier chapters, but there is further information than this that plants gain from sensing light. Plants live in a changing environment, with day/night changes, seasonal changes, weather changes and habitat changes. This means that plants need to be able see their surroundings and then need to be very flexible in their behaviour to respond to these changes. Even photosynthesis has to be modified continually to cope with changing illumination. The sun should be brightest at midday, but few days are without clouds that can temporarily block out the sun. This leads to huge variations

in light intensity with which a plant needs to be able to deal. The main response of plants to varying light is to have an excess capacity in the electron transport chains in the thylakoid membranes and in the enzyme RUBISCO. In fact, it has been shown that there is four to five times more RUBISCO present in some plants than is actually required. Some of this excess activity can be utilized when there are huge variations in the photosynthetic rate as a result of changes in light intensity.

There are several pigments in plants other than chlorophyll which can detect light:

- Protochlorophyllide.

- Phytochromes.

- Cryptochromes.

These influence many different ways in which a plant responds to light.

Protochlorophyllide

Protochlorophyllide is a precursor molecule for the production of chlorophyll (Figure 13.3). Chlorophyll cannot form in the absence of light and the pathway for synthesis arrests at the protochlorophyllide molecule. This influences the whole structure of the plastid. The thylakoid membranes do not form and the plastid remains as an etioplast. Only when light hits the pale yellowish protochlorophyllide pigment does the final conversion to the green chlorophyll pigment occur and a chloroplast forms from the etioplast. Thus, protochlorophyllide acts as a sensor for light and triggers the plant to begin making pigments for photosynthesis.

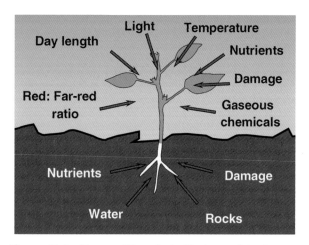

Figure 13.1 The sensitive plant. Plants need to be very sensitive to their environment in order to survive. This figure summarizes the different stimuli a plant needs to be sensitive, to in order to respond in the correct way at the right time.

Phytochromes

The plant *Arabidopsis* possesses one of the smallest genomes known in higher plants. In this simple plant, five different forms of phytochrome have been identified, so in more complex plants there are likely to be more

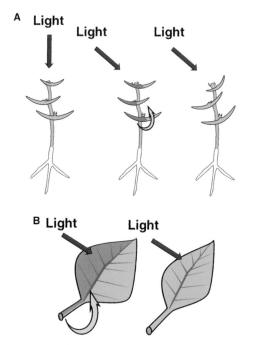

Figure 13.2 Leaf movement in response to light. Light is crucial for photosynthesis, so it comes as no surprise that plants move their leaves to maximize the incident light on the leaf surface (A). The blade of the leaf itself does not move, but the petiole can rotate (B) in order to maximize the light on the leaf surface. The light is sensed by the petiole for this rotation.

forms (Figure 13.4). The presence of these different forms has made the task of unravelling what they do very complicated. In *Arabidopsis* the phytochromes are all very similar; they involve a light-sensitive pigment (chromophore) linked to a protein (apoprotein). The pigment is identical in each phytochrome but the structure of the protein varies and this gives the molecule its specific properties.

Phytochromes have two forms, Pr and Pfr. The Pr form absorbs light of the red end of the visible light spectrum and is converted into Pfr, whereas Pfr absorbs light in the far-red end of the light spectrum (longer wavelengths of red light) and is converted to Pr (Figures 13.5, 13.6). The plant can sense these changes and responds to them. In most studies, Pfr has been shown to be the form that elicits a response from the plant. Now all that needs to be understood is under what circumstances the red and far-red light levels in a plant's environment vary. In Table 13.1 the ratios of red to far-red light are shown under a number of different conditions.

Seed germination

Most seeds require light for germination, and provided the seed is not dormant and has sufficient moisture, it will begin to germinate. The presence of light informs the seed that it is near to the surface of the soil, and the high moisture content of its surroundings informs the seed it is in a suitable place to germinate (Figure 13.7). The ability to germinate in the light is under the control of phytochrome, and lettuce (*Lactuca sativa*) seeds are usually quoted as an example that illustrates this. Lettuce only requires a very short pulse of light in order to germinate. It is red light that triggers germination, and if the seeds are illuminated with red light for a couple of minutes then they will germinate within days. However, if the seeds are subsequently illuminated with far-red light, just after a pulse of red light, then the seeds will not germinate. The Pfr form of the active phytochrome triggers germination and the Pr form inhibits it. Therefore, the sequential illumination of the seeds can reversibly convert the active phytochrome from the Pr to the Pfr form.

Shade-avoidance response

Once a seed has germinated, the seedling uses light signals in its environment to dictate how it grows. If the seedling is covered in soil, the seedling needs to grow rapidly to reach a light source (Figure 13.7). Seeds only have a limited amount of stored food reserves and they therefore need to become autotrophic as soon as possible. Evidence suggests that phytochrome A (Phy A) is responsible for this response. Phy A is degraded rapidly in light but in the dark it is more stable. In a seedling in the dark, Phy A

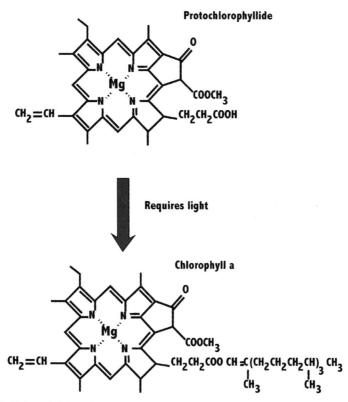

Figure 13.3 Protochlorophyllide and chlorophyll. When chlorophyll is synthesized, the pathway stops at a pale yellowish-green coloured compound called protochlorophyllide. The next step of the process requires light in order to make chlorophyll. If a plant is kept in the dark, it will look etiolated, with yellow leaves. This coloration is due to the build-up of protochlorophyllide in the leaves.

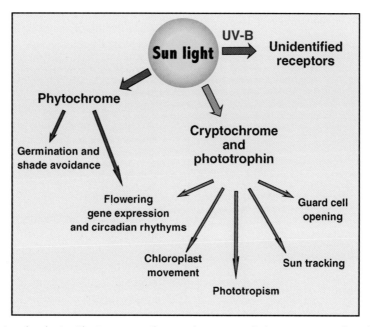

Figure 13.4 Photoreceptors in plants. Plants possess three major groups of photoreceptors; phytochrome (red light sensor), chryptochrome and phototrophin (blue light to UV-A receptors) and an as-yet unidentified UV-B photoreceptor. The figure is based upon that of Pruitt *et al.* (2003).

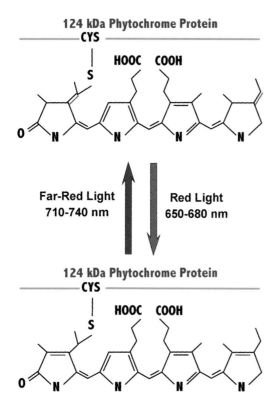

Figure 13.5 Interconversion of phytochrome, a pigment made up of a bilin chromophore and a protein. The complex interacts with red (wavelength 650–680 nm) or far-red (710–740 nm) light. Phytochrome changes its light absorbtion properties upon light absorbtion. In the ground state, the complex is termed P_r. In this form, the pigment absorbs red light most strongly and changes to the P_{fr} form of the pigment. In the P_{fr} form, the complex absorbs light most strongly in the far-red end of the light spectrum and converts to the P_r form, which is usually the physiologically active form of the pigment.

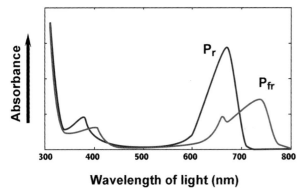

Figure 13.6 The absorbtion spectra of the P_r and P_{fr} forms of phytochrome, showing the difference in the peaks in absorbance of light of the two forms.

predominates and this causes the seedling to adopt an etiolated form. The hypocotyl elongates rapidly and the plumule forms a pointed hook to give low resistance as the seedling pushes through the soil. Leaf expansion is inhibited at this point. As the hypocotyl breaks the soil surface, Phy A is destroyed by the light, the rapid extension ceases, the plumule moves into an upright position, and the leaves begin to unfold and expand.

At this point other phytochromes (B–E) begin to take over the role of interpreting light signals. If a plant is in an open position with nothing limiting its illumination, the plant will tend to have short internodes and form a short bushy plant. If the plant is shaded by other plants, for example in a forest, the plant will respond by trying to get out of this shaded position and grow over competing plants. This response is known as a shade-avoidance response, and involves the plant forming long internodes and few leaves, which maximizes the ability of the plant to grow high rapidly. As can be seen in Table 13.1, under shaded conditions the ratio of red to far-red light falls. Plants in the upper part of the canopy absorb more of the red light for photosynthesis than far-red light, and therefore the ratio falls for light getting through the canopy. In the plants below the canopy, the active phytochromes will be converted to the Pr form. The presence of Pr then initiates the shade-avoidance response of growing very tall rapidly. Plants growing very close together also initiate a similar response. Although light hits plants in the open directly and therefore there is no change in the red to far-red ratio, plants nearby reflect more far-red light than red light. Hence, if plants are close together, it will provoke a shade-avoidance response and the plants will grow taller more rapidly to compete.

What has just been described is the response of what is known as 'a sun plant' to light quality. However, there are also plants that are adapted to a shaded environment, such as perennial dog's mercury (*Mercurialis perennis*). These plants grow in an identical manner whether they are grown in a shaded or unshaded environment (Figure 13.8).

How do phytochromes work?

Phytochromes are mainly located in tissues with high meristematic activity, such as the shoot tip or the root tip. There are three groups of physiological responses controlled by phytochrome, dictated by the amounts of light required to initiate them. The lettuce seed example mentioned earlier requires a short pulse of light to get a response from the seeds and this is termed a low fluence response (LFR). In addition, there are responses referred to as very low fluence responses (VLFR), which require low

Table 13.1 Ratios of red to far-red light in different environments.

Canopy	R:Fr light ratio
Daylight (no canopy)	1.05–1.25
Wheat	0.5
Maize	0.2
Oak woodland	0.12–0.17
Spruce woodland	0.15–0.33
Tropical rainforest	0.22–0.30
Twilight (no canopy)	0.65–1.15
In soil	0.8
In water	1.15+

Data based on those reported by Hopkins (1999).

light, and others known as high irradiance responses (HIR), which require long periods of illumination to initiate a plant response. An example of a VLFR is *Arabidopsis* seeds, which will germinate if illuminated with very low levels of red light. An example of an HIR is the inhibition of stem elongation when a plant is subjected to continuous far-red light.

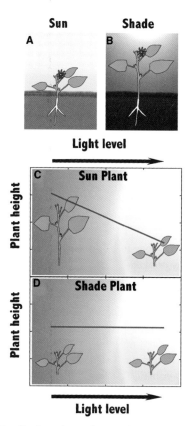

Figure 13.8 Shade and sun plants. When plants are shaded by rocks or other plants, the high far-red:red light ratio of light rises. Phytochrome detects this and some plants respond by increasing their internode length and growing tall with few leaves (A, B). These are termed 'sun plants' (C). Other plants that are accustomed to shaded conditions grow the same way, whether they are in sun or shade (D); they are termed 'shade plants'.

Figure 13.7 The low photon-fluence response in plants. Seeds can sense the ratio of far-red to red light around them. This can influence whether they germinate. If they do, a shade plant or a seedling undergound will experience a high far-red:red light ratio. This will initiate a low photon-fluence (shade-avoidance) response. The hypocotyl of the seed will stay in a hook form and the leaves will not expand. The stem of the plant will be elongated to get the seedling out of the ground. Only when the far-red:red ratio falls (signalling emergence from the soil) will the leaves expand and chlorophyll begin to form.

Phytochromes have such a diverse range of functions in plants as a result of the apoproteins with which they associate. The absorption of light by the chromophore induces a conformational change in the molecule. This also leads to a change in the structure of the protein bound to the chromophore. Phytochrome is thought to be located in the cytosol and migrates to the nucleus on activation. The N-domain of the protein binds the chromophore, and the C-terminal end contains protein sequences associated with signal transmission. Therefore, a conformational change in the protein can lead to activation or deactivation of a specific signal triggered by the light quality. Thereafter, a signal transduction pathway is activated, which leads to a plant response. These responses take one of two forms: first, a rapid response, which utilizes ion gradients across membranes; and second, a slow long-term response, involving the specific expression of particular genes.

Phytochromes and ion movement

Phytochromes influence membrane permeability and ion flows in plants exhibiting a nyctonastic response. In many species of the Fabaceae, at the end of the day their leaves move and adopt a rest position (see Chapter 11). The time for completion of the movements is rapid, and leaf closure is performed within 5 minutes. Phytochrome influences this process by stimulating H^+/ATPase pumps and K^+ pumps in the dorsal and ventral motor cells of the pulvinus. However, it is as yet unclear how phytochrome influences these pumping mechanisms.

Phytochromes and regulation of gene expression

In order to change the form of an etiolated plant to that of a deficiency-etiolated plant, significant changes in metabolism within the plant are required. In an etiolated plant, changes in the state of phytochrome influence gene expression. The phytochrome signal pathways involve transcription factors, which control the expression of batteries of other genes, which then dictate the response of the plant. For example, in the instance of lettuce seed germination, red light causes the formation of Pfr, which in turn activates a signal pathway, which leads to the production of enzymes associated with gibberellin biosynthesis. Gibberellins then stimulate the mobilization of food reserves in the seed to fuel germination. For lettuce seeds, treating the seeds with gibberellic acid can negate the requirement for red light.

Cryptochrome

Plants also respond to UV-A and blue light, and this is detected by photoreceptors known as cryptochromes (Figure 13.4). The cryptochromes are involved in the inhibition of hypocotyl extension when a shoot is subjected to blue light. They have only been studied through mutant analysis in *Arabidopsis* and relatively little is known about their mode of action and the range of roles they play in plant responses to light. There are also other blue light receptors in plants, e.g. for the perception of light for the phototropic response and the movement of stomata (these have been discussed in earlier chapters).

Sensing time

The surroundings of plants change on a daily cycle with day and night (Figure 13.9). This means that the primary means of plant nutrition, photosynthesis, is denied to them at some point during a 24-hour period. This has obviously had a significant influence on the evolution of how plants behave towards their environment. But it is not just the day/night cycle that changes; for most plants their surroundings change dramatically between the seasons. As a consequence, plants have evolved an internal biological clock, so not only does a plant know where it is in the day/night cycle (circadian clock) but also it knows where it is in the year (photoperiodism).

Circadian clock

A series of simple experiments using microarrays (a differential gene expression scan) has demonstrated that over 6% of genes (i.e. over 1000 genes) in *Arabidopsis* exhibit circadian behaviour and vary in expression between day and night (Schaffer *et al.*, 2001). This suggests that a large number of genes need to be expressed at different points over a 24-hour period for a plant to function optimally. This rhythm in plant behaviour is thought to be important to the plant, because a recent study has provided evidence to suggest that the circadian rhythm in *Arabidopsis* plants improves the fitness of the plant, reducing plant mortality and improving the ability of the plant to respond to its changing surroundings (Green *et al.*, 2002). An instance of such an advantage has been observed with genes that encode for pathways for phenylpropanoid biosynthesis. These frequently have their highest expression rates just before dawn, when it would be advantageous to have photoprotective flavonoids present (Harmer *et al.*, 2000). Therefore, gene expression can anticipate the beginning of daylight and provide maximum protection against any stresses to which the dark/light transition may expose the plants (Eriksson and Millar, 2003).

The circadian clock is set by external environmental signals, such as light and temperature, and this process is called entrainment. In nature, day length is the major signal that entrains the clock, but in the laboratory the circadian clock can be set artificially. One curious feature of the plant circadian clock is that plants do not possess a central clock mechanism (Thain *et al.*, 2000). It has been shown that different parts of a plant or different sections of the same organ can be forced to express different rhythms at the same time. Despite most organisms possessing a circadian clock of some sort, the actual mechanism varies widely between different kingdoms. Our current understanding of the proteins involved in the circadian clock in *Arabidopsis* is summarized in Eriksson and Millar (2003). Phytochromes and cryptochromes are known to be involved in the sensing of light for entrainment of the circadian clock in addition to an array of proteins; however, we are still a long way from under-

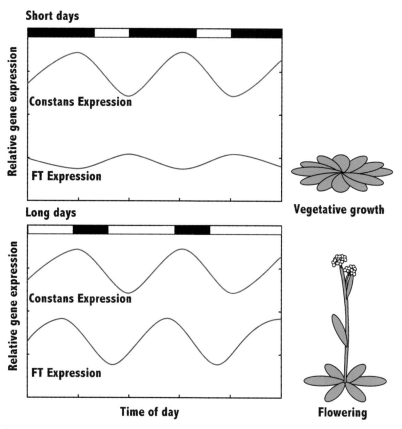

Figure 13.9 Sensing time in plants. Sensing time is very important in plants; it is crucial that they can sense what season they are in and whether it is the right time to flower. We have some understanding of how plants sense time, using the model plant species *Arabidopsis*, which favours long days for flowering. This is controlled by the expression of two proteins that vary in expression across the day: the protein constans is expressed in a cyclical manner across a 24 hour period, but this cycle is constant; the second protein, flowering locus-T (FT), accumulates only during the day. Therefore, if the day is short, little of this protein accumulates. However, as the days get longer more and more of the FT protein accumulates. There is a threshold at which, if constans and FT expression are simultaneously high, flowering in the plant will be triggered.

standing how these components interact and how plants sense the passage of time.

Photoperiodism

For a plant to interpret changes in day length to judge where it is in terms of seasons, a plant must be able to determine the length of the day or night. In an environment with defined seasons, it is crucial for a plant to be able to sense where it is in time (Figure 13.10). Many processes, such as flowering, tuber formation, seed shed and growth, need to occur at the correct time. For example, a plant such as scabious (*Knautia* sp.) flowering in the Northern European in winter time would be disastrous. In this example, the plant relies on insects such as bees to pollinate the flower; in the winter such pollinators are unlikely to be abundant and therefore seed set will be low. In addition,

winter temperatures will be too low for effective growth and development of any fertilized seeds and the flower is likely to be exposed to harsh winter conditions, such as wind, rain and frost. Thus, it is important that a perennial such as scabious can sense whether it is summer time and an appropriate time to flower.

A suitable time for a plant to flower will obviously vary between different regions of the world and thus plants in different habitats will have evolved to behave in different ways. The plant *Arabidopsis* has been used as a powerful tool to understand the mechanism of photoperiodism. *Arabidopsis* is what is known as a long-day plant, i.e. it flowers early if exposed to 16 hours of light followed by 8 hours of dark (Eriksson and Millar, 2003). Under these conditions, the plant will flower once it has reached the size of eight leaves. However, if *Arabidopsis* is exposed to short days (e.g. 8 hours light and 16 hours dark) the plant will produce vegetative growth for much longer, producing up

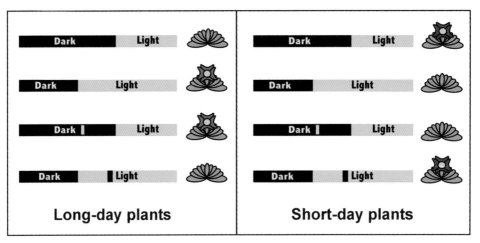

Figure 13.10   Photoperiodism. Some plants flower in response to day/night length. Long-day plants flower when the night period falls below a certain threshold. Short-day plants flower when the night length is greater than a certain time threshold. For both types, night length is the critical period and interruption of the night with a brief flash of light can disrupt or trigger flowering.

to 30 leaves, and then will flower. In its natural environment, *Arabidopsis* is an annual which germinates in late autumn/early winter, when the day length is short. The apical meristem of the plant then develops leaves up to late winter/early spring, when the lengthening day length triggers a conversion of the vegetative meristem into an inflorescence meristem and the plant flowers. As a consequence, the plant flowers in the spring and has the remainder of the spring and summer to develop and release its seeds. By sensing the day length, *Arabidopsis* optimizes flowering and seed release to maximize the chances of survival.

The photoreceptors used to detect light for photoperiodism appear to vary between different species. In *Arabidopsis*, both phytochromes and cryptochromes are used. The best model for how photoperiodism works in *Arabidopsis* is called the external coincidence model, first proposed by Bünning (1936). This model has been further refined by evidence which now suggests that the leaves of *Arabidopsis* produce a protein called Constans, whose concentration follows a circadian rhythm (see Figure 13.9). The Constans protein is relatively unstable and is entirely dependent on gene expression to maintain high cellular concentrations. The expression of Constans rises after 8 hours of illumination and then peaks between 14 and 20 hours of illumination (Eriksson and Millar, 2003). This circadian behaviour is followed regardless of day length. If Constans expression is high (14–20 hours after dawn) and the plant is still illuminated, then the gene *Flowering Locus T (FT)* is expressed. This in turn triggers the expression of a range of other genes, such as *Leafy* and *Apetala 1*, which are all associated with the switch of a vegetative meristem to an inflorescence meristem (see Chapter 10). Genetic manipulation has been used in *Arabidopsis* to alter the concentrations of Constans in transgenic plants and this has

confirmed that Constans has a crucial role in controlling flowering (Suarez-Lopez *et al.*, 2001). Thus, the model follows the following pattern. The level of Constans varies following a circadian rhythm. An *Arabidopsis* plant exposed to short days (around 8 hours) will be delayed in flowering, since Constans will never be in a high enough concentration in the light to trigger FT expression. However, as the days lengthen in late winter, Constans will be present at a progressively higher concentration in the light and ultimately this will trigger FT expression, and then a cascade of events ultimately resulting in flowering. It still remains to be discovered how perception of these signals in the leaves is transmitted to the growing meristem.

Arabidopsis is an example of a large group of plants known as long-day plants, i.e. plants that flower in response to long days. Other examples in this group are certain varieties of wheat (*Triticum aestivum*), spinach (*Spinacea oleracea*) and many of the garden annuals, e.g. sweet peas, petunias, etc. However, there are other groups, such as short-day plants, i.e. plants that require short days to flower. Examples of such species are soy bean (*Glycine max*) and cocklebur (*Xanthium strumarium*). For short-day plants there is a critical night period that needs to be exceeded in order to promote flowering, and for long-day plants there is a critical light period that needs to be exceeded to promote flowering. The exact period varies considerably between different plant species and depends on the habitat they have evolved to occupy. In experiments with both groups, the crucial feature of the 24-hour period appears to be the length of the night time. A short-day plant exposed to a long day and a short night will not flower (see Figure 13.10). If the light is switched off during the day for a very brief period, the plant still does not flower. If the same plant is exposed to a short day and long night, then flowering is promoted. However, if this plant is exposed to a brief period of light in the night then flowering is inhibited. A similar observation has been made with long-day plants. Interrupting the night time with a brief period of light can promote flowering in such plants, even when exposed to short days. Plants appear to be sensitive to the night length rather than the day length. There is also good evidence in a range of species that phytochromes are involved in the sensing of the length of the dark period. If a flash of red light is given in the night time, then the sensing of this period is disrupted.

There are also other groups of plants which respond to progressive changes in the day length, such as long-short-day plants (flowering after a sequence of long days followed by short days) and short-long-day plants (flowering after a sequence of short followed by long days). Finally, there is a group of plants that flower regardless of the day length, known as the day-neutral plants, e.g. kidney beans (*Phaseolus vulgaris*), desert paintbrush (*Castilleja chromosa*) and *Craterostigma plantaginuem*. For these species, usually factors other than day length promote flowering. The kidney bean is native to areas around the equator and is therefore subjected to very little variation in day length throughout the year, and as a consequence flowers independently of the day length. The other two species are examples of plants that live in arid environments and are opportunists. Both of these species need to flower rapidly and reproduce when there has been sufficient rainfall, and therefore responding to daylight signals would not be an advantage in these examples.

There is a range of further plant responses to day length other than flowering, such as leaf fall in autumn and the development of dormant buds, the development of tubers in temperate orchids or bulbs, and the switch from C3 photosynthetic metabolism to CAM in species such as *Mesembryanthemum crystallinum*. Little research has been performed on the mechanisms behind these behaviour patterns in plants.

Sensing touch (feeling)

Sensing wind and vibrations

Plants have already been shown to detect movement and respond to stimuli (Chapter 11). However, it is also clear that plants can detect movements in the air or in their environments through vibrations. The response of plants

Figure 13.11 Wind damage of plants. Plants are very sensitive to damage by wind and this damage tends to make them respond by growing very slowly. Here is shown the damage caused by wind to an apple tree (*Malus* sp.; A, B) and an *Acer* tree (C, D). The apple leaves have been damaged by salt spray from the sea, 1 km away, which has dried the leaves and killed large sections of them. The *Acer* tree shows damage due to leaves being thrown together in the wind and shattering the structure of the leaves.

Figure 13.12   Wind pruning of plants. When trees are exposed to continuous winds, such as this hawthorn (*Crataegus laevegata*) tree on the South Downs, UK, then the whole structure of the tree changes in response to the winds. More branches die on the windward side, so the tree becomes bushier on the opposite side. In addition, the tree will lean away from the wind because of the continual forces involved as the tree grows. Such trees tend to be shorter and bushier than those grown in less windy habitats.

to wind is one example. A plant subjected to strong winds exhibits a range of responses, from stunting of growth to physical deformation of the whole plant or wind damage (Figure 13.11). Trees will develop branches on the side away from the wind (Figure 13.12). There can be many reasons for this. Wind will dry out tissue on the exposed side of the plant which may result in the desiccation and death of buds or tissue growth will be severely restricted. The tissues on the windward side of the plant may also be damaged by salt (in coastal areas) or particulate damage as a result of small particles like sand being picked up by the wind and striking the plant. The greatest limitations on plant growth when exposed to strong winds will be the evaporative lost of water through transpiration, and this will favour growth of the plant on the more shaded side. The shaking effect of the wind is another factor that needs to be considered. An interesting experiment conducted on *Liquidamber* trees demonstrated that by shaking the plants for 30 seconds a day the growth of the plants could be reduced by 70 % (Neel and Harris, 1971). Not all plants respond this dramatically, but plants appear to be able to sense the mechanical stimulation by becoming shorter and sturdier (Jones, 1992). At present it is unclear how plants sense movement. Genes have been identified in *Arabidopsis* which are increased in expression as a

result of mechanical stimulation, but as yet the sensory mechanisms have not been identified.

Sensing chemicals (taste)

Taste and smell are two very similar senses and rely only on the medium through which they are sensed to distinguish the two. For plants, taste could be interpreted as being the detection and response to chemicals in an aqueous environment. Roots are the major structures of plants that are in close contact with an aqueous environment and there is a great deal of evidence that they grow in response to different chemicals in the soil. One of the prime functions of roots (see Chapter 4) is the harvesting of nutrients from the soil and roots are highly adapted to this function. Plants can detect concentration gradients of compounds containing two of the major nutrients they need for growth, namely nitrogen and phosphorus. To some extent, root systems are chemotaxic and will grow towards a nutrient supply but also the presence of high concentrations of nitrogen or phosphorus causes the proliferation of lateral roots to maximize the exploitation of the volume of soil (Figure 13.13). Roots also release enzymes and organic molecules in response to the detection of certain nutrients in the soil, such as insoluble

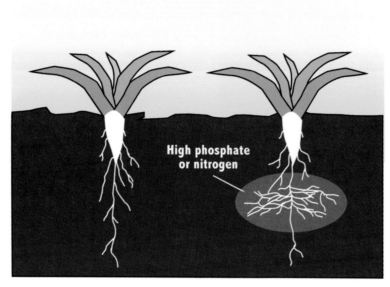

Figure 13.13   The sensing of nitrogen and phosphorus by roots. The schematic diagram shows the response of a plant to the presence of a localized high concentration of nitrates or phosphates in the soil. The root rapidly initiates a large number of lateral roots to take advantage of the nutrients as rapidly as possible.

phosphates and iron (see Chapter 5). The leaves of *Drosera* species can detect, and move in response to, minute amounts of phosphorus and nitrogen on the leaf surface (see Figure 13.14). Seeds are also capable of detecting compounds in the soil. For example, the parasitic plant *Striga* can detect minute quantities of strigol in the soil, which initiates germination. In addition, seeds which germinate in response to cues from smoke are also capable of detecting minute quantities of specific compounds in the smoke, which again trigger germination (Figure 13.15).

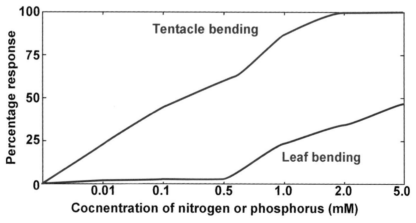

Figure 13.14   The sensing of nitrogen and phosphorus by leaves of sundew (*Drosera capensis*); the graph summarizes a series of experiments. Varying concentrations of nitrates or phosphates were pipetted onto an individual tentacle of the plant. The response of the leaf was in two phases. First, low concentrations of the ions were added; this involved a series of tentacles bending to maximize the use of the added nutrient, the response becoming stronger and stronger with more added nutrient. Water alone elicited no response from the tentacles. However, once the concentration of nitrate or phosphate exceeded 0.5 mM, the leaf began to respond by bending at the point where the nutrient was added, this bending becoming more apparent with higher concentrations of the ion. These data suggest that the sundew leaf can sense these ions on the leaf surface. If the concentration is low, the plant just uses the tentacles to absorb the ions, but if the concentration is high, the whole leaf bends to maximize the leaf area used in absorbtion. Therefore, in a sundew the tentacles will be used only if it captures small prey, such as a gnat, but if a larger insect is captured, the plant will sense the large release of nutrients and initiate folding of the leaf to maximize the rate of prey digestion and nutrient absorbtion.

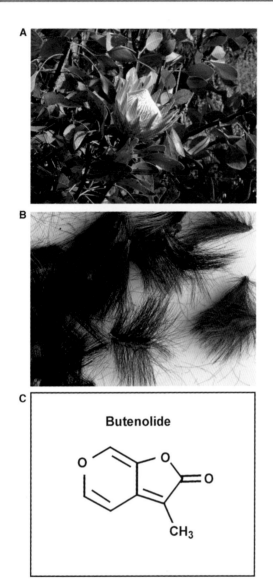

Figure 13.15 Smoke as a trigger for germination in *Protea* spp. For some plants, smoke can be a potent trigger for germination. This is the case in the *Protea* spp. from South Africa (A, B) and little germination can be achieved without it. The active ingredient of smoke that the plant senses is butenolide (C).

Communicating (smell)

The ability to sense smells and to taste is based on the ability of a plant to detect compounds in a gaseous or aqueous environment. There are many examples of plants detecting chemicals in an aqueous environment but there are only three known gaseous compounds that plants can detect and respond to. These are ethylene, jasmonic acid and nitrous oxide.

Ethylene

Ethylene is listed as one of the major plant hormones. Its influence on plant development was first noted where plants grew near to leaks in coal gas pipes. Trees grown near these leaks tended to lose their leaves and become twisted. It was discovered that the cause of this leaf fall was the compound ethylene. It was later discovered that ethylene was naturally produced by plants (see Figure 13.16) and was employed as a signal for certain developmental pathways.

Most tissues in a plant produce ethylene but it is made mainly in the active growing regions, such as the apical meristem. Ethylene production also increases in specific tissues at certain times in the year. For example, relatively large amounts of ethylene are released during the ripening of many fruits, and during autumn as leaves fall from deciduous trees. Plants also produce ethylene in response to stresses such as wounding, cold, drought and disease.

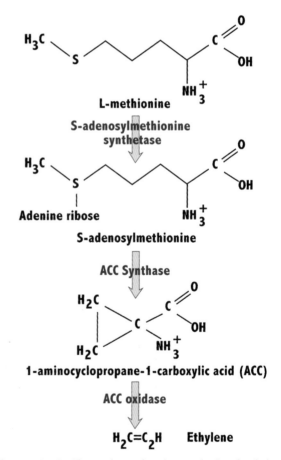

Figure 13.16 The pathway for the synthesis of ethylene. Ethylene is a gas produced by plants as a signal compound. It diffuses through the air and can be detected by leaves on the same plant or neighbouring plants. Therefore, plants can signal to one another and pass on limited information.

Ripening fruits

Climacteric fruits exhibit a rise in ethylene production just prior to fruit ripening (Reid, 1995). The term 'climacteric' describes the sharp rise in the rate of respiration during the ripening of the fruit as a result of rises in ethylene gas (Figure 13.17). The ripening process is followed by a degradation of the cell walls of fruit cells, a conversion of starch and organic acids into soluble sugars and an accumulation of anthocyanins and carotenoids, which colour the fruit. These colours act to advertise and display the fruit to potential seed dispersers and pass on the information that the fruit is now ready to eat. Examples of climacteric fruits are bananas, apples, avocados, tomatoes and melons. With unripe specimens of any of these fruits, the simple application of ethylene gas to the fruit can trigger ripening. The ethylene is generated by the fruit itself and allows the fruit to ripen rapidly and uniformly. As a consequence, not only does an individual fruit ripen uniformly but also gas released from one fruit can influence neighbouring fruits on the same or on separate plants. Thus, plants can communicate their developmental state with one another. This allows fruits on some plants all to ripen in a synchronous manner and this can be advantageous, as certain fruit predators, who may not disperse the seeds, will suddenly be exposed to a glut of ripe fruit which cannot all be eaten or stored, and therefore some fruit will fulfil its reproductive purpose.

Commercial growers of these fruits have exploited this feature of climacteric fruit, since fruit can be picked when

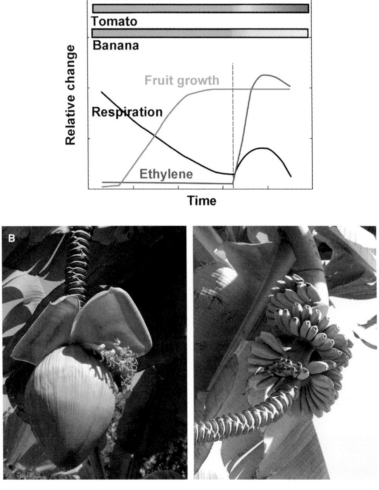

Figure 13.17 Ethylene gas affects the rate of ripening of certain fruits. As the fruit comes to maturity, the ethylene synthesis pathway is activated and the gas is released (A). This triggers a series of physiological responses that lead to the final stages of the ripening process. The ripening of tomatoes and bananas are shown. As the ethylene is generated the rate of respiration in the fruit rises and the starch content falls as sugars accumulate. It is ethylene that triggers these changes in the fruit. (B) Photographs of developing bananas are shown.

ripe and firm and shipped around the world. However, once at the site of sale and consumption, the fruit can be treated with ethylene gas to promote uniform fruit softening and ripening and then sold in prime condition. Not all fruits are climacteric and fruits such as cherries, grapes and strawberries do not ripen when treated with ethylene. In one of the first examples of plant genetic manipulation to reach the commercial market, a tomato variety modified in its ability to form ethylene was made. This variety exhibited a greatly reduced ability to ripen on the plant, and thus tomatoes could be left on the plant for longer to acquire a greater flavour without the worry of fruit softening and ripening. Ethylene has also been demonstrated to promote the senescence of flowers. In carnations, inhibitors of the pathway of ethylene synthesis (silver salts) have been used to extend markedly the longevity of carnation flowers (Reid, 1995). How plants sense ethylene is shown in Figure 13.19.

Pathogenic defence and stress responses

It has been known for some time that wounded plants and plants exposed to pathogenic attack produce ethylene in response. However, the role of this response was never determined. With the production of *Arabidopsis* plants that were insensitive to ethylene, it has been possible to investigate this topic. In some instances there is now evidence that ethylene can act as a defence signal. Certain infections by bacteria in *Arabidopsis* can lead to the release of ethylene, which in turn triggers a defence response from the plant in an effort to combat the pathogen.

Wounding and other plant stresses also elicit a release of ethylene from the plant. The whole plant responds to this by changing its development. The growth rate of the plant and the internode length are reduced and there is increased lignification in the stem of the plant. These responses act to protect the plant from the stress. However, since ethylene is gaseous, neighbouring plants can also be warned of a particular stressor. For instance, a herbivore such as a caterpillar attacking one plant will wound the tissue as it feeds. The plant can then respond by releasing ethylene, which instructs the apical meristem to grow in a defensive manner. In addition, neighbouring plants can also benefit from the ethylene, since their defences can also be primed without actually being attacked.

Abscission

In many plant organs there are layers of cells which form an abscission zone. These usually form at the boundaries between organs and the rest of the plant, e.g. leaves, fruits and flowers. One of the first effects of ethylene on plants discovered was the phenomenon of leaf fall or abscission

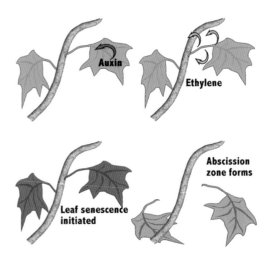

Figure 13.18 Ethylene leaf fall. Ethylene gas is released by plants as they approach any period where leaf fall is initiated. Over most of the year, high auxin levels keep dormant a layer of cells that form the abscission zone on a leaf. As autumn approaches, however, a rise in ethylene intiates the breakdown of the cell walls in the abscission zone. The attachment of the leaf to the plant becomes weaker and eventually the leaf falls off.

(Figure 13.18). Ethylene is now known to promote the release of cellulase and polygalacturonase into the cell wall region of the cells in the abscission zone. The cells become increasingly weakly attached to each other and ultimately fall apart. During the summer, in the leaves of deciduous trees, auxins are constantly transported out of the leaves and this inhibits ethylene production. As autumn approaches, auxins are no longer transported out of the leaves and they begin to senesce. This leads to a rise in ethylene production and release from the cells near the abscission zone, and cell wall degradation is initiated.

Detection of ethylene

The detection method for ethylene used by the plant lies in the presence of an ethylene receptor in the cell membrane (Figure 13.19). This protein uses a copper ion as a cofactor for ethylene binding. In *Arabidopsis* five different ethylene receptors have been identified (ETR1, ETR2, ERS1, ERS2 and EIN4). The different forms appear to be unnecessary, and if a plant is mutant in any particular form it will still grow normally. Only when all five are eliminated is the plant affected (Hua and Meyerowitz, 1998). Evidence suggests that the receptors inhibit the ethylene response pathway until ethylene binds to the protein; the receptors are then deactivated and the response pathway is permitted to occur. The ethylene response pathway then activates a series of transcription factors, which activate a battery of

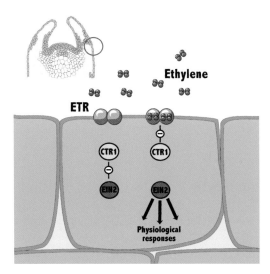

Figure 13.19 Ethylene perception by plants. Ethylene gas is detected by ethylene receptors (ETRs) on the surface of the plant. Once ethylene is bound to the receptor, the complex inhibits the function of CTR1, a copper transporter. This inhibition further relieves the inhibition of a protein called ethylene insensitive 2 (EIN2). Once active, EIN2 initiates a number of physiological responses in the plant associated with ethylene.

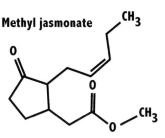

Figure 13.20 Jasmonates in plants. Jamonates form a group of plant growth regulators that help control growth and development in plants. Two of these compounds are shown. Methyl jasmonate is volatile and can act as a gaseous signal in plants.

genes that lead to alterations in the development of the plant.

Jasmonic acid

As discussed in Chapter 16, jasmonic acid and methyl jasmonate are involved in the signalling pathways for some plant defence mechanisms (Figure 13.20). Jasmonic acid is water-soluble, whereas methyl jasmonate is volatile and is released as a gaseous signal. As a consequence plants, must detect methyl jasmonate in the air in order to trigger the defence pathways (Klessig *et al.*, 2000).

Nitric oxide

Nitric oxide (NO) has been shown to be a redox-active compound which signals the activation of a range of different defence responses in mammals (Klessig *et al.*, 2000). Recently there have been reports of NO playing a similar role in plants (Durner and Klessig, 1999). Three examples have so far been identified in which NO is released following pathogen infection of plants. One example is in tobacco plants (*Nicotiana tabacum*) which

have been infected with tobacco mosaic virus (Klessig *et al.*, 2000). Susceptible tobacco plants do not release NO, but in resistant lines nitric oxide synthase activity was stimulated and NO released. NO appears to play a role in the post-infection defence pathway; however, the pathways involved are as yet unclear. NO has also been identified as being involved in the pathway for stomatal closure as a result of abscisic acid signalling (Neill *et al.*, 2002).

Sensing sounds (hearing)

Hearing is the sensing of compressions and rarefactions in the air as a result of objects moving. Communication through sound is very important to mammals, birds and certain insect species, but can plants detect sound and respond? There has been surprisingly little serious research into the area of whether plants can sense sound. There are many spurious anecdotal reports of plants sensing and responding to sound, but there is little rigorous science to test this. With this in mind, a series of experiments were set up to investigate whether plant growth or germination was influenced by playing music or speaking to plants. The results are summarized in Figure 13.21. At first sight, music would appear to increase the rate of germination

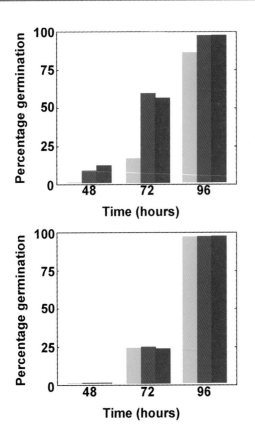

In conclusion, plants possess numerous senses which allow them to constantly monitor their environment and respond to it. Plants can sense light, touch, chemicals, gravity and several airborne gases. Their response is, however, usually slow when compared with animals. Even in dormant states, such as seeds, plants need to be aware of their surroundings, and that is the topic of the next chapter.

References

Bünning E (1936) Die Endogene Tagesrhythmik als Grundage der photoperiodischen Reaktion. *Berichte der Deutschen Botanischen Gesellschaft*, **54**, 590–607.

Durner J and Klessig DF (1999) Virulence and defense in host–pathogen interactions: common features between plants and animals. *Current Opinions in Plant Science*, **2**, 369–374.

Eriksson ME and Millar AJ (2003) The circadian clock. A plant's best friend in a spinning world. *Plant Physiology*, **132**, 732–738.

Green RM, Tingay S, Wang ZY and Tobin EM (2002) Circadian rhythms confer a higher level of fitness to *Arabidopsis* plants. *Plant Physiology*, **129**, 576–584.

Harmer SL, Hogenesch JB, Straume M and Chang HS *et al.* (2000) Orchestrated transcription of key pathways in *Arabidopsis* by the circadian clock. *Science*, **290**, 2110–2113.

Hopkins WG (1999a) Phytochrome under natural conditions. In *Introduction to Plant Physiology*, Wiley, Chichester, UK, pp. 460–461.

Hopkins WG (1999b). In *Introduction to Plant Physiology*. Chapter 18 Phytochrome under natural conditions. Wiley, Chichester, UK, pp. 460–461.

Hua J and Meyerowitz, E.M. (1998) Ethylene responses are negatively regulated by a receptor gene family in *Arabidopsis thaliana*. *Cell*, **94**, 261–271.

Jones HG (1992) *Plants and Microclimate. A Quantitative Approach to Environmental Plant Physiology*, Cambridge University Press, UK, pp. 296–300.

Klessig DF, Durner J, Noad R and Navarre DA *et al.* (2000) Nitric oxide and salicylic acid signalling in plant defence. *Proceedings of the National Academy of Science of the United States of America*, 97, pp. 8849–8855.

Neel PL and Harris RW (1971) Motion-induced inhibition of elongation and induction of dormancy in liquidamber. *Science*, **173**, 58–59.

Neill SJ, Desikan R, Clarke A and Hancock JT (2002) Nitric oxide is a novel component of abscisic acid signalling in stomatal guard cells. *Plant Physiology*, **128**, 13–16.

Pruitt RE, Bowman JL and Grossniklaus U (2003) Plant genetics: a decade of integration. *Nature Genetics*, **33**, 294–304.

Figure 13.21   Plants and sound perception. There is surprisingly little data on the response of plants to sound. The two bar charts summarize a simple experiment in which maize seeds (*Zea mays*) were sown and then the rate of germiantion was measured. Some seeds were germinated in silence (less than 20 dB); other batches were subjected music in the form of *A Bat Out of Hell* (Meatloaf) or Mozart's *Symphonia Concertanze* continually for 96 hours (at around 100 dB). The upper graph shows the first result of this experiment; playing music did increase the germination rate of the seeds. However, having a speaker system so close to the plants had a heating effect, so a fan was placed near to the seeds to completely remove this effect. The experiment was repeated and the lower graph was obtained. This time there was no response to the music and all treatments were identical. This work has been repeated with seedlings and under no circumstances have plants been shown to be sensitive to sound.

in seeds but this work also found that placing a speaker near plants has the effect of causing local heating around the plant. It is this local heating that stimulates germination and plant growth, and once the heating effect is removed sound has no effect on germination of plant growth. Therefore, as would be expected, plants are deaf and cannot sense noises; however, they can sense vibrations made by noises (see earlier, under touch).

Reid MS (1995) Ethylene in plant growth, development and senescence. In *Plant Hormones: Physiology, Biochemistry and Molecular Biology*, 2nd edn. (ed. P.J. Davies), Kluwer, Dortrecht, The Netherlands, pp.486–508.

Schaffer R, Landgraf J, Monica A and Simon B *et al.* (2001) Microarray analysis of diurnal and circadian-regulated genes in *Arabidopsis*. *Plant Cell*, **13**, 113–123.

Suarez-Lopez P, Wheatley K, Robson F and Onouchi H *et al.* (2001) CONSTANS mediates between the circadian clock and the control of flowering in *Arabidopsis*. *Nature*, **410**, 1116–1120.

Thain SC, Hall A and Millar AJ (2000) Functional independence of circadian clocks that regulate plant gene expression. *Curr. Biol.*, **10**, 951–956.

14

Seed dispersal, dormancy and germination

There are no examples of flowering plant species that can physically uproot themselves and move from one site to another. Asexual reproduction allows limited translocation but in most instances this movement is severely restricted. It is through the dispersal of seeds (mainly formed via sexual reproduction) that plants can move from one site to another. Furthermore, unless a plant uses asexual reproduction, a plant cannot persist in a site without seed germination and subsequent growth to sexual maturity. Thus, seed dispersal, dormancy and germination are three crucial components of the survival strategies of flowering plants.

In this chapter we shall look at how plants maximize the dispersal of seeds, and how plants sense their environment to maximize the likelihood of seedlings establishing and reaching maturity.

Seed dispersal

Seed dispersal is important for expanding the range and population size of a plant species (Baskin and Baskin, 2000). In addition, it is part of the process of maintaining a species within a habitat. Seed dispersal occurs in four dimensions (Figure 14.1). The first two dimensions are accounted for by the distance and direction over which seed is dispersed. In addition, once the seed lands, it needs to be dispersed to different depths in the soil. Finally, seed is dispersed over time (Fenner, 1985). Seeds mature and are released at different times and thus are potentially scattered in different directions. Also, the longer the time between each release of seeds from the parent plant, the greater the probability of the seed being dispersed over a larger area or attaining a greater depth in the soil (Howe and Smallwood, 1982).

Plants use various mechanisms to allow the dispersal of seeds, including gravity, wind, water, animal mediation or explosive propulsion.

Gravity-mediated dispersal

Many plant species have no elaborate mechanism for seed dispersal. As their seeds offer little wind resistance, on release seeds fall to the ground in the close vicinity of the parent plant. Such species are reliant upon animal or human intervention for further dispersal.

Wind-mediated dispersal

Many plant species use wind as a means of dispersal of seeds, and this has resulted in the evolution of many extraordinary structures to take advantage of air currents (Augspurger, 1986; Azuma and Yasuda, 1989; Figure 14.2).

Some seeds use wind agitation of the parent plant to cause dispersal, e.g. the shaking of the dried ovaries on the inflorescence disperses seeds of poppy species. This action acts in a similar way to a pepper pot, where shaking of the specialized seed pod throws the seeds in various directions.

The seeds of orchids are small and lightweight. In addition, there are large air spaces in the testa of the seeds which make the seeds very buoyant in the air, so that they can be carried over very long distances. For instance, there are cases of seed of the tongue orchid, *Serapias lingua*, being carried by wind current from France or Spain to the UK. As this species is not native to the UK, its arrival is easily spotted. However, *Serapias* species cannot survive the cold, damp British winters, and usually colonies do not persist for long. Thus, air-filled lightweight seed is an effective mechanism for dispersal over long distances.

During falling, the longer a seed stays in the air, the greater the probability of being dispersed over a large range. Some species have taken advantage of this and have evolved elaborate mechanisms to hold the seeds in the air for longer. For example, winged seeds of *Acer* species keep the seeds in the air for longer and thus increase the chances of air currents carrying the seeds further from the parent

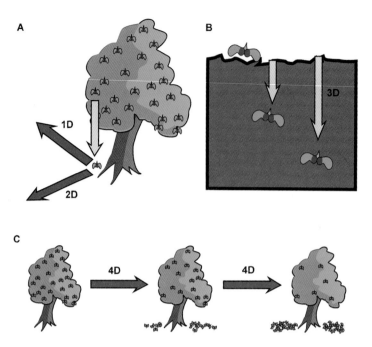

Figure 14.1   Seed dispersal. This schematic diagram summarizes the dispersal of seeds in four dimensions. Seeds can be blown in two dimensions over a large area (A) . Once a seed lands (B) it can acquire a specific depth in the soil (third dimension). In addition (C) , seeds are released at different times and will germinate at different times. Thus, the dimension of time is also a factor in seed dispersal (fourth dimension).

Figure 14.2   Wind dispersal of seeds. The wind is very important in the dispersal of seeds. In the figure are photographs of seeds that use wind to varying degrees for dispersal. (A) Wild carrot (*Daucus carota*) has light spiky seeds; the spikes offer some wind resistance as the seeds fall and may allow a slightly broader dispersal of the seeds from the parent plant. (B) *Acer* spp. use winged seeds (each called a samara). When the samaras fall from the parent plant they spin rapidly and offer great air resistance, allowing them to stay in the air for a long period of time and offering the opportunity for the seeds to move away from the parent plant. (C) Goatsbeard (*Aruncus dioicus*) seed head. The sepals of the flower are modified to form a feathery structure called a pappus, which is held by a column of tissue called a beak. The pappus offers a great resistance to the air and carries the small seeds over great distances. (D) Traveller's joy (*Clematis vitalba*) also uses a feathery appendage to the seeds in order to maximize dispersal.

plant. A more effective mechanism is seen in the family Asteraceae, where feathery appendages are produced on the seeds, acting in a similar fashion to a parachute. This feathery tissue originates from tissue made in the flower (pappus), which unfurls on senescence of the petals and bracts. In the Asteraceae, species such as the dandelions and thistles use this method of dispersal. However, plants such as rose bay willow herb and clematis also produce feathery appendages. Such seeds can stay airborne for prolonged periods, as long as the wind blows, and are capable of really long-distance travel.

Seeds of some species use air currents on the ground for dispersal. Plants such as *Kochia* form a complex ball structure which rolls along the ground, dispersing seeds during its travels. In plants such as tumbleweed, the whole plant can be broken off from the roots after senescence and then roll, following the wind and dispersing seeds. However, seeds of many plant species without these specialized structures are easily blown along the ground by air currents and then tend to clump together near obstacles or in cracks.

Water-mediated dispersal

Seeds from plants species in and around water can also use water currents to facilitate seed dispersal. Species such as the yellow flag iris (*Iris pseudoacorus*) produce seeds that fall onto the water surface and float. Water currents can then carry the seeds over long distances downstream until they get trapped on the banks of the body of water. Species that take advantage of this mode of dispersal usually have lightweight seeds which contain air pockets or cork structures to aid flotation.

Animal-mediated dispersal

There are numerous ways in which plants can use animals for dispersal of seeds. First, the seeds or the parent plant may possess elaborate hook or claw appendages, which allow them to grip onto the fur of passing mammals and birds (Figure 14.3). Plants such as greater burdock (*Arctium lappa*) and wood arvens (*Geum urbanum*) produce small hooks which grip very firmly to hairs. The seed head then slowly breaks up during travelling or through agitation/scratching so that the seeds are slowly released. The seed pod of *Ibicella lutea* produces two large hooks to grip onto the feet of passing animals.

Seeds of some species are dispersed by rewarding animals with food. During maturation of the seed, the ovary swells and forms a fruit. When the seed is ready for dispersal, the fruit ripens and acts as a strong attractant to animals. The fruit may be eaten and carried so that the seeds in the fruit are dispersed further. Alternatively the seeds may be eaten accidentally and, as a result of the thick testa, fail to be digested and are defecated; the seeds can then germinate in their own ready-made compost. The range of dispersal can depend on the behaviour of the animals, e.g. their migratory or territorial behaviour.

Sometimes the seeds themselves are part of the reward to animals. High carbohydrate, protein and fat contents make seeds a major part of the diets of many animals (granivory). Although many seeds are eaten and digested by animals, some are either missed, not digested, or are stored and lost; thus, the seeds are dispersed. Animals such as squirrels and ants are excellent examples of storers of seeds which frequently get lost and thus can germinate. The best strategy for plants to cope with large losses of

Figure 14.3 Animal-mediated seed dispersal. Many plants use the seeds or fruits as a means of attracting animals to disperse the seeds for the plant. (A) Acorn from *Quercus laevis*. Although acorns are eaten by many mammal species, not all will be; mammals such as squirrels attempt to store them in the ground and some of these stores will be eaten but others forgotten. Thus the seeds become dispersed and planted. With brambles (*Rubus fruticosus*; B) a fruit is formed around the seeds to attract mammals to eat them both. The seeds have highly resistant seed coats and therefore survive the digestion process and are excreted. By this time the seeds have been dispersed from the parent plant.

seeds due to granivory is to produce huge quantities of seed in a synchronous manner, so the appetite of the animals is sated, with seed to spare for the next generation.

Importance of post-dispersal mechanisms

Once seeds have hit the ground, the parent plant has dispersed the seed, but there is still a great deal of post-dispersal that can occur. Wind, water and weathering can form part of this process. In addition, dormancy of seeds plays a part in seed dispersal. The longer a seed is dormant, the longer these post-dispersal forces can act to increase dispersal. Since

seeds can be dormant for considerable periods, dormancy can play an important role in dispersal of seeds.

There is evidence that some of the family Poaceae have evolved complex mechanisms for post-dispersal, with some seeds possessing hygroscopic awns that twist in high humidity (Peart, 1979). The awns move around, and this twisting behaviour is thought to push the seed into the soil and bury it. Evidence for the existence of mechanisms for the self-burial of seeds is rare. Two other species known are ivy-leafed toadflax and cyclamen (Figure 14.4). Toadflax grows in very inaccessible places, such as cracks in vertical walls. The chances of seeds falling into another crack are small, but the plant has evolved a mechanism by which at

Figure 14.4 The seeds of some plants are planted by the parent plant, such as the autumn cyclamen (*Cyclamen cyprium*) shown here. Once the flower (A) is pollinated, the seeds begin to develop. At this point, what was the inflorescence stalk begins to coil (B, C) and pulls the seed down close to the soil (D). This means the cyclamen will form quite dense colonies.

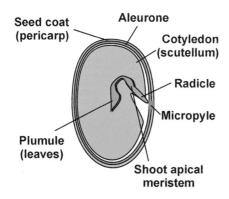

Figure 14.5   The structure of a typical dicotyledonous seed.

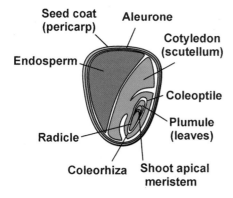

Figure 14.6   The structure of a typical monocotyledonous seed.

flowering the flower stalk is phototropic and grows towards a light source. Once pollinated, the flower stalk becomes negatively phototropic and pushes the developing seeds into the cracks in the wall to establish further plants. Cyclamen plants are alpine plants and tend to live in shallow soil on rocky outcrops. The plant does not release the seeds on maturation; instead, the inflorescence coils up, pulling the seed to the base of the parent plant, where the seed will germinate (Figure 14.4). No other such mechanisms for self-burial have been described in other seeds from other

plant species, but they will certainly exist. Examples of the general structure as a dicotyledon and monocotyledon seed are shown in Figures 14.5 and 14.6.

Dormancy and germination

Recalcitrant seeds

On maturation many seeds can become dormant but a group of seeds known as recalcitrants cannot achieve dormancy, e.g. English oak (*Quercus robur*) and the sugar maple (*Acer saccharinum*) (Figure 14.7). Such seeds are typified by a high water content on maturity (40–60% water), whereas dormant seeds can be almost completely dehydrated without affecting viability (around 5% water). Recalcitrant seeds also have a very short life span and will die if dried, but even if kept moist, their life span is still only a matter of months (Bewley and Black, 1992). Thus, in order to persist in an environment, such seeds must germinate immediately and can therefore only survive in moist environments. Recalcitrant species are generally found in moist or tropical and subtropical environments, where sensing of the environment by the seed is of little evolutionary importance.

A limited number of plants form seeds that are viviparous. Such seeds do not enter dormancy, but instead germinate while still on the parent plant. Examples of viviparous species are some *Allium* species and mangroves. For mangroves, the seedling develops a long, spike-like root. When the seedling is released, it plummets into the ground root first. There is a real advantage to vivipary in mangroves, since the seedling can establish itself in the water very rapidly and begin growing. However, a seed under such conditions would have great difficulty establishing itself, since the silt in the brackish waters inhabited by mangroves tends to be very low in oxygen and would be unfavourable for growth.

Figure 14.7   Non-orthodox or recalcitrant seeds. Most seeds can remain dormant for long periods of time but some are non-orthodox and cannot go dormant and are called recalcitrant. These seeds usually possess a high water content and have a very limited lifespan, such as the two species shown here, an *Acer* sp. (A) and a Turkey oak (*Quercus laevis*; B).

Dormant seeds

Seed dormancy occurs as a result of internal conditions of the seed that impede germination, despite external conditions of water, heat and gases being optimal for supporting germination (Bouwmeester and Karssen, 1992). Many different forms of dormancy have been described in the literature for seeds but in this account we will restrict discussion to the three principal forms:

1. *Primary (innate or true) dormancy.* Here the embryo is incapable of growing until it is relieved by drying or stratification. This state is usually acquired while the seed is still attached to the mother plant.

2. *Coat-, testa- or endosperm-imposed (whole seed) dormancy.* In this case, the embryo is not dormant but the seed coating prevents germination and if de-coated the embryo germinates.

3. *Secondary (induced) dormancy.* Once primary dormancy is broken, if the conditions are optimal, a seed will germinate; however, if the conditions are suboptimal, another state of dormancy known as secondary dormancy can be entered.

Non-recalcitrant seeds, on dehydration, can enter two states, primary dormancy or a non-dormant state (see Figure 14.8). In the non-dormant state the seed will germinate if certain factors are present. For some seeds this may be simply the presence of water, e.g. pea seeds. However, for other species it may be the presence of light/dark, temperature fluctuations or certain components in the soil (Figure 14.9). Without these factors the seeds enter a state of pseudo-dormancy, i.e. the only reason they are not germinating is that the environmental conditions are not favourable. If this state is prolonged the seed can enter the state of secondary dormancy. In the primary dormant state seeds will not germinate even if conditions are ideal for germination and seedling establishment. To recover from primary dormancy usually requires a period of desiccation, imbibition or a particular temperature regime (such as stratification). Secondary dormancy is similar to primary dormancy in most respects. There can be considerable cycling between the states of dormancy and non-dormancy during a seed's existence in the soil. Depending on the environmental conditions, viable seeds can persist for many years (the actual longevity varies from species to species).

Dormancy has several roles in the life strategy of plants. First, the dormant seed can survive very harsh conditions, such as prolonged drought and severe changes in temperature, which could kill an established plant. Some seeds

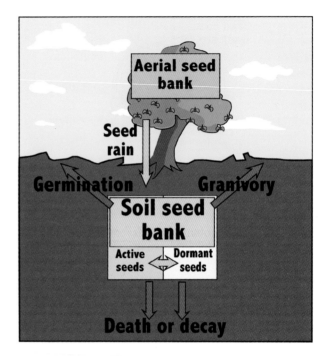

Figure 14.8 *In situ* seed banks. This schematic picture shows the different natural seed banks there are. Seeds can be held in the soil or in the air by the parent plant and will then cycle between being dormant or active. Seeds will leave the seed bank by either germinating, dying or being eaten.

have been reported to tolerate temperatures as high as 150 °C and as low as −196 °C with little damage. Thus, during a wildfire, established plants can be killed but certain plant species produce seeds that can survive the high temperatures of bush fires and are triggered to germinate after the event (see Figure 14.15). Second, dormancy allows a greater opportunity for seed dispersal and for a seed to reach the optimum conditions for germination. The longer a seed takes to germinate, the greater is the probability of the seed being moved away from the parent plant, through either animal intervention or weathering. In addition, dormancy can allow the seed to achieve the optimum depth in the soil.

Throughout the dormant period, the seed senses its environment and waits for signals to initiate germination (Figure 14.10). For some seeds this may take many years. In the Kew Gardens Millennium Seed Bank, UK, seeds are stored to preserve the diversity of flowering plants. In the bank, most of the stored seeds will remain viable for an estimated 200 years. Thus, under optimum conditions seeds can remain viable for considerable periods of time. The oldest verifiable record for seed germination is the sacred lotus (*Nelumbo nucifera*), which gave a rate of 75% germination after 1300 years of storage (Shen-Miller *et al.*, 2002), during which time some of the seeds had dispersed

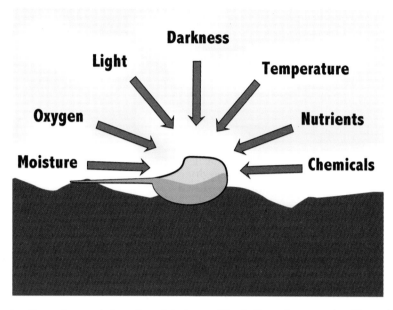

Figure 14.9 Seeds sensing the environment. A seed needs to be sensitive to its environment, just like a whole plant. It needs to be able to sense a whole range of factors, shown here, that influence whether or not the seed germinates.

from China to the USA! Reports of seeds such as wheat and alfalfa being viable in Egyptian tombs for several millennia have never been substantiated and are more likely to be a result of modern-day contamination than a real phenomenon. Wheat under normal circumstances has a dormancy half-life of between 5 and 15 years and thus is unlikely to survive for thousands of years, even under optimum conditions. After years of seeds being shed in a particular site, a large in situ bank of seeds will accumulate. Many of these seeds will die without ever coming out of dormancy,

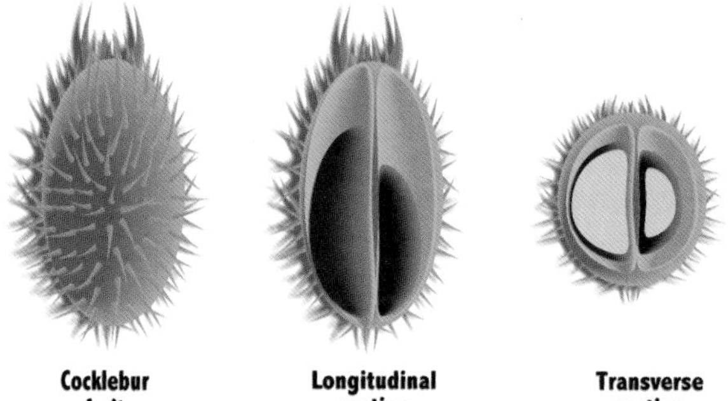

Figure 14.10 Dimorphic seeds. Seeds may look identical but each seed has features different to those of other seeds of the same species. This helps to broaden the time of germination, which could be important in a very variable habitat. The seeds of *Xanthium* spp. are dimorphic. The fruits produce two sizes of seeds, one large and one small. The large seed has a thin testa and germinates soon after falling to the ground. The smaller seed possesses a much thicker testa and takes much longer to germinate.

Cocklebur fruit   **Longitudinal section**   **Transverse section**

but some will survive and wait for the ideal conditions for growth.

Factors governing dormancy and germination

As a seed matures and enters dormancy, abscisic acid accumulates in the seed. During this accumulation, water is lost from the seed and primary dormancy is entered. There are several factors which dictate the length of dormancy in seeds:

1. Presence of germination inhibitors.

2. Presence of inhibitory tissues around the embryo axis which restrict germination.

3. Presence of germination promoters.

Presence of germination inhibitors

The main inhibitor known to restrict seed germination is abscisic acid (ABA). ABA has been shown to be essential for the establishment of primary dormancy (Hilhorst and Karssen, 1992). In *Arabidopsis*, ABA was shown to allow seeds to develop dormancy which dry storage or cold treatment could break. In mutant plants that are ABA-deficient or ABA-insensitive, seeds do not become dormant and instead germinate before release. There is good evidence that ABA levels in seeds reach a peak at maturity and then decline during dormancy. Certain signals, which vary from species to species, can then affect the ABA concentrations in the seeds. ABA can be leached from the seed and thus diluted; it can be metabolized; the seed may lose its sensitivity to ABA; or there may be other competitive stimulatory compounds that accumulate as a result of other environmental factors. All these factors contribute to tipping the balance during dormancy and triggering germination.

Other inhibitory factors have been identified that help maintain seed dormancy but there are fewer examples of these. Jasmonic acid and its methyl ester have been implicated in the establishment and maintenance of dormancy in seeds. Other compounds have yet to be identified but there is good evidence for their presence. For example, species such as *Alysicarpus rugosus* exhibit germination behaviour in which the seeds can sense the presence of high densities of seeds of the same species nearby (Murray, 1998). A leachate from the seeds has been found to be involved in the process which, when present at sufficiently high concentrations, can inhibit seed germination. The active agent in the leachate has yet to be identified. It is likely that the leachate is potent at exceedingly low concentrations and thus may prove very difficult to identify. However, this could be an example of a more widespread phenomenon, whereby seeds avoid clumping. Seed clumping can be created by ants or other insects collecting seeds, vertebrate-dispersed seeds in feces, and footprints or troughs in the ground. Thus, any strategy that allows seeds to avoid competition for identical soil resources could be important as a survival strategy.

There are likely to be many more inhibitors of germination in seeds that have yet to be identified. The major factor hampering their discovery is the probable low concentration for potency of these factors. However, it is clear that for seed germination, despite uptake of water by the embryo, dormancy can still occur, as inhibitors can block metabolic events that allow germination.

Presence of inhibitory tissues

In many instances the embryo of the seed may be non-dormant but the tissues surrounding it may impose dormancy on the seed. A good example of this is the seed of the Devil's claw plant (*Ibicella lutea*). The seed case is lignified and around 1 mm thick. Such a seed case can restrict oxygen getting to the seed and the process of imbibition (water entering the embryo). The restriction of oxygen to the embryo reduces the metabolic rate within the embryo and keeps the embryo dormant. The presence of water in the embryo allows swelling of the embryo axis, which allows the radicle to push through the micropyle,

and thus germination. Preventing water entering the seed can keep the seed dormant. With the seed coat present, *Ibicella* seeds remain dormant for years. However, on removal or breaking of the seed coat, the radicle extends within days if placed in a moist environment, and hence the seeds germinate.

In *Xanthium* species (cocklebur) this mechanism has been taken a step further. Cocklebur produces dimorphic seeds, one large seed and one small seed in separate pods (Figure 14.10). The upper seed possesses a thicker testa than the seed in the lower pod. This thickened testa restricts gaseous exchange with the embryo and insufficient oxygen concentrations reach the embryo to support germination. The lower seed, which lacks this thickened testa, can germinate readily within a matter of months in soil but the second seed only germinates after a lengthy period in the soil. In this respect, the cocklebur is using the second seed as an insurance policy; if the first seed fails to establish a seedling, the second seed can germinate much later, when perhaps conditions are more favourable.

It is a common practice in horticulture to use scarification (scratching of the seed coat to reveal the seed endosperm) as a means of promoting germination of particularly large or lignified seeds. Under natural conditions, scarification of the testa is effected by weathering, decay, damage caused by animals, or passage through the digestion tract of an animal. These factors can take a long time to wear the testa sufficiently to permit germination. In some seeds the testa may not be the tissue restricting germination but the principle is the same.

Presence of germination promoters

It has already been mentioned that oxygen and water can promote seed germination. But there is a range of other factors that encourage germination.

Light

There are many ways in which light can influence seeds. For some seeds, the presence of light is inhibitory to germination. Seeds such as those of the UK marsh orchids (*Dactylorhiza* spp.) will not germinate if in the light, despite all other conditions being suitable. For such seeds this requirement is likely to be a result of the seed sensing its environment and judging whether or not it is at a suitable depth in the soil. There are many other examples of seeds that require darkness for germination; however, there are many other examples of seeds which will show poor germination if left in the dark and require light; a few examples are wild thyme, marjoram, common poppy and common centaury. Such seeds are usually small and the light requirement is likely to be a reflection of the seed

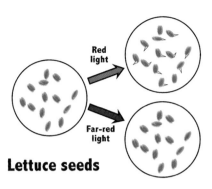

Lettuce seeds

Figure 14.11 Red-light sensing in lettuce seeds (*Lactuca sativa*). Lettuce seeds are slow to germinate in the dark, but will germinate readily if given a brief flash of red light. This converts phytochrome (P_r) into P_{fr}, and this triggers germination. However, far-red light inhibits germination. Under natural conditions this would signal that the seeds are not shaded or deep underground.

sensing its depth in the soil. Light will not penetrate beyond 1 cm in soil, so a seed at any depth greater than this will be in complete darkness. A small seed is unlikely to possess sufficient reserves to reach the surface of the soil. Hence, it is important for small seeds to be able to detect light in their environment in order to be able to establish a seedling.

Lettuce seeds display a requirement for light for germination (Figures 14.11 and 14.12). Some research has been performed on this species to understand how this sensing functions. Lettuce seeds use a chromaphore

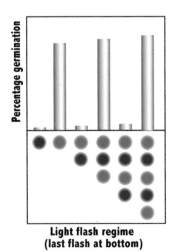

**Light flash regime
(last flash at bottom)**

Figure 14.12 Repeated flashes of red and far-red light can switch germination on and off. Shining red light on lettuce seeds triggers germination but far-red light inhibits this process. The seeds respond to the last flash of light they were given, so the trigger for germination can be switched on and off repeatedly by giving the seeds flashes of the appropriate light.

pigment called phytochrome. In the dark the phytochrome is predominantly in the Pr form and this inhibits germination. However, on the seed receiving the correct quality of light, Pr is converted predominantly to the Pfr form and germination is promoted. This whole mechanism has been studied further to show that shining light in the far-red wavelengths on seeds inhibits germination and shining light in the red wavelength promotes germination. This is shown in Figure 14.12. Through this mechanism a lettuce seed can sense some information about its environment. In the soil either the seed is in darkness or Pr is the dominant form of phytochrome (the soil absorbs red light more than it does far-red light near the surface). However, very near the surface there comes a point where the seed receives the required amount of red light and, if other factors (temperature and water) are suitable for germination, then germination will proceed. It is likely that seeds from many species use this mechanism to sense light. In a similar fashion, the seed can also sense whether it is beneath a canopy of neighbouring plants. Leaves absorb large quantities of red light to fuel photosynthesis, but a small quantity is transmitted through the leaf. However, more red light is absorbed than far-red, so the phytochrome in the seeds will be mainly in the Pr form and thus, although a higher light intensity is detected by seeds, they will be able to sense that they are shaded by other plants and germination will be inhibited.

Temperature

Seeds of different species give optimum rates of germination at particular temperatures (Stokes, 1967). Most species will not germinate at temperatures lower than 5 °C, but this low temperature can be the trigger that converts the seed from a state of primary or secondary dormancy into a non-dormant seed (stratification). At higher temperatures germination may be promoted up to around 35 °C, and then for most species high temperatures are inhibitory (except for those seeds primed by signals associated with fire).

Oxygen

The presence of oxygen can stimulate germination in seeds from some plant species. The restriction of oxygen supply by the tissues surrounding the embryo has already been discussed, but there are other factors that may influence oxygen availability. If a seed is deep in the soil or in an animal's gut, oxygen levels can be low and thus inhibit germination. However, on movement of a seed nearer the surface, or on defecation, the seed is released from these constraints and oxygen availability increases, so germination can be promoted.

Soil conditions

The components of the soil can also be an important cue for germination, e.g. the presence of nitrate in the soil, or the soil pH. These effective signals are probably very variable between different species, but it is important that a seed can sense the environment in which it has landed, to detect ideal growth conditions for seedling establishment. For the seed there will always be a fine balance between maintaining dormancy with the hope that the seed may be moved to a more suitable environment, with the associated risk of never coming out of dormancy, and germinating in a poor environment with little probability of seedling establishment.

Gibberellic acid

The presence of gibberellic acid in seeds is one of the best-understood signals for promoting germination. In many respects, gibberellins can be viewed as having the opposite effect to ABA. Over 120 different gibberellins have now been identified, and the potency of different gibberellins is likely to vary considerably between different plant species.

A dormant seed contains high levels of ABA, but as the seed remains in this state environmental factors trigger gibberellic acid levels in the seed to rise. For most seeds, the relative levels of ABA and gibberellic acid are important in triggering germination. Just as with ABA, the levels of gibberellic acid change with the environmental conditions, and the sensitivity to gibberellic acid of the tissue within the seed also changes with time. Gibberellic acid is thought to play a particularly important role in seeds which germinate after a period of stratification (Figure 14.13). A cold winter period causes a rise in gibberellic acid concentration in the embryo and a fall in ABA levels.

There are two processes in dormant seeds that have been shown to be influenced by gibberellic acid: the mobilization of food reserves and the breakdown of cell wall structures within the seed:

1. *Mobilization of reserves.* There is good evidence that in the seeds of some plant species, gibberellic acid promotes the mobilization of the food reserves in the endosperm (Fick and Qualset, 1975). The release of gibberellic acid from the embryo of the seed promotes the transcription of α-amylase genes in the aleurone layer in cereal seeds. The α-amylase then begins to hydrolyse the starch reserves in the endosperm to release glucose (Figure 14.13). This glucose then fuels the growth of the embryo and therefore germination. However, this is not the complete story, since simply adding glucose or sucrose to a dormant embryo will not induce germination; indeed, in many instances the presence of sugars inhibits germination. Therefore, gibberellic acid must play other roles during this process. ABA acts antagonistically with gibberellic acid and has been shown to inhibit the induction of α-amylase, thus preventing germination under conditions where ABA concentrations are high.

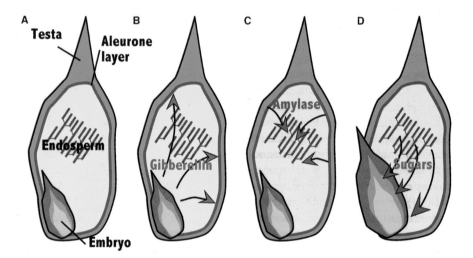

Figure 14.13   The role of gibberellic acid in the mobilization of seed food reserves. Abscisic acid keeps the seed dormant (A) but for many seeds during dormancy gibberellic acid concentrations rise (B) and trigger germination. This rise is often associated with a period of chilling of the seed, as would occur over winter. The rise in gibberellic acid, produced by the embryo of the seed, stimulates the aleurone cell layer in the seed to release amylase enzymes (C) . These break down starch reserves in the seed endosperm, which releases carbohydrate (D) to fuel seed germination.

2. *Loosening of the cell wall.* In *Datura ferox* seeds, dormancy is imposed by the tissues surrounding the embryo; these act as a mechanical barrier to the emergence of the radicle. Extension of the radicle is a turgor driven process. Hydrolases such as endo-β-mannanase have been proposed to be involved in the breakdown of cell wall barriers and the testa (outer coat of the seed) around the radicle to release the constraints on root extension (Bewley, 1997). In *Datura* seeds there is evidence that on triggering of germination there is a dramatic increase in endo-β-mannanase activity in the micropylar region of the seed, where there is a high level of galactomannans in cell walls (Figure 14.14). There is reasonable evidence that there is a link between this cell wall-loosening process and the emergence of the radicle (Figure 14.13). Chen and Bradford 2000 have shown that expansins are expressed in the micro-

pylar region of tomato seeds just before emergence of the radicle. The expression of the expansins (cell wall-loosening enzymes) is stimulated by gibberellic acid and occurs rapidly after imbibition.

Cytokinins

There are several studies that have demonstrated that the application of cytokinins to seeds can promote germination, especially in the presence of gibberellic acid (Thomas, 1992). However, far fewer studies have demonstrated a change in cytokinin levels in seeds leaving their dormant state. As cytokinins are involved in the cell cycle, it certainly comes as no surprise that they can be involved in the germination process of seeds.

Ethylene

The role of ethylene in germination of some seeds is less clear. In seeds of some species ethylene can promote germination, and in others it can inhibit it. For example, the inhibitory effect of high temperature on germination of lettuce seeds can be negated by treatment of the seeds with ethylene (Abeles, 1986; Matilla, 2000). In seeds where ethylene promotes germination, it is unclear whether ethylene is generated as a consequence of the initiation of germination or whether it is a signal that induces it.

Smoke signals

In certain environments of the world, wildfires or bush fires are common occurrences (Brown *et al.*, 1993; Figure 14.15). In fact, they should be regarded as a natural part of certain ecosystems. It is unsurprising that certain plants have evolved to use fire as a cue for flowering, growth and germination. A fire signals an event in which a large quantity of surface plant material is removed, and this presents opportunities for seeds in the ground to germinate and establish seedlings. Plants have evolved to take advantage of wildfires in several ways. Some plants flower in response to fire, others release their seeds, while yet others possess seeds that respond to the temperature shock of a fire or chemicals in ash and smoke and use these as a signal to germinate.

The effect of a fire can be quite variable, depending upon its intensity (Figure 14.15). Fire intensity is measured by kW/m, with some wildfires reaching a maximum of 60 000 kW/m. For practical purposes, a fire possessing energy of around 2000 kW/m is at the limit of control with human interference. Once a fire is initiated, it will not self-propagate unless the intensity reaches around 150 kW/m. Beyond this value, a bush fire is

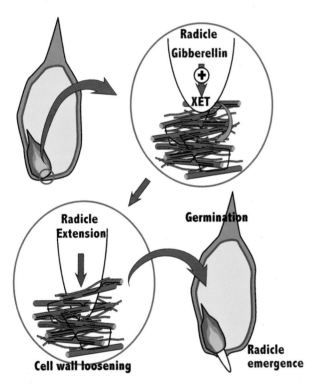

Figure 14.14 The role of gibberellic acid in cell wall loosening in seeds. The supply of carbohydrates to seeds frequently does not permit them to germinate. Therefore, when gibberellic acid concentrations rise there must be more occurring in a seed than just the release of carbohydrate to fuel growth. There is evidence suggesting that in some seeds gibberellic acid also stimulates the release of endo-β-mannanase and xyloglucan endotransglycosylase (XET), which breaks down galactomannans in the micropylar region of the seed. This weakening of the cell walls in this region of the seed permits the radicle to break through the seed coat.

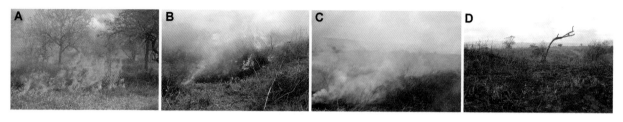

Figure 14.15   For some plants, fire plays a role in germination of the seeds. Bush fires (A–C) are frequent in some habitats and plants have evolved to cope with them and even use them as a signal. Some seeds are released only after a bush fire (D), or only germinate once the seeds have been exposed to high temperatures or the chemicals present in smoke.

started which will spread unless blocked. Morrison (2002) defined the scale of bush fires roughly around the following intensities:

Low intensity: <500 kW/m, flame height <1 m
Medium intensity: 500–2500 kW/m, flame height 1–3 m
High intensity: >2500 kW/m, flame height >3 m

The intensity of a bush fire will have a great impact on the temperature and the chemicals to which seeds are exposed. A wildfire generating 60 000 kW/m is likely to be very damaging, and such a fire is only likely to occur when fire has been absent for a considerable period of time and dry plant material has built up. Thus, in the management of many forest reserves, prescribed fires are initiated periodically to prevent high-intensity wildfires occurring. A prescribed fire would burn at around 500 kW/m and thus would be self-propagating but well under control.

For a majority of plant species, treatment of seeds at 120 °C for a few minutes markedly decreases germination. The exposure of seeds for a couple of minutes to temperatures of greater than 150 °C is usually lethal. However, in *Leucospermum* species the treatment of seeds with such high temperatures causes breakage of the testa and a dramatic increase in the germination rate. The breakage of the testa results in an increase in oxygen to the embryo and then subsequent synchronous germination of seedlings after fires (Brits *et al.*, 1999).

Pyrogenic flowering

For some plant species flowering is triggered by fire. This is called pyrogenic flowering and is exhibited by *Doryanthes excelsa* and *Telopea speciosissima*. Such species will not flower under normal circumstances, but after a bush fire flowering is triggered to occur within 19 months. In this way, seeds are released into a habitat containing less plant ground cover, which enables them to access resources better.

Serotiny

Serotiny is the behaviour of some plant species that retain their non-dormant seeds in a cone or woody fruit for up to several years prior to release. The cones protect the seeds from granivores and the heat generated by bush fires. However, during a bush fire the heat melts resins in the seed that once held the cone or fruit tight shut, which then allows the structures to open and release the seeds. Such survival strategies allow for seeds to be released after fires, which signal the clearance of competitor plants from the environment. Examples of species that use this method of seed release are species of the genera *Banksia*, *Isopogon*, *Conospermum*, *Pinus*, *Eucalyptus* and the family Proteaceae (Bell, 1999; Keeley, 1987; Brits *et al.*, 1999; Tieu *et al.*, 2001). Serotinous species are thus characterized by the possession of an aerial seed bank. Seeds on the ground have a limited life span and are vulnerable to bush fires, whereas in the air the seeds can remain viable for several years and are protected from fires. Fire triggers the release of seeds from these fruits and then they are released into an ash bed resulting from burning of the vegetation (Bell *et al.*, 1993; Bell,). Ash after a fire raises the pH of the soil, which can also be a trigger in the germination of serotinous seeds.

The role of heat in fires

There is good evidence that the heat generated by a bush fire influences the plant population in a particular habitat. Species that cannot tolerate heat soon die out in habitats that are frequently exposed to fire. Research on species around Sydney, Australia, has revealed that certain monocotyledon species and members of the family Fabaceae are more common in areas exposed to high-intensity fires rather than low-intensity fires (Probert, 1992). In contrast, members of the families Proteaceae and Rutaceae are more frequent in habitats exposed to middle-intensity fires and members of the family Epacridaceae are more common in areas of low-intensity fires.

Figure 14.16   Regeneration after fire. Fire is often seen as a destructive force but it can be a natural part of the cycle of life in some habitats. (A) The Arrabida region of Portugal in 2005. In the summer of 2004 a large fire burned much of the area and many shrubs were killed (B). However, many other shrubs and seeds survived and the area rapidly recovered from the fire. Pictures taken in 2006 (C) and 2007 (D) bear testimony to this. Many other plants, such as this wild tulip (*Tulipa tarda*; E) and orchid (*Orchis palustris*; F) show that being dormant at the time of the fire was a potent factor in their survival of the fire.

Evidence suggests that the role of heat in bush fires is to desiccate seeds and thus scarify them, causing splitting of the testa or melting waxy/resinous coverings. Since seeds have a range of coverings, it is not surprising that different species have different preferences for heat exposure.

Chemicals in smoke that trigger germination

It is not only the heat generated by fire that assists germination in certain plant species, but also the chemicals released in the smoke and ash (Brown *et al.*, 1993; Clarke *et al.*, 2000). For a few species there is evidence that nitric oxide is the chemical released in smoke that contributes to triggering germination of dormant seeds. However, for most species the signal chemical has yet to be identified (Brown, 1993). What is known is that it is generated by fires with temperatures in excess of 175 °C. The chemical is stable in the laboratory, is a component of smoke but is also water-soluble. Thus, after a bush fire of the correct intensity, seeds that are primed to germinate in response to smoke will receive a signal and will be able to take advantage of the change of conditions. Species that frequently use chemicals in smoke to trigger germination are members of the families Proteaceae and Ericaceae.

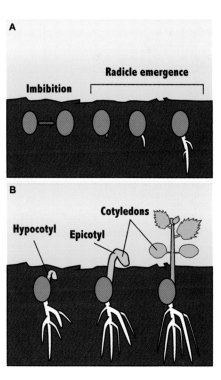

Figure 14.17   The sequence of events in germination of a seed, from the imbibition of water into the seed (A) all the way to establishment of the seedling (B).

For many plants, dormancy in the soil is an important part of ensuring survival of different plant species (Figure 14.16). No two seeds are identical and thus they will have different dormancy periods. This heterogeneity in dormancy acts as an insurance for species survival. Interestingly, in many crop plants, dormancy has been selected against in successive generations. It is important in crops that seeds germinate at roughly the same time and attain a similar size. Seeds failing to germinate would rapidly become shaded by neighbouring seedlings and thus would fail to set seed or survive. Gradually, over centuries, crop plants have become largely non-dormant.

Once a seed has been subjected to the necessary triggers for germination, dormancy is lost and a seedling begins to grow (Figure 14.17). Once the radicle has protruded from the seed coat, germination is said to have occurred and the process has reached the point of no return (see Chapter 10 on plant growth).

References

Abeles FB (1986) Role of ethylene in *Lactuca sativa* cv. 'Grand Rapids' seed germination. *Plant Physiology*, **81**, 780–787.

Augspurger CK (1986) Morphology and dispersal potential of wind-dispersed diaspores of neotropical trees. *American Journal of Botany*, **73**, 353–363.

Azuma A and Yasuda K (1989) Flight performance of rotary seeds. *Journal of Theoretical Biology*, **138**, 23–53.

Baskin CC and Baskin JM (2000) Ecology and evolution of specialized seed dispersal, dormancy and germination strategies. *Plant Species Biology*, **15**, 95–96.

Bell DT 1999 The Turner Review No. 1. The process of germination in Australian species. *Australian Journal Botany* **47**: 475–517.

Bell DT (2001) Ecological response of syndromes in flora of southwestern Australia: fire versus reseeders. *Botanical Review*, **67**, 417–441.

Bell DT, Plummer JA and Taylor SK (1993) Seed germination ecology in southwestern Australia. *Botanical Review*, **59**, 24–73.

Bewley JD (1997) Breaking down the walls – a role for endo-β-mannanase in release from seed dormancy? *Trends in Plant Sciences*, **2**, 464–469.

Bewley JD and Black M (1992) *Physiology and Biochemistry of Seeds in Relation to Germination*, Springer Verlag, New York.

Bouwmeester HJ and Karssen CM (1992) The dual role of temperature in the regulation of the seasonal changes in dormancy and germination of seeds of *Polygonum persicaria* L. *Oecologia*, **90**, 88–94.

Brits GJ, Calitz FJ and Brown NAC (1999) Heat desiccation as a seed scarifying agent in *Leucospermum* spp. (Proteaceae) and its effects on testa, viability and germination. *Seed Science and Technology*, **27**, 163–176.

Brown NAC (1993) Promotion of germination of Fynbos seeds by plant-dervived smoke. *New Phytologist*, **123**, 575–583.

Brown NAC, Kotze G and Botha PA (1993) The promotion of seed germination of Cape Erica species by plant derived smoke. *Seed Science and Technology*, **21**, 573–580.

Chen F and Bradford KJ (2000) Expression of expansin is associated with endosperm weakening during tomato seed germination. *Plant Physiology*, **124**, 1265–1274.

Clarke PJ, Davison EA and Fulloon L (2000) Germination and dormancy of grassy woodland and forest species: effects of smoke, heat, darkness and cold. *Australian Journal of Botany*, **48**, 687–700.

Fenner M (1985) *Seed Ecology*, Chapman and Hall: London.

Fick GN and Qualset CO (1975) Genetic control of endosperm amylase activity and gibberellic acid responses in standard-height and short-statured wheats. *Proceedings of the National Academy of Science of the United States of America*, **72**, 892–895.

Hilhorst HWM and Karssen CM (1992) Seed dormancy and germination: the role of abscissic acid and gibberellins and the importance of hormone mutants. *Plant Growth Regulation*, **11**, 225–238.

Howe HF and Smallwood J (1982) Ecology of seed dispersal. *Annual Review of Ecology and Systematics*, **13**, 201–228.

Keeley JE (1987) The role of fire in seed germination of woody taxa in California chaparral. *Ecology*, **68**, 434–443.

Matilla AJ (2000) Ethylene in seed formation and germination. *Seed Science Research*, **10**, 111–126.

Murray BR (1998) Density dependent germination and the role of seed leachate. *Australian Journal of Ecology*, **23**, 411–418.

Peart MH (1979) Experiments on the biological significance of the morphology of seed-dispersal units in grasses. *Journal of Ecology*, **67**, 843–868.

Probert RJ (1992) The role of temperature in germination ecophysiology. In *Seeds: The Ecology of Regeneration in Plant Communities*, (ed. M Fenner) CAB International, Wallingford, UK, pp. 285–325.

Shen-Miller J, Schopf JW, Harbottle G and Cao R *et al.* (2002) Long-living lotus: germination and soil γ-irradiation of centuries-old fruits, and cultivation, growth and phenotypic abnormalities of offspring. *American Journal of Botany*, **89**, 236–247.

Stokes P (1967) Temperature and seed dormancy. In *Encyclopedia of Plant Physiology*, vol XV/2, (ed. W Ruhlan) Springer-Verlag, Berlin, pp. 746–803.

Thomas TH (1992) Some reflections on the relationship between endogeneous hormones and light-mediated seed dormancy. *Plant Growth Regulation*, **11**, 239–248.

Tieu A, Dixon KW, Meney KA and Sivasithamparam K (2001) The interaction of heat and smoke in the release of seed dormancy in seven species of southwestern Australia. *Annals of Botany*, **88**, 259–265.

15

Interactions with the animal kingdom

It is easy to see plants as victims in the struggle for survival. Plants are constant targets for herbivores and there is a perpetual fight to survive. But the situation need not always be seen in that light, as plants, under many circumstances, manipulate the situation to their advantage. Plants are just as capable of using animals as animals are of using plants, but often it is achieved more through subtle ways. In some examples, such as pollination, plants and animals work together to their mutual advantage. In other examples, plants have used mimicry to fool and control members of the animal kingdom. In this chapter, the means by which plants manipulate the animal kingdom will be discussed.

Animal-mediated pollination

From the fossil record, it appears that flying insects predated flowering plants on the evolutionary scale by many millions of years. Therefore, the ability of plants to use insects for pollination is primarily a matter of plants evolving to take advantage of the circumstances presented. The use of air currents as a mediator for plant sexual reproduction is effective but also restrictive. There are environments where wind-pollinated plants could never thrive. Such plants are completely dependent upon wind to carry many small pollen grains over long distances and on a second plant catching the pollen grains on its stigma. Wind pollination is a very risky strategy and can only succeed through the production of large quantities of pollen.

However, there are whole armies of insects that would make perfect targets to carry pollen from one flower to another and thus achieve fertilization for sexual reproduction (Figure 15.1). A mechanism needed to evolve to attract the insects to flowers and ensure that the pollen could be transferred from one flower to another. Insects require carbohydrate and protein in their diet, and the latter is very limited as a natural resource. Plant material such as

a leaf is generally not a concentrated source of protein. Thus, many flowering plants have succeeded in attracting insects for sexual reproduction by producing flowers that are attractive to insects, through colour and scent, and offer a reward of carbohydrates and amino acids. In addition, numerous insect species are also interested in gathering pollen to provide extra nutrients for their diet. A second important consideration in flowering is the production of a floral signal that allows a pollinator to distinguish a particular flower from others. It is these two factors that have shaped the evolution of flower form and have given rise to the diversity of flowers we see today (Chittka *et al.*, 1997). In contrast to wind-pollinated species, the flowers of animal-pollinated species possess attractive colours and patterns, large forms to act as a focus and stages for insect landing, volatile scents, and fewer but larger pollen grains. However, insects searching for pollen visit even wind-pollinated species, such as oak, maple and corn. Insects are not the only animals to visit flowers; mammals, birds and reptiles are also known to do so. However, insects pollinate the majority of flowers.

The advantages of animal-mediated pollination are that the pollen is directly transferred from one flower to another and less pollen is wasted. Thus, for the expense of a small amount of carbohydrate and amino acids, plants gain a very effective reproductive system. However, one of the crucial aspects of insect pollination is that the animal needs to visit a second flower of the same species to achieve the pollen transfer. To guarantee this, flowers produce a limited supply of nectar that requires the animal to visit several flowers before becoming sated.

The relationship between plant pollination and bees has been studied in great detail and this serves as a good model to focus on in the relationship between plants and animals. Bees that have newly hatched prefer flowers that reflect wavelengths of light in the ranges 400–420 nm (violet) and 510–520 nm (blue/green). However, the most common wavelength of light reflected by flowers is longer than

Physiology and Behaviour of Plants   Peter Scott
© 2008 John Wiley & Sons, Ltd

Figure 15.1   Insect-mediated pollination, used by more plant species than any other method of pollination. Examples are: (A) yarrow (*Achillea millefolium*) and a soldier beetle; (B) marsh woundwort (*Stachys palustris*) with bumble bee; (C) *Orobanche minor* with ant; (D) Asteraceae species with small tortoiseshell butterfly.

these wavelengths. Despite the difference in experimental preference of bees and the range of flower colours plants produce, bees forage from a range of different flower colours in the field. During foraging, bees tend to visit a wide range of flowers initially, and during this process learn which plants provide the greatest reward. Bees then remain faithful to a few species for the rest of their foraging career, a phenomenon known as flower consistency. Thus, to ensure visitation and pollination, a flower must provide a good reward for the insect and a clear signal to be easily distinguished from other species. The range of sight for distinguishing colours is limited for bees and only at short distances from a flower can the actual colour be distinguished. It appears that contrast with the environment is more important than colour for getting a bee to approach a flower. It is estimated to take between three and seven visits to a flower for a bee to relate the floral signals to the reward obtained. Thus, if after these visits a flower has proved to be poor as a source of pollen and nectar, the bee will learn and begin to avoid flowers of this type. After visiting a flower, scent and colour help bees to recognize the flowers again. The scent of a flower that is recognized

by a bee as being from a rewarding flower initiates the landing behaviour of the bee. Thus, a bee soon develops recognition of flowers of particular species of plant. Relationships between other animals and flowers generally follow similar lines to those discussed for bees.

The fidelity of animals to a particular flower type could be very important to a plant, since masses of pollen arriving on the stigmatic surface from a range of different plant species has been shown to be inhibitory to the process of fertilization. Therefore, the flower consistency of a pollinator is crucial in maintaining high seed yield.

Difficulties for pollination arise for rare plants, because there is a decreased likelihood of the pollen being transferred to a second flower. In order to achieve a higher incidence of animal visitation, rare plants have two strategies: either they can produce larger and more attractive flowers than rivals, or they can produce a greater reward for visiting animals. In most studies the latter appears to be the strategy of rare plants. This begs the question of whether plants vary their nectar supply in response to the number of visitations a flower receives. However, as yet there are no data on this topic.

Mimicry in plants

Mimicry can take many forms in nature. In the animal world mimicry is frequently used as an anti-predatory device, where one species copies the behaviour or mimics the colour of another species and in so doing gains some survival advantage. There are examples of mimic species copying other animals and plants. The latter mimics have taken on the appearance of a plant in order to gain camouflage that can be useful in predator avoidance and prey capture. In the plant world, too, mimicry is used to avoid predators, but it is also used to attract and deceive insects.

There are three forms of mimicry:

1. *Batesian mimicry*. This form of mimicry was first described by the British biologist Henry Bates in 1852. It occurs when an organism that is palatable has evolved a resemblance to an unpalatable model species. In this instance, only the mimic obtains a survival advantage. Hence, the mimic gains greatest benefit when it is rare and the model is common.

2. *Müllerian mimicry*. This form of mimicry was first described by the German zoologist Fritz Müller in 1878, to account for certain paradoxes in the Batesian form of mimicry. It was observed that frequently there are distantly related species that resemble one another, but both species are unpalatable. This means that if a predator eats a specimen of one species and finds it unpalatable, it is likely that the predator will avoid similar-looking species. Therefore, individuals act as sacrifices so that the predators can be educated concerning the unpalatability of the organisms sharing similar appearances. Thus, in this example of mimicry both species benefit. The survival benefits of Müllerian mimicry increase with rising population sizes of the organisms involved.

3. *Aggressive mimicry*. With this form of mimicry, an organism mimics a signal or appearance in order to deceive or attract another organism.

There are examples of plant species that exhibit each one of these forms of mimicry.

Batesian mimicry

Mimicry of toxic or poisonous species

This is a very common form of mimicry in both in the animal and plant kingdoms. Possibly one of the best examples of mimicry in plants is that of the Urticaceae (stinging nettles). Stinging nettles have a very wide distribution and are very common plants. Their leaves carry a powerful sting if touched and can be very unpleasant. Some species of nettle in India are powerful enough to kill in extreme cases. The leaf shape of a nettle is instantly recognizable by anyone who has ever been stung by one. However, there are many plant species which possess leaves closely resembling those of nettles, e.g. white deadnettle, mint and other members of the family Lamiaceae (Figure 15.2). These benefit from this similarity and are thus avoided by potential predators.

Rewardless flowering

Several species of plant have evolved flowers that require insects for pollination but do not give any nectar reward to the pollinators. In order to achieve this, a plant needs to use only visual signals to deceive a pollinator into believing that

Figure 15.2 Example of Batesian mimicry in deadnettles and other species. Nettles (*Urtica dioica*) (A) use a combination of formic acid, serotonin and histamine to make them unpalatable to herbivores. However, white deadnettles (*Lamium album*; B) and yellow archangel (*Lamium galeobdolon*; C) both possess leaves that resemble those of nettles and in consequence are avoided by herbivores cautious of being stung.

it is visiting a rewarding flower. This form of mimicry works best when the plant resembles a neighbouring plant that offers nectar secretions to visiting insects. Over 60 % of orchids have been shown to produce non-rewarding flowers (Figure 15.3). The orchid genus *Orchis* is one such group of plants. Some of the dynamics of rewardless flowering have been studied in *O. boryi* (Gumbert and Kunze, 2001). It is essential for the flowers of this species to be visited by bees for pollination, and the primary pollinators for this species are *Apis mellifera* and *Bombus* species. The bees visit the *O. boryi* flowers between visits to other flowers. The bees display avoidance learning, whereby bees that have visited an *O. boryi* flower are less likely to visit a second flower. As a consequence the greater the density of *O. boryi* plants in the region, the less likely the bees are to visit them. As mentioned earlier, bees, on average, will visit a flower between

three and seven times before they begin to learn to associate a particular type of flower with a reward. The *Orchis* orchids rely on this naive period in the bee's foraging career to be pollinated. Thus, the strategy of non-rewarding flowering is very risky and can restrict population size. Since the pollinia of an orchid are removed on the first visitation by an insect, and then deposition of pollen on the stigmatic surface occurs, it has been possible with the orchid *Calypso bulbosa* to demonstrate the likelihood of multiple visits by insects to flowers (Procter and Harder, 1995). In field measurements, 95 % of the *Calypso* flowers had no pollinia and thus had been visited once by an insect. However, only 65 % of the flowers had been fertilized, and therefore visited by an insect that had already visited another flower. Thus, it is clear that insects can learn to avoid non-rewarding flowers. The orchids can offset this problem of recognition by displaying

Figure 15.3 Non-rewarding flowering in *Dactylorhiza* orchids, which have all of the appearances normally associated with flowers offering a nectar reward to insects. The flowers are attractive (coloured), possess nectar trails and also have spurs at the back of the flowers, all signals typical of flowers that attract bees. However, there is no nectar in the flowers. Bees rapidly learn that the flowers are mimics, but by then the flowers have achieved pollination. Over 70 % of orchid species are non-rewarding.

Figure 15.4   Attachment of orchid pollinia to insects. Orchids have a flower form like no other plant family; the anthers are fused and hidden to form pollinia (A, D). When an insect visits the flower, the pollinia become attached to the insect (B, C, E) via a sticky body known as the caudicle. The sequence of events that occur when a hover fly visits a giant helleborine (*Epipactis gigantea*) is shown here. The pollinia have been attached to the fly's back (F) during the visit and, despite the fly's best efforts, it cannot be removed. A section of an orchid flower is also shown (D–G).

a high degree of variation in scents produced by the flowers and in shape, colour and pattern.

To discriminate between flowers, insects use mainly a combination of colour and scent. Only known floral mimics that copy the coloration of a nectar-rewarding plant species are known, and there are no known floral mimics that mimic the scent of another flower. The advantage to the orchid for using this survival strategy is that it does not have to invest carbohydrate and valuable nitrogen reserves in nectar secretions. However, such orchids only do well in areas where there is a high density of plants that reward insects, and there is some evidence that the floral signals for non-rewarding orchids need to be stronger than those of rewarding plants in order to attract insects to them.

The pollination system of the orchids also makes the pollen reserves inaccessible to visiting insects (Figure 15.4). Instead of producing anthers that rub on the back of an insect visiting the flower, an orchid produces two pollinia (Figure 15.5). These are complex aggregations of pollen attached to stalks. These stalks bear an adhesive on the base and thus when an insect visits a flower, the orchid sticks two pollinia onto the insect's back or head. This targets the pollen very precisely on the insect and reduces wastage of protein reserves in the pollen. Insects such as bees cannot gather the pollen from an orchid flower and therefore orchid flowers are doubly rewardless. However, to achieve pollination, the pollinia will burst or rub off onto the next flower that the insect visits and achieve pollination.

Figure 15.5   Pollinia structure and function. Here the detail of the positioning and structure of the pollinia (singular = pollinium) in a *Dactylorhiza* orchid (A) is shown. An individual yellow pollinium has been removed from the flower. The pollinium is a mass of pollen attached to a stalk known as the caudicle (B–D). The caudicle dries once the pollinium is removed from the flower. This causes the pollinium to flip forwards and then it touches the stigmatic surface of the next flower the insect visits.

Although orchids appear to take a very risky strategy with reproduction by being rewardless, seed set in many plants is good and does not limit their reproductive capacity.

Lithops

In 1811 the botanist William Burchell, on an expedition to South Africa, by chance picked up an interesting stone but found to his surprise it was actually a plant! The plants have been called stone faces, stone plants or belly plants, but it was not until 1922 that they were first named *Lithops*, which is derived from the Greek meaning 'living stones'. Since that time, 36 species of *Lithops* have been identified and they are found only in the southern-most tip of Africa. There are a number of plant species that resemble stones, but none more so than *Lithops* plants (Figures 15.6, 15.7). This mimicry of the natural environment is really best described as camouflage.

The plants are succulent and use Crassulacean acid metabolism (CAM) to reduce water losses during photo-synthesis. Their structure is an adaptation to drought-stressed environments and they are only found in locations with very low rainfall. The plants are succulent in nature and store water in the leaves. In its most basic form, the plant is made up of a single pair of leaves attached to a short stem and a long root system. As the plant matures, more and more leaf pairs develop, resulting in a patch of what look like pebbles on the ground. The pebble or stone appearance results from only a small upper surface of the leaf being visible above ground. This small surface is pigmented and acts as a window for the rest of the leaf. A majority of the leaf is held underground, including the cells that contain chlorophyll. The photosynthetic cells are placed around the cone that forms the leaf, with the centre of the leaf being made up of succulent cells that are crystal clear. Light entering the leaves through the upper window is scattered through these succulent cells onto the photo-synthetic cells. This scattering permits more of the photo-synthetic cells to use light from a range of angles. Thus, the plant is adapted to photosynthesize underground. The pigmentation of the upper leaf window forms the

Figure 15.6 The stone plants (*Lithops* spp.) possess very little surface growth and most of the plant remains below ground, the upper parts of the leaves (A, B) being coloured to look like stones on the ground. Most herbivores fail to see the plants because of their camouflage.

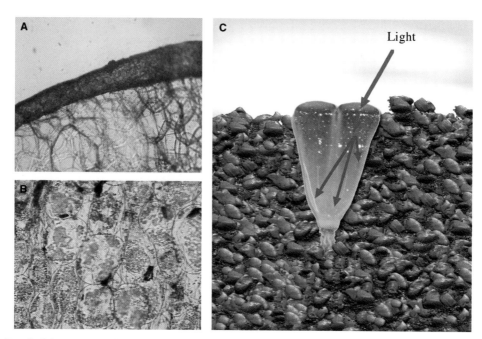

Figure 15.7 Detail of the structure of *Lithops* stone plants. They have a heavily pigmented upper surface to their leaves to hide them from herbivores (A). In order to survive below ground the plants need a number of adaptations. The photosynthetic cells are places at the sides of the conical leaves (B) and the cells in the centre of the leaves are clear. Thus, light hits the leaves and is scattered onto the photosynthetic cells (C). This means that the leaves are protected not only from herbivores but also from the intense heat of the desert environment in which they live.

camouflage and mimics the surface of the ground around the plants. These plants are easily mistaken by herbivores for stones and thus overlooked.

Grass species as crop weeds

It has already been mentioned in Chapter 7 that parasitic plants often are perpetuated in agriculture through the similarity between the parasite seed and that of the crop plant e.g. *Striga*. The first documented case of mimicry of a crop species by another plant occurs in Matthew's Gospel in the New Testament. In Matthew Chapter 13 Jesus tells a parable of 'The wheat and the tares'. Wheat is sown in a field but a weed is also sown at the same time. The weed and the wheat plant are indistinguishable until the time of harvest when the weed can be removed. With the extent of agricultural monocultures on the Earth's surface there is considerable evolutionary pressure on wild species to attempt to inhabit agricultural fields (Figure 15.8). If a weed is dissimilar to the crop plant then it can easily be spotted and removed. However, the more similar a weed is to a crop, the harder it becomes to spot, and thus it can easily contaminate seed stores at harvest time. Even in agricultural systems today, invasive grass species, such as darnel ryegrass (*Lolium temulentum*), can be aggressive weeds of wheat (*Triticum aestivum*) and barley (*Hordeum vulgare*) fields.

Müllerian mimicry

Many plants possess an appearance which they share with related species, which benefits all of the plants. Plants in the family Labiaceae (mints) are frequently aromatic and contain toxins which assist the plant in resisting herbivores (Figure 15.2). The common appearance of the plants in this family means that, although the toxic compounds in the plants vary markedly, all of the species benefit from the similarity. Furthermore, species in the mint family also bear a remarkable resemblance to species in the nettle family (Urticaceae) as a result of leaf shape. Thus, an encounter with a nettle is likely to give any herbivore an aversion to other nettle species and members of the mint family. Similarly, the aromatic compounds produced by the Labiaceae could help to ward herbivores off nettle plants. The situation is slightly more complex, since some members of the mint family are palatable, and thus benefit from a similar appearance to the nettles (Batesian mimicry, see above).

Aggressive mimicry

Egg mimics

Herbivory is a major problem for plants and, as has been discussed earlier, plants have evolved many means by

Figure 15.8 Wheat and tares. In agriculture, large areas of crop plants are planted in order to obtain a product. This is a huge advantage to a plant species as it achieves its objective of reproduction with relative ease. However, to take advantage of this situation some plants attempt to colonize the same land as the crops. Many grass species have seeds of similar size to those of crops such as wheat (*Triticum* spp.; A). This makes it difficult to separate the seeds and therefore such plants can be propagated by accident with the crop plant. The agricultural weed field cow wheat (*Melampyrum arvense*; B) has seeds of similar size to those of wheat and was once a major problem as a parasitic weed of wheat plants. However, with improvements in seed collection and cleaning, this weed has been almost completely eliminated as an agricultural pest and is now seriously endangered in the UK and the rest of Europe.

which to deter predators. However, in a small number of species this has taken the form of elaborate changes in the structure of the plants. Passionflower species are the best example of this. The major herbivores of passionflower plants are *Heliconius* butterfly larvae. These butterflies recognize the passionflowers and lay their eggs on the leaves. However, the larvae are omnivorous and will attack and eat other larvae that they encounter. As a consequence, the females lay eggs where there is no evidence of the presence of other larvae or eggs. Around 2 % of the passionflower species have evolved pigmented protuberances on the structure of the plant that mimic *Heliconius* eggs. These structures can be found on the stem, leaves and growing shoot. To a passing butterfly, the plant will appear to be infested with *Heliconius* eggs and thus it is more likely to by-pass that plant and lay eggs on another passionflower plant. The plant therefore deceives the butterfly into believing it is an unsuitable food source. A similar mechanism has been observed on *Streptanthus* plants (Brassicaceae).

Passionflowers use several other different mechanisms to deter herbivory by heliconid butterflies. First, some plants, such as *Passiflora foetida*, generate an odour which fools butterflies into believing that the plant is unfit and rotting (Echeverri *et al.*, 1991). Again, the butterflies avoid laying eggs on such plants and hunt out a more suitable food plant. Second, passionflower species display a remarkable variation in leaf shape, with *P. suberosa* reportedly being the most extreme example. Thus, insect pests have great difficulty in recognizing a

particular passionflower species for egg laying. Third, the plants form extrafloral nectaries, which attract ants (Wirth and Leal, 2001; Labeyrie *et al.*, 2001). The ants in turn protect the plants from insect larvae, such as those of the heliconid butterflies. Fourth, there is evidence in *P. foetida* that plants produces carnivorous bracts around the flowers, which act rather like the tentacles in sundews (Radhamani *et al.*, 1995; see Chapter 8). As a consequence, the genus *Passiflora* displays the most remarkable series of aggressive actions to deter insect predation known for any plant genus.

The bee orchids

While passionflowers use mimicry to deter potential predators, certain orchids use mimicry to attract potential pollinators. Possibly the most remarkable form of mimicry in the plant kingdom is that found in the genus *Ophrys*, the bee orchids. This genus is the largest of the European terrestrial orchids, comprising 40–200 different species (the actual figure depends on the classification authority consulted). Bee orchids prefer warmer climates and are more frequently encountered in the southern European countries. The flowers of these orchids have evolved to mimic the appearance of a bee or wasp and the plants use three means to fool insects. First, the lower petal, known as the labellum, bears a remarkable resemblance to a bee in size, shape and coloration (Figure 15.9). The labellum also bears a patterned reflective yellow or white mark, which resembles the wings of the visiting insect. The upper petals

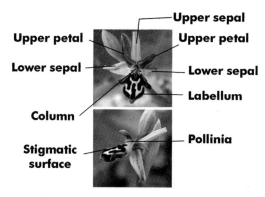

Figure 15.9   Structure of the flower of the orchid *Ophrys cretica*.

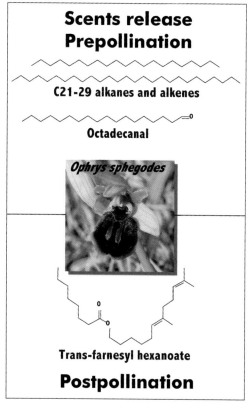

Figure 15.10   Volatile compounds released by the flowers of the early spider orchid (*Ophrys sphegodes*). Over 20 such volatile compounds have been shown to elicit sexual behaviour in *Adrena* bees. The flower superficially resembles a bee and this, combined with its scent, fools male *Adrena* bees into believing that the flower is a female bee. The males attempt to copulate with the flower and end up with pollinia attached to their heads or tails. If a flower is pollinated, it then releases a pheromone identical to that released by a female bee that has been fertilized, therefore other bees avoid this flower.

and sepals are usually coloured and therefore the whole structure resembles a bee visiting a flower. Second, the flower does not stop at just using visual signals to attract insects; it also uses tactile stimuli, with the presence of hairs on the sides of the labellum. Third, the flower has evolved to produce fragrances that mimic sex pheromones of the target insect (Figure 15.10). These pheromones can be volatile and effective at attracting pollinators from great distances. However, these sex pheromones are very specific for particular species of bee or wasp. Thus, the plants rely wholly upon a single species, or a very few, for pollination. Different species of bee orchid are adapted to attract different species of insect. Thus, there is great variability in the appearance of the flowers and the pheromones released (Figure 15.10). Only a few species of *Ophrys* orchid have been investigated better to understand the intimate relationship between bee orchids and their pollinators.

One such species that has been investigated is the early spider orchid (*Ophrys sphegodes*), which is pollinated by the solitary bee species *Andrena nigroaenea*. The labellum of the plant has been reported to produce more than 100 different compounds (Ayasse *et al.*, 2000). By separating and purifying these compounds released by the spider orchid, then investigating the neurone activity in the bee species subjected to the different compounds, it has been possible to identify compounds released that act as insect pheromones (Schiestl *et al.*, 1997,1999). The spider orchid releases 24 alkanes and alkenes that elicit copulation behaviour in *A. nigroaenea*. These compounds make up the sex pheromones released by the plant. These compounds are not very volatile in this instance, but this helps mimic the normal copulation process in the bees. Under normal circumstances the female bees leave odour trails at which male bees congregate and search for the female. Once the males congregate, the female returns to the site to copulate. The spider orchid, using visual, tactile and olfactory stimuli, fools the male bees into believing the flower is a female bee. The male bees then try to copulate

with the flower, which inevitably involves the insect brushing against the pollinia of the flower. These pollinia are packages of pollen produced by the orchids that attach to the insect. The frustrated bee then visits another flower and the same process is repeated, and the pollen is transferred onto the second flower, thus achieving pollination. The bees recognize the cocktail of chemicals (its bouquet) produced by an individual flower and avoid that plant from then on (Ayasse *et al.*, 2000). The insect appears to realize it has been fooled and takes steps to avoid a second encounter with the same plant. But the plants are one step ahead and, although the pseudo-sex pheromones produced by an individual *O. sphegodes* are constant, the compounds produced by the labellae of other individuals do vary. Thus, the bee fails to recognize the next flower it visits as a mimic, since it is likely to produce a slightly different bouquet. There is also some evidence that flowers on the same inflorescence produce different scents; thus, a bee can be fooled by other flowers on the original plant. In addition, the visual appearance of the labellum on a flower varies markedly within a population of *Ophrys* orchids; this can be in the form of different shapes of labella or in altered white patterns on the labella (Figures 15.11, 15.12). Thus, flowers on different plants possess not only different bouquets but also, to a certain extent, a different appearance.

In a bizarre twist of events, the mimicry does not stop there. Schiestl and Ayasse 2001 have also identified the production of farnesyl hexanoate by the flower after pollination (Figure 15.10). This compound is produced by the Dufour's gland in female bees to inform male bees that the female has already been inseminated. The production of the farnesyl hexanoate by the spider orchid instructs the bees not to visit the flower. The evolutionary advantage of this to the plant is not immediately clear, but it probably serves to protect the fertilized flower from the over-amorous attentions of a male bee, which may cause the flower damage during pseudocopulation! In experiments, unpollinated flowers treated with farnesyl hexanoate were markedly less likely to be visited that untreated flowers. The mirror orchid (*O. speculum*), which is pollinated solely by the scoliid wasp *Campsoscolia ciliata*, has been investigated to a lesser extent, but evidence suggests that again the flower attracts the wasps using pheromones to fool the insect into copulating with the flower. After fertilization, the flower reduces its attractiveness to the wasps by reducing the pheromones produced.

Some other orchid species use similar systems to attract insects, such as the Australian orchid genus *Chiloglottis*, e.g. *Chiloglottis formicifera*, the ant orchid, which is pollinated by thynnine wasps (Bower, 1996). As with the bee orchids, the associations between the different species are restricted and only certain species of thynnine wasps will visit specific *Chiloglottis* species.

Such an evolutionary process does, however, have its hazards. The relationship can become so specific that the plants' reproductive success is easily perturbed by the relative abundance of the mimic insect. In the worst-case scenario, the plant could easily become extinct. The British species of bee orchid (*O. apifera*) has encountered just this problem. The wasp or bee species that pollinated the flower has possibly become extinct, which may be why *O. apifera* now only uses self-pollination for reproduction (Figure 15.11). Other UK *Ophrys* species still use insects for fertilization, e.g. the fly orchid (*O. insectifera*) mimics the solitary wasp species *Gorytes mystaceus*.

Figure 15.11 The self-pollination of the UK bee orchid (*Ophrys apifera*). The UK bee orchid is a member of an orchid genus that has evolved to attract specific species of bees or wasps. However, for the UK bee orchid the flower (A) is usually self-pollinated. The pollinia within the flower flick across the flower to touch the stigmatic surface and thus achieve self-pollination (B–D). This bee orchid has evolved to flower later than the other bee orchid species and beyond the hatching time of *Andrena* bees, the usual pollinators of *Ophrys* spp. Perhaps this late flowering time has meant that it is essential that the plant usually self-pollinates.

Pollination in the Araceae

The *Arum* lilies, like the *Ophrys* orchids, use a scent to attract insects to the flowers. The flower possesses a large upright structure known as a spadix (Figure 15.13). When the female flowers are mature, the spadix heats up through a stimulation of respiration in the tissue. This allows the volatilization of odorous compounds (Seymour *et al.*, 1999). Usually these scents resemble rotting flesh and attract a large number of flies. The flies are fooled into believing that the flower contains a feast, and enter a chamber containing the ripe female flowers. If the insects are bearing any pollen, this will be transferred to the stigmas. However, the flies cannot escape the chamber and are trapped. Once the female flowers cease to be sexually receptive, the male flowers shed their pollen onto the trapped flies. Then hairs that once trapped the insects in the chamber wither away, thus releasing the pollen-dusted flies to visit a second *Arum* flower. Like bees, the flies will learn that the *Arum* lilies offer little in the way of reward

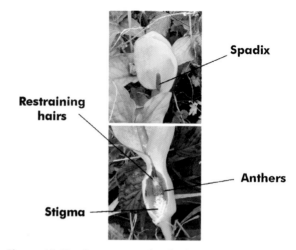

Figure 15.13 Insect attraction in Araceae. Flowers in the family Araceae possess flowers that act as insect lures and traps. The flower emits volatile scents that resemble the odour of rotting flesh. Insects are attracted to the flowers and enter the structure shown. Once in, it is difficult to get out. If the insects possess any pollen on their bodies, they pollinate the flowers; if not, the female parts of the flower cease to be receptive and the anthers shower the trapped insects with pollen. The hair traps to the flower whither and the insects are then released. If the insects are then trapped by another flower, they will pollinate it. The flower shown is lords and ladies (*Arum maculatum*).

Figure 15.12 Variability of *Ophrys* flowers. Here a selection of flowers from a colony of late spider orchids (*Ophrys holoserica*) is shown in the Bourgogne region of France. Each flower has a slightly different appearance, so any bee visiting one flower and becoming frustrated with it may visit a second flower, fooled again into believing that it is in fact a female bee. The scents of individual orchids also vary, making it very difficult for the bees to learn quickly that the flowers are in fact mimics.

with respect to a feast of rotting flesh, but by then many *Arum* flowers will already be pollinated!

Seed dispersal and food reserves: the role of humans

Plants are significant food supplies for herbivores and omnivores, and humankind has obviously had a major

Figure 15.14 Edible fruit production. Through the production of fruits, animals are attracted to plants as a food source. In exchange for the food, the animals disperse the seeds of the plants. Here brambles (*Rubus fructicosus*; A) and plums (*Prunus sp.*; B) are shown.

Figure 15.15 Humankind has taken the relationship between crop plants and animals a stage further. Through the selection of plants that give high yields of fruit or seeds, large areas of the Earth's surface have been planted with particular plant species. Therefore the plants are successful and humankind can produce enough food to feed large populations. Tomatoes (*Lycopersicon esculentum*; A), oranges (*Citrus sinensis*; B) and squashes (*Cucurbita* sp.; C) are shown as examples.

impact on plant life on this planet (Figures 15.14–15.16). Through interaction with humans, many species have become agricultural crops and, as a result of this, probably cover more of the Earth's surface than any other plants. Using the area of the Earth's surface covered by a particular plant species is a reasonable index to measure the success of a plant. Through this index, crop plants represent the most successful species on Earth today. The two species that take up the largest measured area of the planet are wheat and rice. Estimates of the total area occupied by these plants comes to around a billion hectares.

With the crop plants, we see an intimate relationship between humans (the consumer) and plants (the producer). Provided the plant produces the large food reserves that we want, we in turn plant out large areas and propagate the crop plants. Thus, both the crop and humankind are mutually served in the relationship. Of the 250 000 higher plant species, only around 2000 species have an economic value as crop plants. The bulk of the human population in the world is supported by only around 30 different plant species (Walters and Hamilton, 1993)

Crop plants were probably first domesticated around 10 000 years ago. Seed crops were the first plants to be used. There is some debate as to what factors seed crops had which made certain species targets for domestication and not others. The most likely hypothesis is that plants such as wheat, rice and barley produced larger seed heads than most grass species. However, that is not thought to be enough, since much of the high productivity we see today in cereal crop plants is a result of years of selective breeding. The other important factor is thought to be mutant varieties of these plants arising, which failed to 'shatter' on seed maturity. When the seeds are mature, grasses tend to form a weakened cell layer between the seed and the plant. The seeds are easily dropped from the parent plant with little agitation, and this process is known as 'shatter'. If a plant underwent shatter, then by the time most seeds were mature, many would have already been shed onto the ground and harvesting a crop would be very difficult. Consequently, there would be a great advantage to early farmers if a species held onto all of the seeds for a longer time. To the plant, the failure of the shatter mechanism is not an advantage, since it will result in a reduction in the capacity for seed dispersal. However, it would also mean that early farmers would be able to gather all of the

Figure 15.16 Mimicry in wild carrot (*Daucus carota*), which has evolved a means of attracting insects to the flowers. Around one in ten individuals possess a central red flower in the umbel, thought to appear like an insect, such as a soldier beetle, on the inflorescence. Other insects are thought to be more likely to visit the inflorescence as it looks as though it is already occupied by a potential mate.

seeds from individual plants. The harvest sizes would increase and the time taken to gather the harvest reduced. This allowed humans to progress gradually from being hunter-gathers to becoming farmers.

References

Ayasse M, Schiestl FP, Paulus HF, Lofstedt C *et al.* (2000) Evolution of reproductive strategies in the sexually deceptive orchid *Ophrys sphegodes*: how does flower-specific variation of odour signals influence reproductive success? *Evolution,* **54**, 1995–2006.

Bower CC (1996) Demonstration of pollinator-mediated reproductive isolation in sexually deceptive species of *Chiloglottis* (Orchidaceae: Caladeniinae). *Australian Journal of Botany,* **44**, 15–33.

Chittka L, Gumbert A and Kunze J (1997) Foraging dynamics of bumble bees: correlates of movements within and between plant species. *Behavioral Ecology,* **8**, 239–249.

Echeverri F, Cardona G, Torres F, Pelaez C *et al.* (1991) Ermanin – an insect deterrent flavonoid from *Passiflora foetida* resin. *Phytochemistry,* **30**, 153–155.

Gumbert A and Kunze J (2001) Colour similarity to rewarding model plants affects pollination in a food deceptive orchid, *Orchis boryi. Biological Journal of the Linnean Society,* **72**, 419–433.

Labeyrie E, Pascal L and Delabie J (2001) Protection of *Passiflora glandulosa* (Passifloraceae) against herbivory: impact of ants exploiting extrafloral nectaries. *Sociobiology,* **38**, 317–321.

Proctor HC and Harder LD (1995) Effect of pollination success on floral longevity in the orchid *Calypso bulbosa* (Orchidaceae). *American Journal of Botany,* **82**, 1131–1136.

Radhamani TR, Sudarshana L and Krishnan R (1995) Defence and carnivory: dual role of bracts in *Passiflora foetida. Journal of Biosciences,* **20**, 657–664.

Schiestl FP, Ayasse M, Paulus HF, Löfstedt C, Hansson BS, Ibarra F and Francke W (1999) Orchid pollination by sexual swindle. *Nature,* **399**, 421–422.

Schiestl FP and Ayasse M (2001) Post-pollination emission of a repellent compound in a sexually deceptive orchid: a new mechanism for maximising reproductive success? *Oecologia,* **126**, 531–534.

Schiestl FP, Ayasse M, Paulus HF, Erdmann D and Francke W (1997) Variation of floral scent emission and post pollination changes in individual flowers of *Ophrys sphegodes* subsp. *sphegodes* (Miller) (Orchidaceae). *Journal of Chemical Ecology,* **23**, 2881–2896.

Seymour RS and Schultze-Motel P (1999) Respiration, temperature regulation and energetics of thermogenic inflorescences of the dragon lily *Dracunculus vulgaris* (Araceae).*Proceedings of the Royal Society of London Series B Biological Sciences,* 266,1975–1983.

Walters M and Hamilton A (1993) *The Vital Wealth of Plants,* Gland, Switzerland, WWF International.

Wirth R and Leal IR (2001) Does rainfall affect temporal variability of ant protection in *Passiflora coccinea*? *Ecoscience,* **4**, 450–453.

16

Plant defences

Plants are autotrophic organisms, fixing carbon dioxide and synthesizing carbohydrate and protein. The vast majority of global food chains have plants at their base. Thus, it is not surprising that an estimated 18% of plant production is devoured by herbivores each year (Cyr and Pace, 1993). Of this damage, an estimated 10% is caused by insects and the remainder by other herbivores. Since plants cannot move to avoid predators, pathogens and neighbouring plants and they lack an immune system that would be the equivalent of that in animals, plants have developed numerous defence systems to protect themselves. These mechanisms include both structural adaptations and chemical defences. It is probably plant defence that has led to the evolution of the phenomenal ability of plants to synthesize a vast range of organic compounds. The possession of this range of organic chemicals in plants has led to the development of the term 'secondary plant metabolism'. These are compounds that are present in a wide range of plant species but have no known metabolic function (Fraenkel, 1959). Plants are estimated to have the capacity to manufacture over 100 000 different organic compounds. Thus, there is a large arsenal of toxic cocktails plants can deploy. It is many of these compounds that have been of interest in terms of medicines from plants. Without these defences, the losses of material to herbivores and pathogens would be considerably greater.

The response of plants against herbivores should not be seen as always passive; in many instances poisons are produced in response to herbivore attack. Plants should be viewed as actively defending themselves through weapons of mass destruction and the enrolment of predators to their assistance, and in a small number of cases they are the aggressive parties.

In this chapter we shall look at the range of defences available to plants, using case studies to illustrate the mechanisms.

Physical defence structures

One of the first lines of defence is the outer structure of the plant (Figure 16.1). If a predator can be kept out or away, then the plant will be safe. These defences can take many forms. Many species have evolved thorns which adorn the stem, branches or leaves. These can be very large and threatening, as in acacia trees, or much softer and less pronounced, as in thistle species. It all depends on the target herbivore species. Acacias in Africa target the larger browsing mammals, such as giraffes and antelopes. The large thorns are a great deterrent to many species, as they hurt the mouth-parts of the browsers. A common phenomenon in plants which use thorns or spikes to protect themselves is that the greater the herbivore damage to the plant, the larger the density of thorns produced on new tissue. Thus, plants can increase their physical defences in response to herbivore damage.

Many grass species use sharp silicates as a defence. These are deposited on leaf edges and form a serrated edge. This forms a potent deterrent to many herbivores from attacking grasses as the silicates give grasses a sharp edge.

As leaves form, they become reinforced with cellulose and lignin and become rigid. The rigidity of the leaf acts to protect it from herbivores. Thus, a herbivore will routinely attack the youngest leaves on the plant in preference to the oldest. One nice example demonstrating this is the holly blue butterfly and holly trees. This butterfly only eats holly leaves in the spring, when they are soft and newly formed, but later in the season, once the leaves have hardened, the butterfly lays its eggs on ivy leaves, as the mature holly leaves are reinforced and unpalatable (Figure 16.2).

One of the easiest places for microbial attack in plants is through the stomatal pores in the leaf surface or the lenticels on the stem. These pores need to stay open for photosynthesis, but they also provide a means for bypassing other physical defences of the leaf. Whether there are

Physiology and Behaviour of Plants   Peter Scott
© 2008 John Wiley & Sons, Ltd

Figure 16.1 Physical defences of plants. The simplest way of defending against predators is to keep them out in the first place and plants have many physical defences that do just that. The figure summarizes a number of physical defences used by plants. A) *Cerinthe major* - leaves are soft and covered in a wax which presents little defence. B) Herb Robert (*Geranium robertianum*) – the plant has firm pigmented leaves which may offer a little in defence. C) *Sonchus* sp.- the leaves possess small soft points along the margin, other plants in the Asteraceae possess reinforced spines on their leaf margins such as *Cïrcium* sp. (D). E) *Lapsana communis* - the cactus-like Euphorbia possesses a profusion of long needle-like spines that are a serious discouragement to any herbivores. F) *Aloe nobilis*. This plant uses thick reinforced leaves with a large number of short spines. G) Ferocious Blue Cycad (*Encephalartos horridus*). The leaves of this plant are extremely rigid and very sharp. Few herbivores attack cycads. H) Scot's pine (*Pinus sylvestris*). The photograph shows a wound to the trunk of the pine tree with a protective resin issuing forth. This resin quickly hardens and plugs up the wound preventing insect or fungal attack. I) Cork oak (*Quercus suber*). The thick reinforced bark of this tree is a potent defence for the tree.

Figure 16.2 The holly blue butterfly (*Celastrina argiolus*; A) preferentially feeds on holly leaves (*Ilex* spp.; B). For the first brood of the butterfly, the larvae are laid on young holly leaves and feed off these. As the leaves expand and become more rigid, they become to resistant to attack by feeding caterpillars, so for the next mid-summer brood the butterfly has to completely change its food source to ivy (*Hedera helix*; C). The tough leaves of holly are a potent defence against attack, but the leaves are very vulnerable in their early stages of development.

specific features of guard cells that prevent microbial infection through the stomatal pore is unknown.

The cuticle

The surface of plant structures above the ground is covered with a structure known as the cuticle (Brett and Waldron, 1990). This acts as the primary barrier to repel pathogens (Figure 16.3). The structure is formed from cutin, a hydrophobic material made up of a complex structure of fatty acids and fatty esters. In grasses silica is also deposited in the cuticle. This has several functions, but the main one is to produce a sharp surface which can act to deter herbivores, and silica is known to be toxic to many fungi.

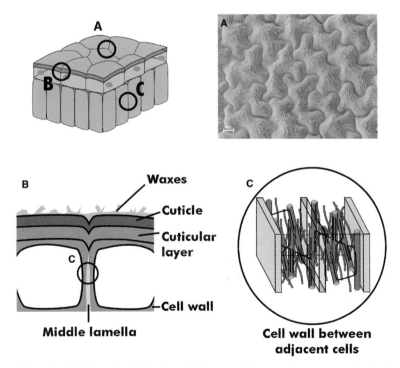

Figure 16.3 Structure of the cell wall. The cell wall of plant cells is one of the most important methods of protecting the plant. This leaf section is shown, with highlighted regions (diagram, top left) identifying the different areas that contribute to defence: (A) The waxy cuticle, secreted by the epidermal cells, acts as a barrier to prevent penetration of cells and to minimize the grip of water and insects on the outer surface of the plant. (B) A cross-section of a leaf surface showing the individual elements of the cell wall on the cell surface and between the cells, including barriers across the leaf surface, wax, cuticle and cuticular layer. (C) The structure of the cell wall between cells.

The cell wall

The cell wall as a passive barrier

There is considerable diversity in the cell wall composition of different cells in separate organs of a plant. However, for most cells the dense network of fibres which make up the cell wall is sufficiently compact to exclude even virus particles (Brett and Waldron, 1990). Only if the cell wall is ruptured in some way can micro-organisms gain entry to the protoplast. Some fungi have adapted to this, and eke out a parasitic existence living in the middle lamella region of the leaf, never needing to penetrate the cell wall structure.

The cell wall as an active barrier

There can be considerable change in the structure of the cell wall in response to penetration by a pathogen. The most likely response is that the cells at the point of penetration will become reinforced with lignin or suberin. Cell death also occurs at this point and this whole process forms part of what is called the 'hypersensitive response' of a plant (Bell, 1981). Thus, the point of infection is isolated from the rest of the plant and progress of the pathogen is severely restricted, if not arrested. The whole process of the hypersensitive response requires signals from a pathogen being recognized by the plant, which then responds. If the pathogen is not recognized, then it can effectively begin parasitizing the infected structure. A further response can occur in the cell wall if the structure is penetrated. Callose is deposited where the pathogen contacts the protoplast. This acts to prevent the spread of the infection into the remainder of the cell. During attack of the cell wall there is considerable digestion of the pathogen and the plant cell wall. The compounds themselves can act as elicitors for phytoalexins, which form another defence mechanism (see later).

Physical defence barriers are effective to an extent, but with larger herbivores the plant needs to make itself unpalatable. One of the most effective methods of achieving this is to tie up a great deal of the carbohydrate in the cells as cellulose, which acts as a barrier but also is indigestible to animals. Cellulose is mainly excreted by insects, and mammals require a specialized process using bacteria to aid in its digestion. But using indigestible materials is only part of the defence story; plants have a whole array of other chemicals to rely on.

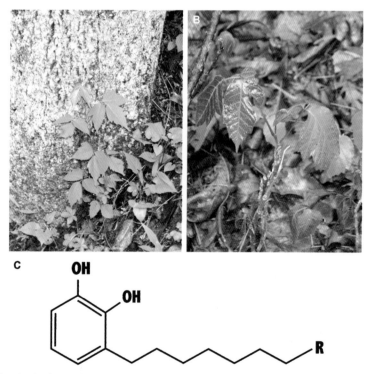

Figure 16.4　Poison ivy (*Toxicodendron radicans*; A, B), which produces a group of toxic chemicals called urushiol (C) on the surface of the leaf. If the plant is touched the urushiol spreads to the skin and diffuses through it. A T cell-mediated immune response is then initiated, in which the body recognizes the urusiol as a foreign body and initiates the body's defences, leading to the development of a rash, although the urushiol is in fact harmless. Only insects such as sawfly larvae (*Arge humeralis*) can attack the plant.

Poisons by injection or touch

The two most familiar examples of toxins elicited by touch must be those produced by poison ivy and nettles (Figures 16.4 and 16.5). The leaves of poison ivy are coated in a chemical called urushiol. On touching or bruising of the leaf tissue, this oil is deposited on the skin and can cause an allergic contact dermatitis. There is a delay in the skin's response to the urushiol, as it interacts with proteins in the skin, which trigger the T lymphoyctes to produce cytokines, which cause a severe inflammatory response. Urushiol is in reality harmless to the body, and it is a failure of the immune system which misreads external signals, believing the body to be under attack. This is a very potent defence mechanism, and anybody who has ever been stung by poison ivy can confirm that these plants are to be avoided.

Nettles are another familiar species which employs a defensive stinging mechanism. The aerial tissues of a nettle are covered in fine hairs, known as trichomes. If touched, these break at the tip and release a cocktail of formic acid, histamine, acetylcholine, serotonin (5-hydroxytrypta-

mine) and other unidentified ingredients. These all combine to give a powerful sting if the plant is touched. The presence of formic acid gives nettles their characteristic smell when damaged. When a nettle is damaged, the plant responds by producing new tissues possessing more defensive trichomes.

Chemical defences

Plants produce three main groups of compounds to act as defence molecules; terpenes, phenolics and nitrogen-containing organic compounds (Figure 16.6). Each species or group of closely related species may produce a cocktail of defence compounds peculiar to them alone. Thus, there is very limited knowledge of the range of secondary metabolites produced by plants. These chemical defences give plants protection against herbivory and parasitism. But a careful balance in the plant needs to be achieved in which the metabolic costs of synthesizing these chemicals is balanced against the likelihood of predation and the danger of the chemicals to the plant itself. The costs to the

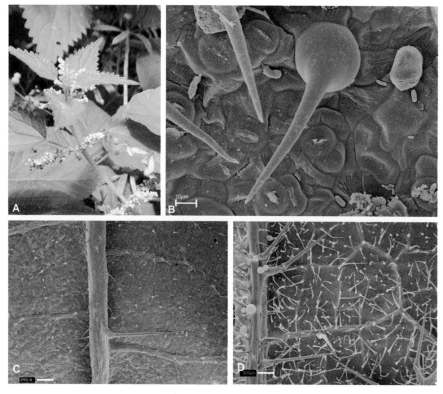

Figure 16.5 Chemical defences in nettles (*Urtica* spp.). Nettles such as *Urtica dioica* (A), release a cocktail of chemicals from silica spines on the leaf surface. The spines surround the leaves and the stem of the plant. If touched, the spines fracture and inject histamine, choline, formic acid and serotonin, causing reddening of the skin and a stinging sensation. If the plant is damaged, new leaves develop many more spines in response. The leaf shown in (C) is a leaf untouched during its development; (D) shows a leaf that has been repeatedly brushed to simulate damage – note the large increase in the number of spines.

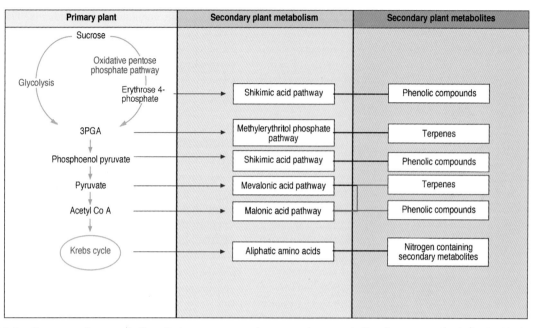

Figure 16.6 Plant secondary metabolism. By its very nature, plant secondary metabolism is very complex indeed. This figure is an attempt to summarize the major pathways involved in plant secondary metabolism and the main compound families formed. These pathways produce a vast bulk of the 100 000 organic compounds identified in plants so far.

plant in deploying these chemicals must be considerable, since in many instances the secondary metabolites are produced in response to attack (post-infectional defences), rather than constitutively (pre-infectional defences). There follow examples of each of the divisions of compounds that act as pre- or post-infectional defences.

Terpenes

Terpenes are synthesized via the mevalonate or methyl-erythritol phosphate pathways from a basic isoprene building block (Figure 16.7). They are usually water-insoluble but are soluble in lipids. This is the most diverse group of secondary metabolites known in plants and as such it is only possible to give a limited number of examples of these compounds acting in the defence of plants.

Pyrethrins

In the plant family Asteraceae, many monoterpenes are produced which act as potent insecticides (Figure 16.8). The most potent naturally occurring compounds are chrysanthemic acid and pyrethric acid, which are produced abundantly in the flowers of *Chrysanthemum cineraria-folium*. These compounds are collectively known as the pyrethrins. They are potent chemicals which act against the

nervous system of insects, paralysing and killing them if they attempt to eat the leaves. However, they have little effect on mammals. Pyrethrins can be extracted from the flowers of *Chrysanthemum* and have been used for many years as a natural pesticide. Chrysanthemic acid and pyrethric acid are short-lived once exposed to the air and are readily biodegraded. These molecules have been used as the basis for numerous synthetic pyrethrins, which last longer and are more toxic to insects. Pyrethrins are an example of a post-infectional defence.

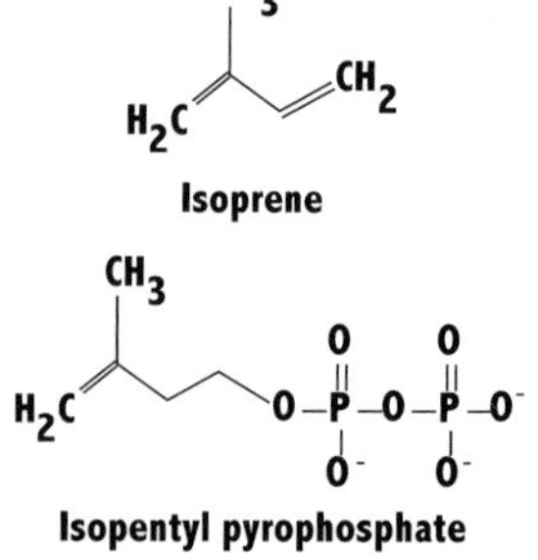

Figure 16.7 The terpenes, the largest group of plant secondary metabolites. They are made from the basic unit isoprene, which is converted into isopentyl pyrophosphate. The name is derived from terpentine, which is a terpene. From these units the mono-, di-, tri- and sequiterpenes, and the steroids, are formed.

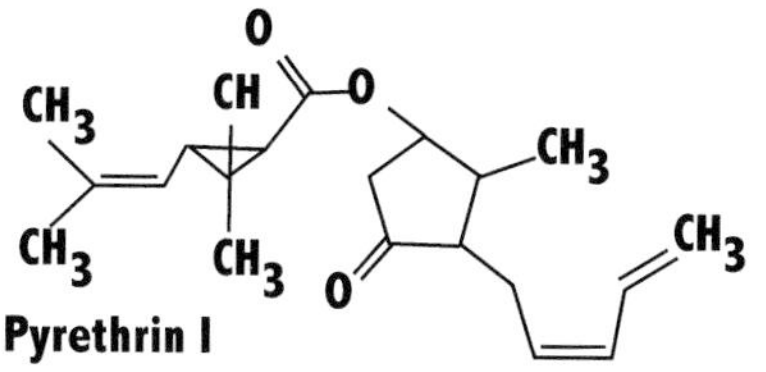

Figure 16.8 Pyrethrins, a closely related group of monoterpenes that act as natural insecticides. Two pyrethrins are shown from *Chrysanthemum cinerariaefolium*. The plant produces these compounds in response to insect attack. The pyrethrins act as neurotoxins that paralyse insects. This group of compounds has been used to produce a whole group of compounds used as insecticides, known as the pyrethroids.

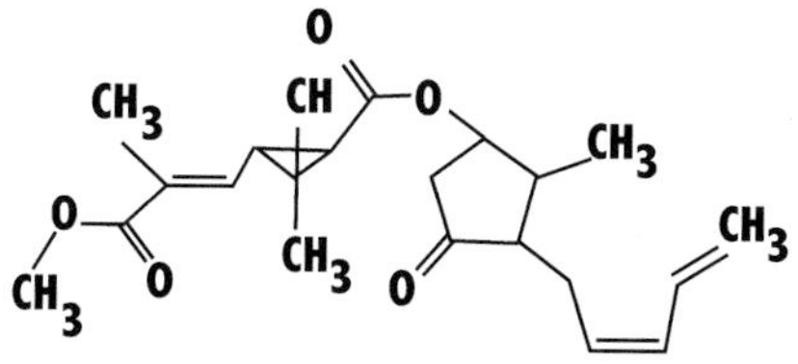

Trees of the genus *Taxus* (yew trees) produce a range of diterpenes which are toxic to insects, birds and mammals if eaten. Birds can only eat the red flesh of the berries of yew and this is used in the seed dispersal mechanism of the tree. These compounds include baccatin, cephalomannine and, in one species (*Taxus brevifolia*, the Pacific yew tree), taxol. There is now great interest in taxol as an anticancer drug (see Chapter 16 on plant medicines).

Plants in the family Lamiaceae (the mints) are known to produce a large range of monoterpenes and sesquiterpenes, which usually accumulate in trichomes (glandular hairs) on the leaves and stems of the plants. Once damaged, the leaf releases these volatile compounds and this gives the plants their characteristic odours. Plants such as mint (*Mentha spicata*) produce menthol, menthone and menthyl acetate. These compounds have been demonstrated to act as insect repellents. Other species in this family produce thymol (thyme, *Thymus vulgaris*), cineole (rosemary, *Rosmarinus officinalis*) and thujone (Sage, *Salvia officinalis*) (Figure 16.9).

The neem tree (*Azadirachta indica*) from India has attracted considerable interest because it produces a triterpene known as azadirachtin. This is related to the limonoid compounds found in the skins of citrus fruit, and gives rise to the bitterness of the rinds of these fruits. Azadirachtin is a potent insecticide which is toxic at very low levels. This plant has recently hit the news as a result of extravagant US patents attempting to exploit traditional knowledge from India that has been known for centuries.

The last group of terpenes that has been shown to have insecticidal activity is the steroids. A group of compounds known as phytoecdysones have been shown to mimic the activity of moulting hormones in insects (Figure 16.10). Thus, the moulting activity in the life cycle of insects is disrupted by the steroids and the insect is killed. It was observed that after locust swarms, certain plants were left relatively untouched. After further investigation, one species, bugleweed (*Ajuga remora*), was discovered to produce a phytoecdysone called 20-hydroxyecdysone. When extracts of the plant were fed to insects, they produced multiple head capsules, which resulted in a failure to produce mouthparts properly, and insects subjected to the extracts died. Other steroids of the cardenolides and saponins are active against mammal herbivores. One excellent example is digitoxin from the foxglove plant (*Digitalis purpurea*). Digitoxin is a glycosolated-cardenolide which interferes with Na^+- and K^+-activated ATPases. Cardenolides are bitter and very toxic; one ATPase particularly susceptible to this compound is in the heart (Figure 16.11). Thus, if plants containing cardenolides are eaten, the result is often heart failure. However, as will be mentioned in a later chapter, digitoxin has become a powerful heart drug. Saponins have been reported in a number of different plant species (Figure 16.11). These compounds are so named since they are water- and lipid-soluble and possess soap-like properties. They are toxic because they interfere with membrane function and cause haemolysis (breakdown of the red blood cells), and they also react with sterols in the blood. In humans saponins are only hazardous if injected into the bloodstream and are relatively harmless if ingested.

Phenolic-based defence chemicals

Phenolics form the second largest division of plant secondary metabolites enrolled into defence. Over 10 000 different plant-based phenolic compounds have been identified. Phenolics are made from the shikimic acid or malonic acid pathways (Figure 16.6). Animals lack these pathways; thus, they are unable to synthesize the aromatic amino acids phenylalanine, tyrosine and tryptophan. Phenolic compounds are also used in the plant for the formation of lignin and coloured compounds such as the anthocyanins.

Coumarins

The simplest plant phenolic toxins are the coumarins. Coumarins are present in high concentrations in clover plants and they contribute to the smell that comes from

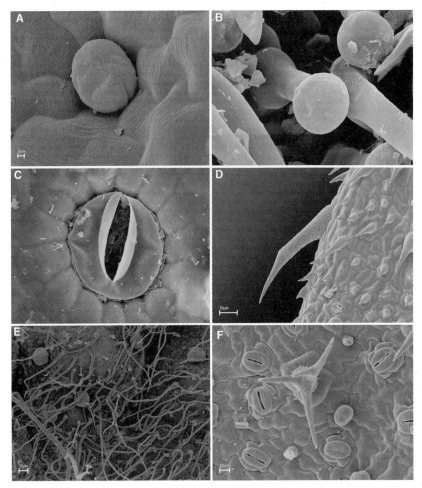

Figure 16.9 The essential oils. The essential oils are plant extracts that are made up of volatile aromatic compounds. Many of these compounds are from the terpene group, and are used in the perfume industry and aromatherapy and as flavourings. The oils are secreted in glands or trichomes on the leaf surface. Shown here are electron micrographs of trichomes of: (A) basil (*Ocymum minumum*); (B) rosemary (*Rosmarinus officinalis*); (C) thyme (*Thymus vulgaris*); (D) lemon balm (*Melissa officinalis*); (E) sage (*Salvia divinorum*), underside of leaf; (F) lavender (*Lavandula angustifolia*).

freshly-mown hay. One coumarin, furocoumarin (a psoralen), is present in the plant families Umbellifereae and Rutaceae (Figure 16.12). When one of these plants is damaged, the furocoumarin is exposed to UV-light and becomes activated into a high-energy state. This causes the molecule to interact with DNA of the herbivore and this can lead to the death of affected cells. In mammals this results in severe rashes on contact with the plant; in extreme cases such contact can be fatal. Plants such as celery routinely possess low levels of furocourmarin, but on infection by the fungus *Sclerotina sclerotiorum* the plant releases xanthotoxin as a defence reaction. This causes photophytodermatitis in farm workers handling the crop. Plants such as giant hogweed (*Heracleum manegazzianum*) possess very high concentrations of psoralen,

bergapten and xanthotoxin and thus are particularly dangerous. Insects have evolved to deal with furocoumarin by either attacking the plants from the stem only or producing nets or webs around the feeding insect to shield off the UV-light. Furocoumarin an example of a post-infection defence in celery, where the amounts of the compound rise in response to infestation by fungi.

Isoflavenoids

The rearrangements of the phenol ring necessary for the formation of isoflavenoids are very complex and thus rare in nature. Isoflavenoids are almost exclusive to the plant family Fabaceae. However, many hundreds of different isoflavenoids have been identified. Synthesis of the

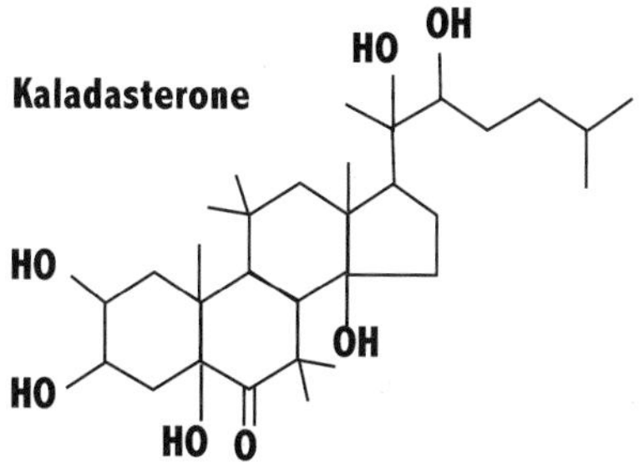

Figure 16.10 Plants also can produce steroids, of which one group are the phytoecdysones. These compounds mimic the moulting hormones produced by insects. Therefore, when insects eat the leaves of the plant the moulting process is upset and the insect is killed. One example of a plant which uses this defence is bugleweed (*Ajuga reptans*). The kaladasterone is shown.

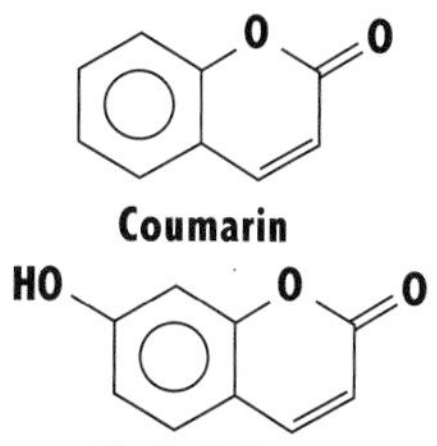

Figure 16.12 Coumarins, phenolic-based compounds found in the families Umbelliferae and Rutaceae. Many coumarins interact with UV light and react with DNA, causing cell death. In most instances this results in a severe rash on mammals or death in the case of insects. To avoid these defences, most insects that attack plant species containing coumarins do so in the dark or under the cover of a defensive web.

molecule is costly in terms of energy and oxygen. These molecules possess antimicrobial activity. Peas (*Pisum sativum*) produce the isoflavenoid pisatin following fungal infection (Figure 16.13). This acts to kill the fungus and stop the spread of the infection. *Trifolium* species produce the isoflavenoids daidzein and coumestrol, which possess

oestrogen activity. These compounds bind to oestrogen receptors and frequently cause fertility problems in sheep grazing on a diet high in clover.

A number of plant species closely related to *Derris elliptica* and *Lonchocarpus utiliz* produce the isoflavenoid rotenone. This compound can act as a potent insecticide if the plant's root system is attacked. Rotenone acts to inhibit oxidative phosphorylation in the respiratory electron transport chain. In mammals the toxicity is low, since it is rapidly metabolized in the gut and rendered harmless. Extracts are still prepared from these plants and applied as insecticides. However, rotenone is quickly rendered harmless by exposure to light and the air. Isoflavenoids also form the basis for the antimicrobial (fungal or bacterial attack) compounds called phytoalexins. These will be discussed later, with other compounds which also act as phytoalexins in other plant species.

Tannins

These are polyphenols produced by plants (Figure 16.14). The name 'tannin' comes from the Celtic word for oak and refers to the ancient source of tannins for treating animal skins to make leather (curing). They are frequently present in fruits (e.g. grapes and bananas), leaves (e.g. tea plants) and bark of some trees (e.g. oak and birch trees). They are

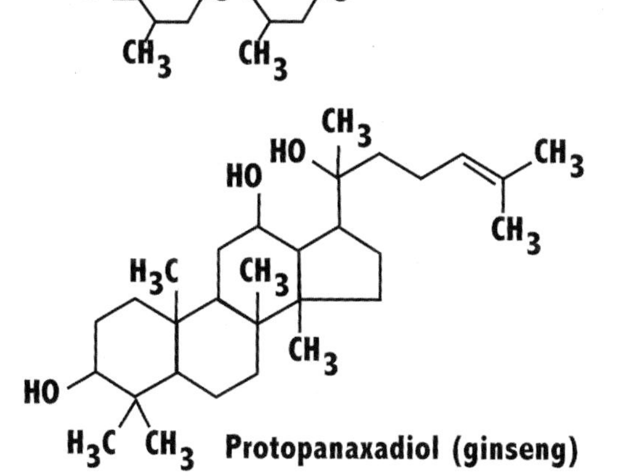

Figure 16.11 Other steroid-derived plant defences. The defence of the foxglove (*Digitalis purpurea*) is digoxin, a saponin (a steroid glycoside) which interrupts normal functioning of the mammalian heart. Protopanaxadiol is a saponin from ginseng (*Panax ginseng*), which has been shown to affect cell division in mammals.

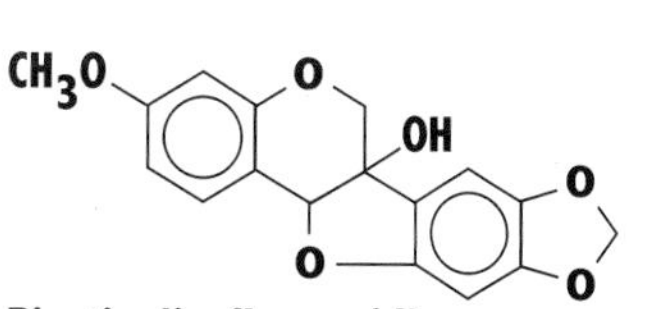

Figure 16.13 Pisatin, an isoflavonoid produced by pea plants (*Pisum sativum*) in response to fungal attack. This compound is known as a phytoalexin.

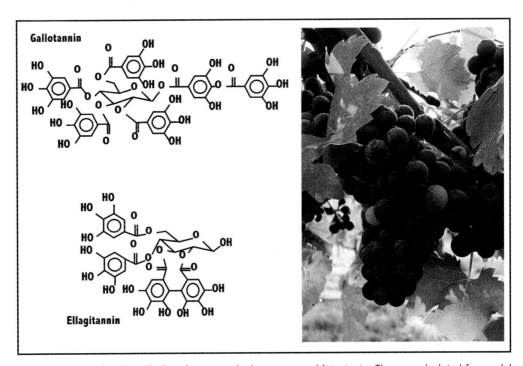

Figure 16.14 Tannins, complex phenolic-based compounds that possess a bitter taste. They were isolated from oak bark by early civilizations to tan leather. Tannins possess the property of interacting with proteins, denaturing them and precipitating them from solution. For herbivores this means that they are unable to digest any plant material ingested and often this results in severe illness if not death. Tannins are often found in fruits, such as bananas and grapes, and some leaves, such as tea, and are an important component of red wine. The tannins in red wine can be astringent and leave a dry feeling in the mouth because the tannins interact with proteins in the mouth associated with taste and denature them. However, if the wine is drunk with a food with a high protein content, such as red meat or cheese, the tannins interact with the protein in the food rather than those in the mouth and cause the wine to taste more palatable. In a similar way, milk is added to tea to reduce its bitterness as a result of the high tannin content.

produced in particularly high quantities in plants of the family Fabaceae. Tannins are deposited on the surface waxes of the plant or are stored in the vacuole. This ensures that they do not interfere with normal plant metabolic processes. Tannins are only released on damage to a cell or cell death. They have an astringent taste and thus organisms can learn to avoid them. Their bitter taste is a result of their interfering with the glycoproteins in saliva, the tannins reacting with proteins to form insoluble complexes that precipitate out of solution. In the digestive tract, tannins interact with digestive proteins, making food indigestible. Thus, arthropods and mammals can be severely affected by a high tannin content in food. In addition, tannins also inhibit microbial attack of plants. Insects have adapted to plants with high tannin contents, having an alkaline gut pH which reduces the effect of tannin interaction with protein. Many animals secrete mucin (a tannin-binding protein), again to mitigate the effects of tannin on food digestion. Many fruits, such as bananas and grapes, lose their tannin content as they ripen, and this allows the plant to carefully control when the fruit is ripe and ready to

be eaten for seed dispersal. Tannins have gained numerous uses in human society, such as in leather production and in certain foods. Tannins produce the bitter taste in tea. To make it more palatable, milk is added to the drink to reduce the free tannin content, because it forms complexes with proteins in the milk. In addition, tannins are an important component of red wine, whose taste is frequently improved by drinking it with foods containing high protein content (cheese and meats) and thus reducing the free tannin content. Tannins protect wine from bacterial attack but begin to polymerize and precipitate with time. Thus, the tannin content of wine decreases with age, thereby improving the taste of the wine.

Nitrogen-containing organic compounds

This is a group containing many compounds which are less closely related than the other divisions already covered in this chapter. The main feature they have in common is that they are all synthesized from common amino acids.

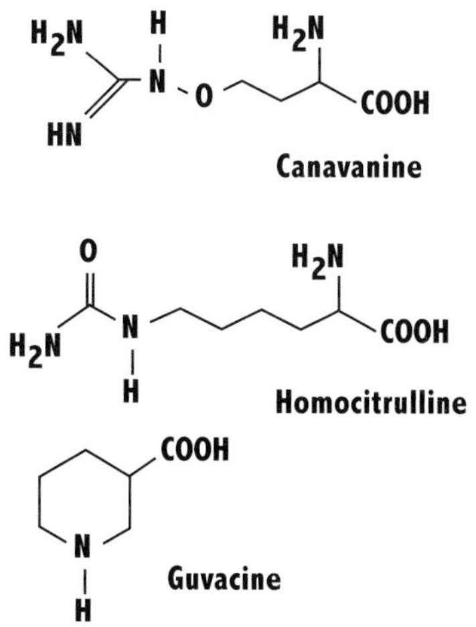

Figure 16.15 Non-protein amino acids. Plants in the family Fabaceae tend to store protein reserves in their seeds. However, to protect the seeds from herbivores the plants then make the proteins alter some of the amino acids. Thus, when the amino acids are eaten they are not recognized by the herbivore and then are either inserted incorrectly into new proteins or inhibit the uptake of normal amino acids. This results in a major disruption of protein metabolism in the herbivore and can result in its death.

Non-protein amino acids

There are 20 different amino acids used to make proteins in animal cells. However, in plants some amino acids are modified within proteins but also non-protein amino acids are produced as defence compounds (Figure 16.15). Around 300 non-protein amino acids are known, and they are generally very similar to amino acids normally used in protein synthesis. Hence, in animal tissues these amino acids are incorporated into proteins by accident and this leads to a failure of function and ultimately to cell death. In some examples, non-protein amino acids have a toxic effect as a result of inhibiting amino acid uptake or the synthesis of amino acids. Non-protein amino acids are made by members of a small number of plant families: Asteraceae, Curcubitaceae, Hippocastenaceae, Fabaceae and Sapindacae.

One such amino acid is canavanine, which is an analogue of arginine. Canavanine was first discovered in jack beans (*Canavalia ensiformis*) but has since been found in seeds of a large number of members of the Fabaceae. It can account for as much as 10% of the dry weight of the seed in some examples. Thus, the amino acid serves as a nitrogen storage compound in the seed and a defence against gramivores (seed eaters), including humans. The species that produce these amino acids have evolved enzymes that can recognize the non-protein amino acids and thus do not incorporate them into proteins, thereby avoiding poisoning themselves.

Proteinase and amylase inhibitors

Many plant species protect themselves by using inhibitors of protein digestion. Hence when the plant is attacked by a herbivore, the proteinase prevents proper digestion of the cell contents and thus starves it. These are effective against both mammals and insects and are most frequently observed in members of the family Fabaceae. For example, the seeds of the kidney bean (*Phaseolus vulgaris*) are filled with 20% protein and 60% carbohydrate, and are thus an attractive meal for herbivores. However, they also contain the proteinase inhibitor phasin (a toxalbumin). Phasin is a potent inhibitor of trypsin and chymotrypsin and thus renders the seeds inedible to most species. For humans to eat the beans, they must first be soaked to imbibe water, then the seeds cooked to denature the phasin protein. Phasin binds to the active site of proteinases in the gut and, instead of being digested, fuses irreversibly. Kidney beans also contain an α-amylase inhibitor, phaseolamin. This protein inhibits enzymes that break down starch. Since a great deal of the carbohydrate content of leaves and seeds is starch, this is an important defence mechanism. The α-amylase of kidney beans is unaffected by the inhibitor. The α-amylase inhibitors are currently receiving a great deal of interest as potential slimming drugs.

Experiments in tomato plants have shown that attack of leaves by the Colorado beetle leads to the accumulation of proteinase inhibitors in both wounded and non-wounded leaves. Thus, inhibitors of digestion can be an active and responsive form of plant defence. However, in this case merely wounding the leaves led to an identical response; thus, the tomato plants did not recognize and react to a specific herbivore and this is a general response to wounding.

Cyanogenic glycosides

Cyanogenic glycosides have been identified in over 200 different plant species, including dicotyledons, monocotyledons, gymnosperms and ferns (Figure 16.16). They are made through the conversion of amino acids to oximes, which are then glycosylated. In the cell the cyanogenic glycosides are contained in separate compartments from the enzymes which break them down. Upon rupture of the cell (e.g. by herbivore damage), the contents of the compartments mix and the glycosides are degraded to liberate hydrogen cyanide. Cyanide is a potent inhibitor of the respiratory pathways of cells in the mitochondria and is toxic in low doses. The release of cyanide has been demonstrated to be a potent protection for plants against both herbivores and pathogens. Cyanogenic glycosides are found in bracken and the seeds of many species, such as peach, apple and apricot (Table 16.1). Apple seeds contain amygdalin (a cyanogenic glycoside).

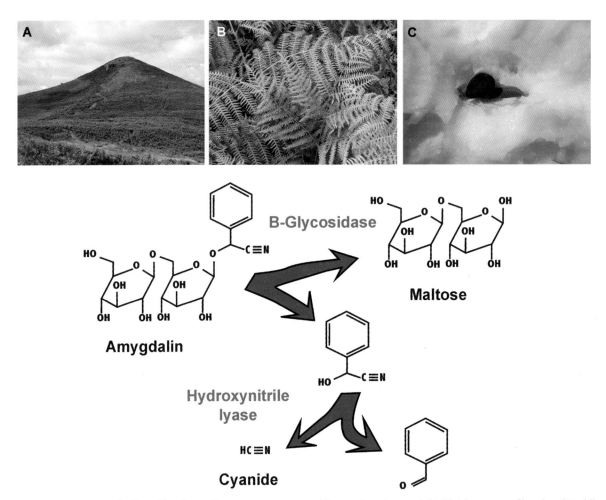

Figure 16.16   Cyanogenic glycosides. Some plants can generate cyanide gas when damaged. (A, B) A large area of bracken (*Pteridium esculentum*) on Roseberry Topping, North Yorkshire. The new shoots of bracken contain cyanogenic glycosides in the vacuole. These compounds are also found in apple seeds (C). When the cells are ruptured as a result of herbivore damage, the cyanogenic glycosides mix with β-glycosidase and hydroxynitrile lyase in the cytosol of the plant cells and the reaction shown occurs. The cyanide gas generated is highly toxic. Most areas of bracken show very little damage due to herbivory as a result of this potent defence.

Table 16.1   Cyanide generation capacity of selected seeds.

Species	Cyanide generation capacity on damaging tissue (mg/g seed)
Apple	0.6
Apricot	2.9
Peach	2.6

The LD_{50} for cyanide is 1.5 mg/kg tissue. Thus, for a child the consumption of a small number of peach or apricot seeds could be fatal. This is only likely if the seeds are mistaken for almond seeds. However, literally hundreds of apple seeds would have to be consumed for toxicity.

Glycosinolates

This group of toxins occur mainly in the Brassicaceae. They take the form of thioesters between D-thioglucose and aglycone. Aglycone spontaneously forms nitrile and isothiocyanate (amongst other compounds) on hydrolysis by thioglucosidase of the glycosinolate molecule. Thioglucosidase is produced by mammalian herbivores but is also present in plant cells. The plants maintain the glycosinolates and thioglucosidase in different compartments to prevent self-toxicity. The presence of these compounds gives the brassicas their characteristic mustard smell and accounts for the flavour of the vegetable brassicas we eat. These spontaneous breakdown products of aglycone are toxic to herbivores and also act as repellents. These breakdown products have the following toxic effects:

isothiocyanate acts as an irritant of mucus membranes and interferes with thyroid function, and nitrile causes liver damage. Plant breeders have attempted to reduce the presence of glycosinolates in oilseed rape in order to allow the leaf material as a forage crop, but this usually has the net result that insect pests become a major problem. Although these compounds are toxic, in low concentrations they can be tolerated and they provide some unique flavours to the human diet. In addition, there is growing evidence that the glycosinolates and their derivatives may have anti-cancer properties.

Lectins

Lectins are sugar-binding proteins that agglutinate cells or precipitate proteins with long carbohydrate conjugates. They are found in plants, animals and bacteria. In plants there is evidence that they are used in defence. The structure of the lectin dictates the type of carbohydrate with which it will interact. In a limited sense, lectins are similar to antibodies; however, they are not inducible and are constitutively expressed. In addition, they are effective against only a certain carbohydrate structure, whereas antibodies can be raised against many different structures. In addition to plant defence, lectins are thought to be involved in the processes of cell recognition in plants and in the storage of proteins in seeds. The French bean (*Phaseolus vulgaris*) produces *Phaseolus* agglutinin I. This lectin is toxic to the cowpea weevil (*Calosobruchus maculatus*). The lectin present in castor beans is discussed in detail in Chapter 19.

Alkaloids

The alkaloids are a huge group of related compounds produced from either amino acids or purines and pyramidines. They are alkaline (hence their name) and freely water-soluble. There are estimated to be more than 10 000 different alkaloids found in around 20% of higher plant species. Most alkaloids are toxic to mammals and are thought to play a role in plant defence.

As a paradox, these toxic compounds should be familiar to most readers as being also used as medicines or drugs. For example, plants of the genus *Nicotiana* produce the alkaloid nicotine in response to herbivory. When a herbivore attacks, the plant signals to the root system using jasmonic acid (which is transported in the phloem). This causes the synthesis of nicotine in the roots, which is then transported again in the phloem to all plant tissues. Nicotine causes vomiting, diarrhoea and respiratory paralysis in high concentrations. However, as will be seen with most alkaloids, nicotine also is a stimulant in lower concentrations and is highly addictive

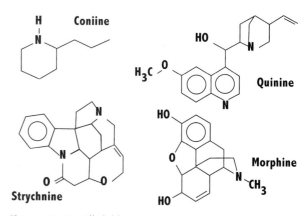

Figure 16.17 Alkaloids are nitrogen-containing organic compounds produced by a range of different plant species. They generally have a bitter taste and plants containing them are avoided by herbivores. Four alkaloids are shown: coniine, from hemlock (*Conium maculatum*), a potent neurotoxin causing paralysis and ultimately death; strychnine, from the strychnine tree (*Strychnos nux-vomixa*), which causes violent muscular spasms and death; quinine, from *Cinchona* spp., a potent toxin of the malarial parasite *Plasmodium*; and morphine, from the opium poppy (*Papaver somniferum*), a potent relaxant of the nervous system, used as an analgesic.

in humans. Nicotine can be delivered to the body in many different forms, but smoking is the usual way it is administered.

Examples of drugs in the alkaloid group are morphine, caffeine, codeine, cocaine, marijuana, strychnine, atropine, curare and quinine (Figure 16.17). All of these compounds potentially incapacitate or kill herbivores. Even small quantities that effectively drug the herbivore are likely to prove fatal, since other organisms are likely to prey on a prone herbivore. Only an organism at the top of a food chain, such as humans, can afford to take these and run little risk of dying, but even then the addictive qualities of these compounds bring further hazards.

Defence reactions and signal pathways

Not all of the toxic defences are present constitutively in plants, and many need to be induced first. The most frequently occurring fast response of plants to infection by a bacterium or fungus is a hypersensitive response, in which a few cells in the local region of the infection die and antimicrobial compounds accumulate, such as those mentioned above. Lignin and callose are also deposited in the cell wall region to prevent further spread. If successful infection is limited or eradicated, defence pathways may also be activated by the release of jasmonic acid, salicylic acid or systemin.

Jasmonic acid

Jasmonic acid has already been mentioned as being involved in the production of alkaloids in tobacco plants in response to herbivory. It is a plant stress growth regulator, which is involved in defence signalling throughout the plant. There are two major forms of the growth regulator: jasmonic acid, which is water-soluble and is transported in the phloem; and methyl jasmonate, which is volatile. These have been found in many different plant species and they have dramatic effects on plant metabolic pathways.

There is a growing list of plant defences which are induced by jasmonic acid. For example, in potato plants jasmonic acid causes the production of proteinase inhibitors in treated leaves and thus protects these leaves from herbivore attack. There is evidence that most of the defence mechanisms quoted in this chapter can be influenced by jasmonic acid.

Salicylic acid

Salicylic acid is involved in systemic acquired resistance, which is associated with a damaged/infected part of the plant signalling to other parts of the plant the presence of the infection and initiating defences in those tissues. These defences usually include chitinases and β-glucanases to attack fungi. The mechanics of this signalling pathway are not fully understood.

Systemin

Systemin is an 18-amino acid polypeptide. When plants are wounded, a number of defence related proteins are produced rapidly. Systemin is thought to be one of the key regulators of this response. Systemin is stored as a larger protein, prosystemin. Upon leaf damage, prosystemin is cleaved by proteinases to release systemin, which then activates transcription of a battery of genes to give the defence response. Systemin can also be transported in the phloem and can induce a response to wounding in other tissues. In tomatoes there is evidence that systemin controls the production of proteinase inhibitors to affect herbivore digestion. There are suggestions now that systemin may be linked to the same signalling pathway as jasmonic acid.

Phytoalexins

Phytoalexins are a diverse group of compounds which accumulate rapidly in response to microbial attack in the vicinity of the attack. The mechanism of resistance is common to many different plant species. Examples of phytoalexins covering all of the divisions of organic compounds discussed earlier have been identified in plants, e.g. isoflavone phytoalexins from the Fabaceae, and terpenoid phytoalexins, mainly from the Solanaceae and Convolvulaceae.

Allelopathy

Allelopathy is the chemical means by which plants exclude other species from their close vicinity. The chemical can be released into the air, water or soil around the plant, and it exhibits some inhibitory effect on neighbouring plants or seeds. It has been remarkably easy to demonstrate that such chemicals are present in plants. A standard test has been developed whereby lettuce seeds are exposed to plant extracts and the rates of germination and growth are assessed. Such tests have permitted the identification of numerous plant compounds that possess an allelopathic effect. However, what is less clear is whether these tests are sufficient to establish allelopathy in the field, and whether the inhibitory concentrations used in experiments are relevant to the situation in the soil around a plant. In order to establish whether an allelopathic phenomenon exists, three issues must be addressed:

1. It must be demonstrated that the observations are not just a direct result of competition for resources.

2. The active agent must be identified.

3. Autotoxicity must be discussed.

Unfortunately in most cases research has not addressed all of these issues.

There are three examples where there is good evidence for the phenomenon of allelopathy. Possibly the best-known example is that of the black walnut (*Juglans nigra*) (Rietveld, 1983; Coder, 1999). The black walnut is primarily grown for its wood and nuts and it has been known for a long time that plants planted near to or under the shade of *J. nigra* tend to turn yellow, wilt and then die. It is now known that *J. nigra* releases the compound hydrojuglone from its roots and leaves. In the soil, hydrojuglone is oxidized to juglone, which is very toxic to a wide range of plants. The black walnut itself survives by actively excluding the juglone from its roots. Thus, *J. nigra* releases hydrojuglone onto other plants through the leaves (leaching out of leaves in the canopy during rainfall through the degradation of fallen leaves in autumn) or from the root system. This then inhibits the growth of neighbouring plants and gives the black walnut room to expand its range. Juglone is not very mobile in the soil and remains in highest concentrations just near the root system of the walnut. Thus, the toxicity only extends to the limits of the root system. Hydrojuglone has been identified in other plants of the walnut family but *J. nigra* possesses the highest concentrations.

A second example is of shrubs growing on the chaparral on the south-eastern coastal plain of North America. It is frequently observed that where the shrub *Ceratiola ericoides* grows there are bald halos around the shrub where no grasses grow. This is a positive advantage to the shrubs, since the grasses cause frequent fires on the chaparral and their presence can cause fires of high enough intensity to cause major damage to the shrubs. The grasses regenerate efficiently after these fires but the shrubs do not; hence exclusion of grasses is an advantage to *Ceratiola*. Studies have been performed to demonstrate that the halos are not a result of nutrient competition between the grasses and the shrubs (Williamson, 1990). Ceratiolin is released from the leaves and leaf litter of *Ceratiola* shrubs. Once released, this compound is degraded to hydrocinnamic acid, which is inhibitory to the germination of grasses in low concentrations (Tanrisever *et al.*, 1987). This degradation occurs as a result of heat, light and soil acidity. Thus, the hydrocinnamic acid is likely to be formed around the plant but not directly under the plant where there is shade. It has been suggested that the plant produces monoterpenes to inhibit autotoxicity. Another possibility is that the hydrocinnamic acid has relatively little effect on a mature shrub but seeds of grasses are highly susceptible to it.

A new approach has been developed in the area of allelopathy, with tissue culture being used as the growth medium instead of soil (Bias *et al.*, 2003). A great problem in this area has always been that it is difficult to detect and purify plant products in soil and thus very hard to identify potential allelopathic compounds. The plant spotted knapweed (*Centaurea maculosa*) has always been suspected to have the ability to kill neighbouring plants; however, it was only in tissue culture that the active compound could be assayed and identified. Spotted knapweed releases catechin in (+) and (−) forms (Figure 16.18); (+)-catechin acts as an antibiotic, killing soil bacteria, and (−)-catechin acts as a herbicide, killing neighbouring plants. (−)-Catechin does not affect the growth of the spotted knapweed plant and it is thought that the knapweed plants can exclude the compound by pumping it out of the root cells before it can do any damage. (−)-Catechin acts as an agent causing the release of free radicals in the root system and upwards in the plant. This triggers cell death, ultimately leading to the demise of the whole plant. (−)-Catechin has proved to

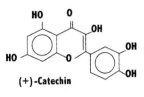

Figure 16.18 Catechin and allelopathy. The +isomer of catechin (catechin+) is released into soils by spotted knapweed (*Centaurea maculosa*) and acts a toxic agent in the soil, killing neighbouring plants. Plants that have lived in the presence of spotted knapweed for thousands of years have evolved to respond to this toxic survival mechanism. Hence, spotted knapweed is not particularly successful in Europe, but in North America, where it has been introduced, it has become very successful indeed. The North American plants do not have mechanisms to cope with the toxicity of catechin+ in the soil and therefore the spotted knapweed has come to dominate in many areas of the USA, where it has become a serious pest. This is one of the few examples where it has been definitively shown that plants do use allelopathy as a method of reducing competition in a particular habitat.

be a potent herbicide and is now being trialled as a potential commercial product. Spotted knapweed is an invasive exotic species in the USA occupying large tracts of land. It was originally introduced from Europe. In Europe spotted knapweed is not a considered an invasive species, and there is growing evidence that, through co-evolution with this plant in Europe, native species have developed a tolerance to the toxin released by the plant (Fitter, 2003). However, in the USA the native species have never encountered the toxin before, and thus the knapweed has become a serious pest.

Recognizing self

In earlier chapters we have discussed how plants recognize pollen from plants of the same allelic group and prevent self-pollination. However, this recognition can be extended further. Plants can frequently have pieces of closely related species grafted onto them. In compatible reactions, the vascular tissue meets and plasmodesmata form between neighbouring cells (Yeoman, 1984). However, in an incompatible reaction the stock plant recognizes the scion as being a foreign body. In this case the vascular tissue shuts off the scion, effectively starving it of nutrients and water and killing it. The mechanism of recognition is still not understood fully, but plants obviously possess further mechanisms for detecting and repelling intruders.

Mimicry and the enrolment of other organisms for protection

The topic of mimicry is covered in greater detail in a separate chapter, but it is worth mentioning here that there is good evidence that plants use mimicry to deter would-be predators. Plants such as *Xanthium trumarium* and *Arisarum vulgare* bear dark flecks on the leaves and stems (Lev-Yadun and Inbar, 2002); these flecks are thought to mimic the presence of ants on the plant and thus act to deter herbivores. In other examples, plants such as *Paspalum paspaloides* and *Pisum fulvum* have evolved structures that resemble aphids and caterpillars, respectively. These accurately mimic plants that are already infested with the herbivores and thus are avoided by other herbivores, since there will be increased competition or the plant appears unhealthy (Figure 16.19).

There are numerous examples of plants that have enrolled insects and other organisms to aid in their defence against other herbivores. Ants are considered to be a predator of plants. In many cases, however, plants have turned this situation around. About 1000 different plant species have been described that give predatory insects (mainly ants) a reward, and in return the insects protect the plants from herbivores. By providing a sugary secretion from extra-floral nectaries, the plant can attract arthropods that are predatory on herbivores, e.g. *Impatiens* species. There is evidence in species such as *Vicia faba* that after leaf damage the number of these extra-floral nectaries increases within a week to act as an attractant to predatory insects (Mondor and Addicott, 2003). In *Ricinus communis*, attack of leaves resulted in a 12-fold increase in the amount of nectar produced by extra-floral nectaries in the damaged leaf and to a lesser extent on neighbouring leaves (Wäckers *et al.*, 2001). Thus, plants can actively enrol insects using a carbohydrate reward and bring them to their defence in direct response to being attacked.

Other plant species release volatile chemicals (herbivore-induced volatile emissions) which are similar to those deployed by arthropod predators and parasitoids in locating arthropod herbivores (Dicke and Sabelis, 1988; Drukker *et al.*, 2000). This has now been demonstrated in a large number of plant species, including maize, cabbage, lima beans, cotton and tomatoes. The maize system has been studied in the greatest detail to date. Maize, when attacked by the beet armyworm (*Spodoptera exigua*), releases volatile chemicals that are known to attract at least two different species of parasitoids (*Cotesia marginventris* and *Microplitis croceipes*) which attack the armyworm (Turling and Tumlinson, 1992). Mechanical damage of the leaf does not induce the same response; only through the addition of the regurgitated products of the armyworm does the volatile compound get released. Further analysis has shown that the plant detects the presence of volicitin in the region of damage and responds by releasing herbivore-induced volatile emissions (Turlings *et al.*, 2000). The presence of volicitin activates a signal cascade that induces the production of indole and terpenoid emissions (Frey *et al.*, 2000; Shen *et al.*, 2000). These in turn attract the parasitoids. This highlights the presence of a very species-specific detection system in plants in response to herbivore damage. Thus, there is growing evidence that plants are able to respond specifically to herbivore predators.

Figure 16.19 Mimicry and leaf damage. One unusual defence used by some plants is a method of making the plant look as though it is damaged when it is not. Here leaves from a Swiss cheese plant (*Monstera deliciosa*; A) and a papaya (*Carica papaya*; B) are shown; both plants possess leaves with holes them or large indentations in them. There is evidence that herbivores seeing these leaves avoid the plants as they look as though there must be another herbivore doing serious damage to the plant and therefore there will be intense competition for the food resources.

Other plants form dwellings for ants as well as offering nectar rewards. At least four species of ant live on *Acacia drepanolobium* (Stanton *et al.*, 2002). As the shoots form on this plant, the thorns swell and the leaves develop nectaries. Ants of the genus *Crematogaster* make homes in these swollen thorns and then act as bodyguards to the *Acacia* tree, attacking any herbivores that settle on the plant. This defence is so effective it has been shown that the ants can deter even large mammals such as giraffes! Many species of *Acacia* trees use ants as a means of defence in this way.

Plants may be fixed to one place but they are far from defenceless. They have a huge variety of defences that can be deployed to protect them. Some of these defences are always present such as thorns and stinging hairs, whereas others are produced in response to a threat. Plants can be regarded as synthetic factories. However, it is the vast array of defensive that make plants some useful as medicines and this is the topic of the next chapter.

References

Bias HP, Vepachedu R, Gilroy S, Callaway RM and Vivanco JM (2003) Allelopathy and exotic plant invasion: from molecules and genes to species interactions. *Science*, **301**, 1377–1380.

Bell AA (1981) Biochemical mechanisms of disease resistance. *Annual Review of Plant Physiology*, **32**, 21–81.

Brett C and Waldron K (1990) *Physiology and Biochemistry of Plant Cell Walls*, Topics in Plant Physiology 2, Unwin Hyman Ltd, London, pp. 137–154.

Coder KD (1999) Allelopathy in trees. *Arborist News*, **8**, 53–60.

Cyr H and Pace ML (1993) Magnitude and patterns of herbivory in aquatic and terrestrial ecosystems. *Nature*, **361**, 148–150.

Dicke M and Savelis MW (1988) How plants obtain predatory mites as bodyguards. *Netherlands Journal of Zoology*, **38**, 148–165.

Drukker B, Bruin J and Sabelis MW (2000) Anthocorid predators learn to associate herbivore-induced plant volatiles with presence or absence of prey. *Physiological Entomology*, **25**, 260–265.

Fitter A (2003) Making allelopathy respectable. *Science*, **301**, 1337–1338.

Fraenkel GS (1959) The *raison d'etre* of secondary plant substances. *Science*, **129**, 1466–1470.

Frey M, Stettner C, Pare PW, Schmelz EA *et al.* (2000) An herbivore elicitor activates the gene for indole emission in maize. *Proceedings of the National Academy of Science of the United States of America, 97*, 14801–14806.

Hiradate S, Yada H, Ishii T, Nakajima N *et al.* (1999) Three growth inhibiting saponins from *Darantia repens*. *Phytochemistry*, **52**, 1223–1228.

Lev-Yadun S and Inbar M (2002) Defensive ant, aphid and caterpillar mimicry in plants? *Biological Journal of the Linnean Society*, **77**, 393–398.

Mondor EB and Addicott JF (2003) Conspicuous extra-floral nectaries are inducible in *Vicia faba*. *Ecology Letters*, **6**, 495–497.

Rietveld WJ (1983) Allelopathic effects of juglone on germination and growth of several herbaceous and woody species. *Journal of Chemical Ecology*, **9**, 295–308.

Shen BZ, Zheng ZW and Dooner HK (2000) A maize sesquiterpene cyclase gene induced by insect herbivory and volicitin: characterization of wild-type and mutant alleles. *Proceedings of the National Academy of Science of the United States of America*, 97, 14807–14812.

Stanton ML, Palmer TM and Young TP (2002) Competition-colonization trade-offs in a guild of African *Acacia* ants. *Ecology Monographs*, **72**, 347–363.

Tanrisever N, Fischer NH and Williamson GB (1988) Calaminthone and other menthofurans from *Calamintha ashei*; their germination and growth regulatory effects on *Schizachyrium scoparium* and *Lactuca sativa*. *Phytochemistry*, **27**, 2523–2526.

Tanrisever N, Fronczek FR, Fischer NH and Williamson GB (1987) Ceratiolin and other flavenoids from *Ceratiola ericoides*. *Phytochemistry*, **26**, 175–179.

Turlings TCJ and Tumlinson JH (1992) Systemic release of chemical signals by herbivore-injured corn. *Proceedings of the National Academy of Science of the United States of America*, 89, 8399–8402.

Turlings TCJ, Alborn HT, Loughrin JH and Tumlinson JH (2000) Volicitin, an elicitor of maize volatiles in the oral secretion of *Spodoptera exigua*: its isolation and bio-activity. *Journal of Chemical Ecology*, **26**, 189–202.

Wäckers FL, Zuber D, Wunderlin R and Keller F (2001) The effect of herbivory on temporal and spatial dynamics of foliar nectar production in cotton and castor. *Annals of Botany*, **87**, 365–370.

Williamson GB (1990) Allelopathy, Koch's postulates, and the neck riddle. In *Perspectives on Plant Competition*, (eds Grace JB and Tilman D), Academic Press, London, 143–158.

Yeoman MM (1984) Cellular recognition systems in grafting. In *Encyclopedia of Plant Physiology*, (eds Linskens HF and Heslop-Harrison J), Springer-Verlag, Berlin, NS, 17, 453–472.

17

Plants and medicines

It is sobering to think that for 80% of the population of the world, the pharmacy is a field or forest. Humankind has always been very dependent upon plants as a source of medicines and even for the majority of modern medicines, their original source was a plant. It is the remarkable ability of plants to synthesize a huge range of compounds that has given them this feature. Since plants cannot move from a fixed position within their environment, they need to be able to cope with stresses and herbivory, and to be self-reliant in terms of synthesizing every compound they need. Thus, plants produce a huge range of compounds. It is estimated that the 10% of plant species so far analysed contain over 100 000 different organic compounds. In this sense, we can view plants as highly capable synthetic factories. These compounds have huge potential as medicines for humankind. It is tempting to view all plants as potential natural sources for beneficial medicines. However, it must always be remembered that plants are, in many instances, using these chemicals as a defence (see Chapter 16), and many of the chemicals are potentially toxic when accompanied by other compounds or at an incorrect dose.

As a starting point for understanding the use of plants for medicinal purposes, we have over 30 000 years of human history and culture to which to refer, and this is what we will look at first in considering plants as medicines. The lore concerning the use of plants as medicines is gigantic and justice cannot be done to this subject in a single chapter. The discussion is restricted to examples of how certain significant plant medicines were discovered.

Doctrine of signatures

In sixteenth century Europe, physicians believed that when God created the world he also left humankind cures for all of the ailments on the planet in plants. With around 250 000 plant species to choose from, this would mean a great deal of trial and error before any simple illness could be cured. However, it was believed that any plant with medicinal qualities would possess some clue in its appearance as to what particular ailment the plant was effective against. This was known as the 'Doctrine of signatures'. Often today we see the traditional medical use of plants reflected in the name of the plant. This can be illustrated with a few examples.

Lungwort (Pulmonaria officinalis)

A clue as to the traditional application for this plant is immediately apparent. The suffix '-wort' was traditionally used for a plant that possessed some medicinal or culinary use. The prefix 'lung-' also suggests that the plant was used for treatment of the lungs or respiratory diseases (coughs, bronchitis and asthma). The 'clue' for this application is believed to come from the spotted appearance of the leaves, thought to resemble a diseased lung (Figure 17.1). Lungwort is still recommended by herbalists as a natural medicine in the form of a tea drink. However, the question has to be asked, as with all such medicines – is it effective? Lungwort has been identified to contain allantoin and the flavenoids quercitin and kaempferol, and it lacks many of the toxic alkaloids present in many other members of the same family (Boraginaceae). Allantoin is an agent used in skin creams, with reputed healing properties for ulcers. Quercitin and kaempferol are powerful antioxidants. These agents may help in lung diseases but there is no evidence to support this and extracts of lungwort are not used in any conventional medicines.

Greater celandine (Chelidonium majus)

Greater celandine is a perennial herb of the family Papaveraceae, frequently found in rough scrubland, walls and disturbed areas. As a result of its yellow flowers and the

Physiology and Behaviour of Plants   Peter Scott
© 2008 John Wiley & Sons, Ltd

Figure 17.1 The doctrine of signatures. With around 250 000 different plant species to choose from, it is difficult to have a systematic method for investigating the medicinal properties of plants. For most of the world's cultures, the most basic method has been to use the appearance of the plant. In Europe this practice was called the doctrine of signatures; for every illness there was a plant that could cure it and a clue was left in the appearance of the plant (during creation). Here are shown: (A) common self-heal (*Prunella vulgaris*), used to treat throat infections (clue, wide mouth to the flower); (B) greater celandine (*Chelidonium majus*), used to treat jaundice (clue, sap and flower are yellow); (C) lungwort (*Pulmonaria officinalis*), used to treat lung infections (clue, the appearance of the leaves); and (D) viper's bugloss (*Echium vulgare*), used to treat snake bites (clue, shape of flower heads and seeds).

presence of an orange-coloured latex produced on damaging the plant, it was reputed to be a cure for jaundice. Indeed, several plants, such as dandelion, agrimony, hawkweed and marigold, were all reputed to be treatments for jaundice because of the colour of their flowers. There is no evidence that celandine can be used to treat jaundice. However, in Gerard and Johnson's *The Herbal or General History of Plants* (1633, reprinted 1975), celandine is described as follows:

. . . the juice of the herbe is good to sharpen the sight, for it cleanseth and consumeth away slimie things that cleave about the ball of the eye and hinder the sight . . .

It is still recommended by many practitioners of herbal and alternative medicines for many uses, such as for stimulating the liver and gall bladder and as an antispasmodic treatment for stomach pains.

Viper's bugloss (Echium vulgare)

Viper's bugloss is a common plant found throughout Europe, particularly on chalky soils. The name is derived from the use of the plant as a treatment for snake bites. The plant gained its usage from the resemblance of the seed to a snake's head. This is further reflected in the Latin name *Echium*, which also means viper. The name 'bugloss' is derived from the Greek term for an ox's tongue, which the leaves resemble in shape and texture. In Gerard and Johnson's *The Herbal or General History of Plants* (1633, reprinted 1975), viper's bugloss is described as follows:

> The herbe chewed and the juice swallowed downe is a most singular remedy against the biting of any venomous beast, and the root so chewed and laid upon the sore workes the same effect.

In reality, the plant extract has no effect upon snake-bites and it is rarely recommended in herbal remedies, owing to the high level of pyrrolizidine alkaloids, which can cause liver failure.

Common self-heal (Prunella vulgaris)

Self-heal possesses many old names but they all suggest that it has remarkable healing properties. It was believed that the flowers resembled a throat; thus, the plant was used to treat throat ailments. In Germany the plant was called Brunella (derived from Brunellen) since the plant was used to treat a mouth ailment called 'die Breuen' or, in English, trench mouth. The name was mistranscibed by Linnaeus and it became known ever since as *Prunella*. In Gerard and Johnson's *The Herbal or General History of Plants* (1633, reprinted 1975), self-heal is described as follows:

> There is not a better wound herbe in the world than that of Selfheale is, the very name importing it to be very admirable . . .

Despite the attractive name, self-heal has never been identified to possess any properties to aid the healing of wounds.

Sage (Salvia officinalis)

The Latin name for sage, *Salvia*, was derived from *salvere*, to heal, and was first used by Pliny. Sage has been used for centuries as a treatment for healing wounds. In addition, it was also used as a means of improving memory and thinking. In Gerard and Johnson's *The Herbal or General History of Plants* (1633, reprinted 1975), sage was described as being:

> . . . singularly good for the head and brain, it quickeneth the senses and the memory, strengtheneth the sinews, restoreth health to those that have the palsy . . .

The English name 'sage' is believed to be derived from the French word *sauge*, which means 'save'. As a consequence of the plant being thought to improve memory, a person that possessed a good memory became known as a sage. Thus, in modern times the name possesses a dual meaning.

The leaves of sage are rich in flavenoids, with these compounds making up 1–3% of the leaf weight, and oils such as thuyone and camphor (Bruneton, 1999). There is some evidence that extracts from sage can act as an antispasmodic or as an antioxidant. However, there is little evidence that it does indeed improve memory or heal wounds.

From these examples it is clear that the Doctrine of Signatures cannot be trusted as a means of identifying medicinal plants. Plant appearance in no way reflects its potential medicinal use. The origin of plant names can often be convoluted and unreliable, but this does not mean that we cannot learn from folklore. A whole scientific discipline, known as ethnobotany, has developed around this area.

Ethnobotany

The term used to describe the area of research into plants and medicines is 'ethnobotany'. In the strictest terms, ethnobotany is the study of the relationship between plants and traditional peoples, but given that the topic of plants and medicine is the subject of most of the research into this area, we shall restrict the definition to medicinal uses of plants by humankind. Ethnobotany is an extension of the Doctrine of Signatures across the whole world, relying on centuries of tradition and wisdom passed on through the generations.

In Europe, written herbals first originated in ancient Greece, and were again documented and extended by the medieval physicians John Gerard (Gerard and Johnson, 1633/1633) and Nicholas Culpeper (Culpeper, 1651). These herbals documented the medicinal use of plants and the folklore remedies for illnesses and diseases. These then, just as they do now, provided the basis for further studies and research into the use of plants in medicines.

One interesting example is a treatment for the disease dropsy (oedema), which has symptoms of fluid accumulation in parts of the body, accompanied by swelling (Cotton, 1996). An English physician, William Withering, uncovered a treatment for the illness in the form of a secret recipe formulated by a woman in Shropshire. The initial treatment contained extracts from more than 20 different plants, but through methodical research Withering identified the active ingredient as coming from leaves of the purple foxglove (*Digitalis purpurea*) and this gave rise to a crude drug known as digitalis (Figure 17.2). Through a great deal

of research, the effective concentration of the drug was assessed. However, it was immediately apparent that the amount of digitalis in leaf material was very variable, and in order to limit this as much as possible, Withering harvested the plants only at certain times in the year. This highlights the great problem with traditional medicines. The variation in amounts of digitalis makes it almost impossible to obtain a consistent dose from the leaf tissue, so the effectiveness of the medicine is compromised. It was

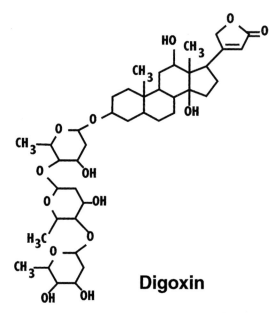

Figure 17.2   Use of *Digitalis purpurea* as a medicine. The purple foxglove contains potent cardiac glycosides and is very dangerous to eat. However, these compounds can also be used as a treatment for oedema or dropsy. If the heart rate falls and is weak, it can lead to the accumulation of fluids in the lower part of the human body. Digoxin from the foxglove can be used to strengthen the heartbeat and thus cure the condition.

only the efforts of Homolle and Quevenne in 1841 that permitted the first isolation and purification of the active ingredients, digitoxin, digitalin and digitalein (Mann, 1994). Once purified, the medicine could be applied in a controlled dosage. However, it was the early work of Withering that formed a milestone in the use of ethnobotany and the subsequent development of medical drugs.

To make the step from traditional plant lore to isolating a medicine, such lore needs to be carefully tested to establish the authenticity of the traditional use of a plant. Then the compounds within the plant that are effective as a medicine need to be isolated and identified. Once identified, and with the molecular structure established, a range of related compounds can be synthesized artificially and used as tests to optimize the effectiveness of the treatment. This will yield a useful medicine that could then be marketed in the Western world. However, for people in the poorer nations, the cheapest and most realistic source of medicines will always be the plants around them.

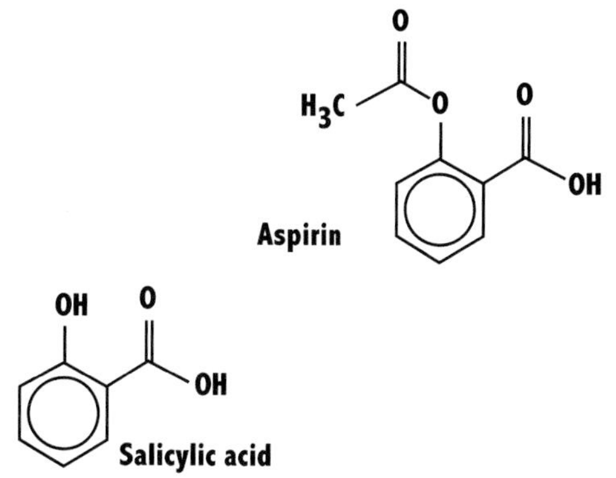

The origins of aspirin

The origins of aspirin illustrate quite nicely how the purified plant product can be used as a basis for the development of further drugs. In the writings of the Greek writer Hippocrates (500 B.C.) there are references to the use of willow bark as containing a bitter substance that could ease aches and pains and reduce fevers. Interest in willow bark was revived in 1763 by the Reverend Edmund Stone, who wrote further concerning the medicinal properties of the plant. Within 50 years the active bitter ingredient was isolated and given the name salicin (after the Latin name for the willow, *Salix alba*). As salicin is broken down to salicylic acid in the body, the acid was synthesized and produced commercially as a medicine. However, salicylic acid has severe side-effects, since it reduces the secretions by the stomach wall which protect the stomach lining from being irritated by acid produced during digestion of food. Deeming the problem of salicylic acid to be due to the acidity of the compound, Felix Hoffmann at the Bayer Company in 1899 produced an acetyl derivative of salicylic acid, acetyl salicylic acid, which was marketed under the name aspirin (Figure 17.3). In the body, aspirin is broken down to salicylic acid by the stomach, liver and kidneys. Hence, aspirin is a safer delivery system for salicylic acid into the body, bypassing most of the adverse reactions of the stomach lining. It is highly likely that the plant product salicin would also serve as a safe delivery system for salicylic acid, but obtaining the product directly from a plant source produces unpredictable dosage, as has been discussed with digitalis.

Figure 17.3 The origins of aspirin. Certain plants accumulate high concentrations of salicilin in their bark (e.g. willow trees, *Salix* spp.) or leaves (e.g. meadowsweet, *Spirea ulmaria*). On extraction, this breaks down to salicylic acid (shown), which causes stomach disorders when supplied to patients to relieve pain; however, when an acetyl group was added these side-effects were reduced. This produced ASPIRIN (A for acetyl; SPIR from Spirea; and IN to finish the name of the compound).

The origins of antimalarial drugs

The credit for the discovery of drugs that can cure malaria belongs to South American Indians. Legend tells of a man suffering from a malarial fever in the Andes, drinking from a pool of water to quench his thirst. The water was bitter and contained extracts from the bark of tree species near the pool. Miraculously, the man's fever disappeared and through some investigation bark from the quina-quina tree was identified as containing the active agent. Thereafter the bark was used by the South American Indians to treat malaria fevers. In the 1600s, Jesuit priests noted the use of the bark to treat fevers and chills and introduced the use of this plant to Europe. Linnaeus classified and named the tree after the Countess of Chinchon, the wife of the Viceroy of Peru at that time, and it has since been known as the chinchona tree (Hobhouse, 1987). In Europe, the active ingredient was isolated and identified and named quinine (Figure 17.4). Seeds of the

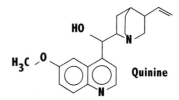

Figure 17.4 The antimalarial alkaloid quinine, which is extracted from the bark of the South American *Cinchona* tree. Quinine inhibits the life cycle of the malarial parasite (*Plasmodium*) in the human body and therefore can treat malaria. It was used as a successful cure for malaria for centuries until it became necessary to make synthetic equivalents during World War II.

chinchona plants were purchased by Dutch traders and plantations of the trees were grown in Java. There are five different species of chinchona tree, but *Chinchona officinalis* contains the largest amounts of quinine. As a consequence, the Dutch had a monopoly on world quinine supplies for several hundred years until synthetic quinine could be synthesized. During World War II the availability of quinine became restricted as a result of Japanese occupation of areas where the chinchona tree plantations were. Thus, there was great need to synthesize a synthetic equivalent to quinine. This led to the development of new synthetic antimalarial drugs that were more effective and easier to make than quinine, such as chloroquine and mefloquine. However, several strains of the *Plasmodium* parasite have since acquired resistance to chloroquine and the use of quinine is experiencing a revival.

Malaria is an infectious disease that is spread through the bite of the female *Anopheles* mosquito infected with the *Plasmodium* parasite. The illness causes severe fevers, which can be followed by organ failure and death. It is estimated that around a million people each year die from malaria and approximately 200 million people are infected by the disease. The discovery of quinine allowed Europeans to resist malaria and thus permitted the colonization of areas of South America and Africa that had previously not been possible as a result of malarial infections. Quinine kills the *Plasmodium* blood parasite in its active blood-infecting form but not in its dormant form, requiring continuous use of quinine.

St John's Wort (Hypericum spp.)

This is a common plant found throughout Europe and was introduced into North America. There are many different species in the genus *Hypericum* but all are ascribed similar properties. St John's wort is described in Gerard and Johnson's *The Herbal or General History of Plants* (1633, reprinted 1975) as a vulnerary:

> … a moste precious remedie for deepe woundes, or wound made with a venomed weapon …

It was used in Europe and by the North American Indians to treat wounds. However, its use for these purposes has declined owing to the high probability of secondary skin diseases. In some traditional medicines, St John's wort has been used as an antidepressant and to treat insomnia and anxiety. In recent years there has been a revival of interest in the plant as a possible cure for depression. Two compounds, called hypericin and hyperforin, are thought to be the active ingredients of the plant extract, but there is little evidence of how these drugs act in the body (Figure 17.5). It is one of the few natural alternatives to Prozac as an antidepressant and is available from many practitioners of herbal remedies. One bonus of St John's wort is that it is free of the side-effects suffered by users of other antidepressants.

Natural alternatives to Viagra

In the 1998 the pharmaceutical company Pfizer released a revolutionary new drug for the treatment of male impotence, called Viagra™. Virtually overnight, the company was making huge profits. This illustrates the tremendous capacity of the pharmaceutical industry to synthesize products and humankind's ingenuity. However, impotence products and aphrodisiacs are nothing new to humankind; virtually every culture has derived some

 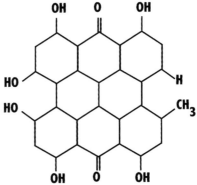

Figure 17.5 Antidepressants from St John's wort (*Hypericum perforatum*), which has been demonstrated to have a medicinal effect upon depression. There are two active compounds present in extracts of the plant, hyperforin and hypericin, which can now be purchased as treatments for mild forms of depression.

remedies which are reputed to have these effects. Many treatments are useless and are sometimes denuding the environment of these plants and animals. For products such as powdered rhinoceros horn and tiger penis, there is no evidence of their efficacy in treating impotence in humans. However, such traditional medicines are driving such fauna to extinction.

The plant kingdom does offer some natural alternatives that have been around for centuries. In Europe, the mandrake (*Mandragora officinarum*) was deemed to possess aphrodisiac qualities but its use has died out almost completely. From the continent of Africa there are some more promising treatments. Three treatments have been heralded as being competitors to Viagra. The first, from Zimbabwe, is called vuka, which is an undescribed herbal blend. This is frequently sold as an 'herbal Viagra' or 'viagre'. The second treatment is used in the central African countries, Cameroon, Gabon and Zaire. The medicine is called yohimbe and is an extract from the bark of the yohimbe tree (*Pausinystalia yohimbe*). This extract has been given approval as an alternative medicine by the US Food and Drug Administration (FDA). In laboratory tests, however, the extract has inconsistent results and can dangerously raise the blood pressure. The active ingredient has been identified to be the alkaloid yohimbine (Figure 17.6). Yohimbine is a selective inhibitor of presynaptic α-2-adrenergic receptors (Bruneton, 1999).

This causes vasodilation of the corpus caverosum and thus offers a treatment for impotence (Riley, 1994). Yohimbine has also been identified in other aphrodisiac plants, such as *Aspidosperma quebracho* (Sperling *et al.*, 2002). From China there is a phytochemical, protodioscin, extracted from *Tribulus terrestris*, which again is reported to act as an aphrodisiac (Adimoelja, 2000).

Probably the most promising herbal treatment comes from the Kwazulu Natal region of South Africa. The Zulu people have a range of impotence-combating herbal remedies, collectively known as 'ubangalala'. Many of these plants have been tested in the laboratory and one species has been shown to be more effective than Viagra in correcting erectile dysfunction. The active ingredient has been isolated and identified. This isolated plant product is now undergoing trials to investigate its safety.

Natural treatments for AIDS

AIDS is estimated to have killed over 25 million people to date, with a further 40 million people that are infected with the virus. Around 95% of the new AIDS virus infections occur in poor countries that cannot afford the current drug treatments. Without treatment AIDS will surpass the Black Death (bubonic plague) as the worst killer disease of all time (bubonic plague killed an estimated 40 million people

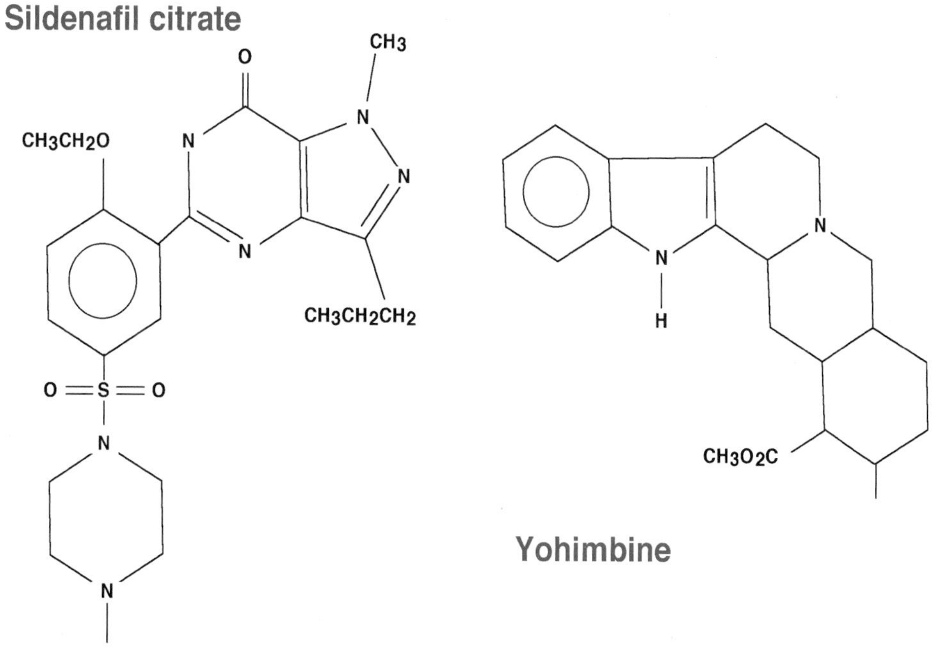

Figure 17.6 Aphrodisiacs and plants. The discovery of sildenafil citrate (Viagra, shown left) by Pfizer was a major boost to the pharmaceutical company's earnings. However, plants have been used for the same purposes for centuries. Here is shown the drug yohimbine, from the bark of the central African yohimbe tree (*Corynanthe yohimbe*).

in Europe and Asia in the fourteenth century). Several life-prolonging drugs have been manufactured to date, but no ethnobotanical-based remedies have thus far been trialled. However, in the near future this will no longer be the case.

Over 40 000 plant extracts have been scanned since 1986 for significant anti-AIDS activity. Only five have exhibited such activity and four are undergoing preclinical trials:

1. (+) Calanolide A and (−) calanolide B, isolated from *Calophyllum lanigerum* and *C. teysmannii* (tree species from Sarawak, Malaysia).

2. Conocurovone, isolated from the Western Australian shrub *Conospermum incurvum* (saltbush).

3. Michellamine B, isolated from *Ancistrocladus korupensis*, a vine from the rainforests in Cameroon. In tests this proved too toxic to human cells to develop further.

4. Prostratin (Figures 17.7 and 17.8), from the mamala tree (*Homalanthus nutans*) from Samoa (Gustafson *et al.*, 1992). Bark from this tree has been used as a traditional Samoan medicine to cure two viral diseases, yellow fever and hepatitis. The bark has been discovered to contain the compound prostratin, which is a phorbol ester. In tissue culture tests, prostratin has been shown to inhibit HIV infection. In addition, the compound does not cause the development of tumours, as many phorbols do. Prostratin has now been shown to activate protein kinase C, which prevents the cells harbouring latent HIV particles and thus allows the body's immune system to eradicate the virus from the body more effectively (Nacro *et al.*, 2000). The drug will be subjected to a human trial in the next few years to test its efficacy. In addition, the drug structure will be

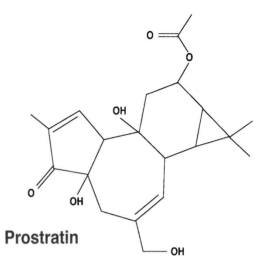

Prostratin

Figure 17.8 Phorbol esters and HIV. An extract from the Samoan *Homalanthus nutans* tree has been traditionally used as a treatment for the viral illnesses hepatitis and yellow fever. Trials have shown that the active compound in the plant, prostratin (a phorbol ester), has potential as an anti-HIV drug. An agreement has been reached with the Samoan people that 20% of the profits from the drug will return to the Samoan government.

further analysed to test whether more effective drugs can be derived, based on the structure of the molecule.

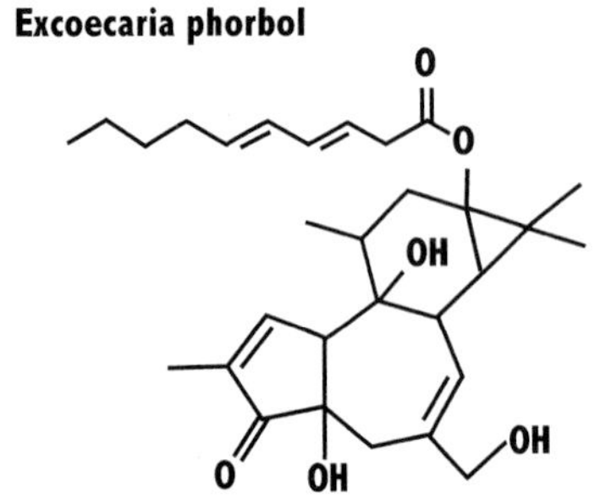

There are further drugs derived from plants that may have anti-AIDS activity, which are in the early stages of development. Two proteins, MAP30 (*Momordica* anti-HIV protein of 30 kDa) and GAP31 (*Gelonium* anti-HIV protein of 31 kDa) possess anti-HIV plant proteins that have been identified, purified and cloned from the medicinal plants *Momordica charantia* (the bitter melon) and *Gelonium multiflorum*, respectively (Leehuang *et al.*, 1995). These antiviral agents are capable of inhibiting infection of HIV type 1 in T lymphocytes and monocytes, as well as replication of the virus in infected cells. They are also not toxic to uninfected cells because they are cannot enter healthy cells. These proteins will shortly be used in further trials.

A cure for certain cancers

It is estimated that one in every three people in the USA and Europe will die from cancer, with lung cancer representing over 25% of cancer cases. This makes cancer the second greatest killer after heart disease in the USA. Since there are over 100 different forms of cancer, each caused by different conditions, it is unlikely that one drug will ever be

Excoecaria phorbol

Figure 17.7 Anti-HIV drugs from plants. HIV is a relatively new disease to medicine and therefore there are no folklore examples to help in its treatment. The drug excoecaria phorbol, isolated from *Excoecaria agallocha*, is being trialled as an antiHIV treatment from plants.

effective against all of these illnesses. However, there are prospects that plant-derived drugs will help in the fight against this disease. In order to identify possible cures for cancers, an effective screening method is required. For cancer cells, such screens have involved the growth of cancerous cell lines in tissue culture. Once predictable growth is achieved, the cell lines can be treated with plant extracts to investigate how the cell division is affected. As potential compounds are isolated, further trials need to be conducted to investigate how the drugs affect the whole organism. Only after passing through all these stages can a medicine enter into the phase of human trials.

Since the 1960s, only seven plant-derived drugs with anticancer properties have received FDA approval. These are taxol, vincristine, vinblastine, topotecan (a camptothecin derivative), irinotecan, etoposide and tenisposide.

Vincristine and vinblastine

Unlike many plant-derived drugs, these were not developed directly based on ethnobotanical observations. In the 1950s, Beer and Noble were investigating local tradition in Madagascar that the rosy periwinkle (*Catharanthus roseus*) was effective as a treatment for diabetes. They isolated an alkaloid called vincaleukoblastine (vinblastine) in their studies (Figure 17.9). A second group, led by Dr Gordon Svoboda in 1957, working for Eli Lilly & Co., used extracts from the periwinkle and injected it into mice with P-1534 leukaemia. In a surprise result, the team found that 60–80% of the mice had a prolonged life span. Eli Lilly & Co. identified the active ingredients in the plant extracts and produced the alkaloid drugs vinblastine and vincristine. Both of these compounds proved to be very effective at inhibiting tumour cell growth of particular types of cancer. The two forms of cancer against which they were particularly effective were Hodgkin's disease and childhood leukaemia. Before this treatment was discovered,

these forms of cancer were particularly deadly, with a very high mortality rate. However, after the 1960s there were many instances of a complete recovery from these cancers. To this day, the structure of these drugs is too complex to synthesize in large quantities by the drug company and the compounds are still isolated directly from plants. The amount of the alkaloids in the plants is around 0.2–1% and there are around 95 different compounds present (Bruneton, 1999). Approximately 3 g of vincristine and vinblastine can be isolated from a ton of plant material. These two drugs act to inhibit cell mitosis by binding to tubulin and preventing the formation of microtubules during the formation of the mitotic spindle (Brossi and Stuffness, 1990). As a consequence, the cell division of cancerous cells is inhibited. As with all alkaloids, they are highly toxic and the correct dosage is essential for effective treatment of cancers.

It was the discovery of these two compounds that sparked a great deal of the interest in plant-derived anti-cancer drugs. There was great optimism that plants could yield many cancer-fighting drugs in the future. To date this has not been realized, because such studies present many difficulties. First, there is little tradition concerning cancer treatment using herbal remedies, so there is little knowledge to use as a basis for studies, and in many respects the screening is nothing more than a blind search for effective compounds. Second, the assays for the anti-cancer activity are crude and may not always be representative of the *in vivo* state of a tumour. Third, many of these compounds may only be effective in combination with other drugs. Such combinations are endless and will be found only by chance, as were vincristine and vinblastine. Fourth, alas, many compounds that do exhibit anti-tumour activity frequently prove to be too toxic for use as treatment in humans. What follows are some of the current plant medicines being explored as cancer treatments.

  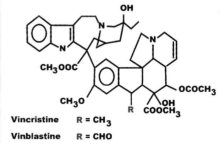

Figure 17.9 The rosy periwinkle (*Catharanthus roseus*) was initially the subject of a trial for a potential drug to treat diabetes. However, quite by accident it was identified as possessing remarkable anticancer properties. It contains two active ingredients, the alkaloids vincristine and vinblastine, which have proved to be particularly successful against childhood leukemias, with a 99 % success rate in treatments! The discovery of these alkaloids sparked a great interest in potential anticancer drugs in plants.

Figure 17.10 The North American mandrake (*Podophyllum peltatum*) was used by native North Americans as a cure for intestinal worms and warts. It has also proved to contain a potent anticancer drug, epitopside, which interferes with the cell division cycle involving topoisomerase. Certain cancer cells are particularly susceptible to the epitopside and therefore the growth of a cancer can be arrested.

Etoposide and teniposide

The North American mandrake plant or mayapple (*Podophyllum peltatum*) has also been used as a source of anti-cancer drugs (Dobelis and Inge, 1989). The plant is a perennial member of the family Berberidaceae and is found in woodlands of eastern North America. The rhizome of the plant contains high concentrations of the compounds podophyllotoxin, etoposide and teniposide, all of which have anti-cancer properties. The roots have a long history of use as traditional medicines by native North America tribes. The roots would be ground into a powder and then eaten as a tea to act as a laxative or rid the body of intestinal worms. The powder was also used as a treatment for warts.

Etoposide and teniposide have been isolated from the plant and used in chemotherapy to treat certain forms of cancer (Vogelzang *et al.*, 1982). It is currently unclear where in the cell cycle they act, but these drugs have been effective against carcinomas, Kaposi's sarcoma, lymphomas and malignant melanomas (Figure 17.10).

Topotecan and irinotecan

In 1958 the anti-tumour activity of extracts of a Chinese tree, *Camptotheca acuminata*, was identified in the course of a screening project for a natural source of steroids. By 1966, the active ingredient was successfully isolated from bark extracts and identified as the quinoline-based alkaloid camptothecin (Figure 17.11). Some clinical trials were then performed using camptothecin, but the drug was too toxic for clinical use. Following these trials, research focused on discovering how the drug killed cells. Through

this work it was discovered that camptothecin was an inhibitor of DNA topoisomerase I and thus prevented cell division in tumours. The anti-tumour activity of camptothecin was eventually harnessed through the production of less toxic derivatives, topotocan and irinotecan. These compounds are some of the only known naturally occurring DNA topoisomerase I inhibitors. They are currently used to treat leukaemias, ovarian and lung cancers and non-Hodgkin's lymphomas.

Taxol

Bark from the Pacific yew tree (*Taxus brevifolia*) has been used in traditional medicines by Native Americans for centuries. However, in the late 1980s the active ingredient, taxol, was extracted and purified and proved to be very successful at destroying ovarian cancer cells (Figure 17.12).

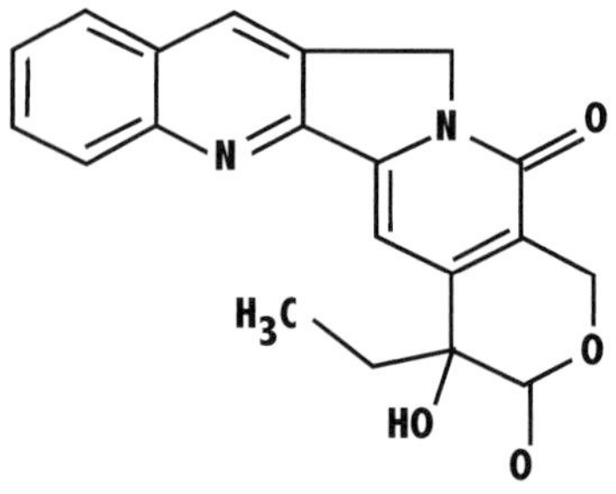

Figure 17.11 Camptothecin, an alkaloid derived from the Chinese tree *Camptotheca acuminata*. It inhibits DNA topoisomerase I, used in cell division, and therefore inhibits the growth of certain tumours and destroys them.

Figure 17.12   Taxol, an alkaloid extracted from the North American Pacific yew tree (*Taxus brevifolia*) and marketed under the name tamoxifen. Initially, taxol needed to be extracted from the bark of yew trees and up to 10 trees were required to treat a single patient, which put incredible pressure on the availability of the drug. Now the drug is synthesized artificially.

The amount of taxol in the bark of yew trees varies greatly, but the Pacific yew possesses the highest concentration of the compound; 0.001–0.01% of the bark fresh weight is made up of taxol, and around 0.7 g of taxol can be extracted from a single tree. Since it takes around 2.5 g of taxol to treat a patient, then around three trees are required for one course of treatment. Initially, this caused problems with over-harvesting of the trees; however, this problem has now been overcome by the creation of a semi-synthetic approach, using leaf extracts of the trees to harvest an intermediate compound. The mode of action of taxol has been found to be unique. Taxol binds onto polymerized microtubules and promotes their synthesis. During mitosis, the disassembly of the microtubules is inhibited and thus cell division is arrested. The novel action of taxol will almost certainly lead to the development of a number of new anti-cancer drugs.

Sustainable development of medicines from plants

The examples presented here only scratch the surface of this topic, but they are enough to demonstrate that plants have a great capacity to synthesize a range of different compounds. Many of these compounds are produced by the plant to make itself toxic to herbivores, so medicines from plants need to be treated with respect (see Chapter 16). For many illnesses, studies of folklore and tradition can be important indicators for possible species that may be useful sources of medicines. However, for more recently discovered illnesses, such as cancer and AIDS, different approaches are required because there will be no folklore concerning potential

cures. Other screening methods for medicines are required. It is becoming increasingly clear that there are not large numbers of plants that will contain cancer-curing drugs, but there are many illnesses known to humankind and each illness will require a separate screening procedure. Screening for the possibility of potential drugs in plants is fraught with problems. The amounts of compounds in plants could vary between the organs from which an extract is derived, the time of year the extract is made, and the growth conditions of the plant material. Thus, there are many different variables that can influence whether a drug is discovered and these need to be considered during any screening procedure.

Many of the compounds that have proved useful as medicines have been all but impossible to manufacture synthetically on a commercial scale. The molecular structures are just too complex to make. Thus, we are totally dependent upon the plant as the source of the drug. This puts great pressure on plant material, and in many tribal cultures over-collection of plants has resulted in many species becoming very rare. It is therefore important to treat medicinal plants as a crop and to introduce them into agriculture.

There are few examples of a developing nation benefiting from the use by Western nations of its folklore to generate medicines. However, there is one example; the Samoan government has signed a contract with the AIDS Research Alliance that 20% of all the profits derived from the use of prostratin (see AIDS research above) will be returned to the Samoan people (Cox, 2001).

From the numbers of drugs that have been isolated which treat major illnesses, such as AIDS and cancer, it is obvious that there will not be huge numbers of these compounds that will be isolated from plants in the future.

However, the larger the number of different plant species available to screen, the greater the likelihood that a new drug will be discovered. The huge synthetic capacity of plants and the tremendous diversity between different plant species means there are large untapped reserves of medicines for present diseases and those of the future. As a consequence, to lose plant diversity could have major implications for human health.

References

Adimoelja A (2000) Phytochemicals and the breakthrough of traditional herbs in the management of sexual dysfunctions. *International Journal of Andrology*, **23** (suppl 2), 82–84.

Brossi A and Stuffness M (1990) Antitumour bisindole alkaloids from *Catharanthus roseus* (L.). In *The Alkaloids*, vol 13, Academic Press, London.

Bruneton J (1999) *Pharmacognosy. Phytochemistry of Medicinal Plants*, Lavoisier, Paris.

Cotton CM (1996) *Ethanobotany. Principles and Applications*, Wiley, Chichester, UK.

Cox PA (2001) Ensuring equitable benefits: the Falealupo covenant and the isolation of anti-viral drug prostratin from a Samoan medicinal plant. *Pharmaceutical Biology*, **39**: 33S–40S.

Culpeper N (1651)English physician.

Dobelis IN (1989) (ed.) *Magic and Medicine of Plants*, Reader's Digest Books, Pleasantville, NY.

Gerard J and Johnson T (1633) (eds) *The Herbal or General History of Plants*, Reprinted by Dover Publications, New York, (1975).

Gustafson KR, Cardellina JH, McMahon JB and Gulakowski RJ et al.(1992) A non-promoting phorbol from the Samoan medicinal plant *Homalanthus nutans* inhibits cell killing by HIV-1. *Journal of Medicinal Chemistry*, **35**, 1978–1986.

Hobhouse H (1987) *Seeds of Change – Five Plants that Transformed Mankind*, Harper and Row, New York.

Leehuang S, Huang PL, Huang PL, Bourinbaiar AS *et al.* (1995) Inhibition of the integrase of human-immunodeficiency-virus (HIV) type-1 by anti-HIV plant-proteins MAP30 and GAP31. *Proceedings of the National Academy of Sciences of the United States of America*, 92, 8818–8822.

Mann J (1994) *Murder, Magic and Medicine*, Oxford University Press, Oxford, UK.

Nacro K, Bienfait B, Lee J, Han KC et al.(2000) Conformationally constrained analogues of diacylglycerol (DAG). 16. How much structural complexity is necessary for recognition and high binding affinity to protein kinase C? *Journal of Medicinal Chemistry*, **43**, 921–944.

Riley AJ (1994) Yohimbine in the treatment of erectile disorder. *British Journal of Clinical Practice*, **48**, 133–136.

Sperling H, Lorenz A, Krege S, Arndt R and Michel MC (2002) An extract from the bark of *Aspidosperma quebracho* blanco binds to human penile α-adrenoceptors. *Journal of Urology* **168**: 160–163.

Vogelzang NJ, Raghavan D and Kennedy BJ (1982) VP-16-213 (etoposide): the mandrake root from Issyk-Kul. *American Journal of Medicine*, **72**, 136–144.

18

Plant tissue culture and the rise of plant biotechnology

A remarkable property of many plant cells is the phenomenon of totipotency; that is to say, single plant cells during differentiation still maintain the ability to form whole plants (Figure 18.1). In animal cells the fate of a particular cell is usually decided very early on in its development, and thus such cells lose their totipotency rapidly, which is why animal cloning has proved such a difficult exercise. This is not so for most plant species. Nice examples of this are the clonal plants mentioned in Chapter 18, where new plants are formed from the root system of a parent plant.

In tissue culture, the totipotency of plant cells can be used to advantage as a new technology. However, in order to achieve this, the cells need to be given signals from plant growth regulators, such as auxins and cytokinins. The best way to give these signals to the plant cells or tissues is through the tissue culture medium. Tissue culture enables all the components of a culture medium to be controlled, thereby permitting optimization of a plant's growth. As tissue culture is sterile and growth occurs in temperature- and light-regulated growth cabinets, plants can be grown in isolation from many stresses to which they are subjected in greenhouse and field culture. In this chapter we shall look at the enormous potential of plant tissue culture and its pivotal role in plant biotechnologies. First, however, we need to look at the history of plant tissue culture and how it arose.

The development of plant tissue culture media

In the late nineteenth century it became clear that plants required nutrients in the soil in order to grow. These became known as essential elements. However, these were not enough to sustain a plant on their own. In the 1940s it was thought that seeds would contain all the growth substances for successful plant growth in tissue culture. Most seeds are quite difficult to extract and use to test this view, but one particular seed, the coconut, presented an elegant way of analysing whether this was so. Therefore, coconut milk was used in combination with sucrose to yield the first tissue-cultured plant, which was a callus culture of *Datura stramonium* (van Overbeek *et al.*, 1941). The use of tissue culture then began to spread and by 1960 orchids were cultured in a coconut-derived medium, and this yielded virus-free material and multiplication of the cultured plants (Morel, 1960, 1964). These early experiments began to highlight some of the exciting possibilities of plant tissue culture. This was the start of micropropagation of plants in tissue culture. However, the use of coconut milk presented a problem; its composition was variable and frequently results could not be repeated between laboratories – something more reproducible was required. This was the reasoning behind the experiments of Murashige and Skoog (1962), who developed the first completely artificial plant tissue-culture medium. Their work allowed all of the components of the culture medium to be controlled, and therefore tissue culture methods could be repeated around the world. The MS medium, developed by Murashige and Skoog (1962), is still the most common medium used for plant tissue culture; however, for particularly tricky tissues, laboratories still have to resort to adding undefined components, such as coconut milk and banana powder, to the medium to get successful results.

Components of the medium

In their natural habitat, plants can gather all the nutrients they need from the soil, the air or rainwater. In a

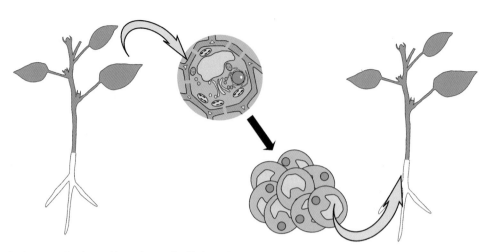

Figure 18.1   Totipotency in plants. Many plant cells, if given the correct instructions, can dedifferentiate and form whole new plants again, an ability called totipotency. It probably rarely happens naturally, but using plant tissue culture the growth regulators can be varied such that the cells can multiply as shown.

sealed culture vessel, a plant is completely dependent upon the medium composition to obtain the nutrients it needs for growth. In plant tissue culture, nutrient components are divided into major or minor elements, depending on the concentration required in the medium (Table 18.1).

Major elements

For optimal growth of plants, fertilizers are used. These contain three major components, nitrogen, phosphorus and potassium. These are the three essential major elements required in the medium:

1. *Nitrogen*. This is supplied in a number of forms: as a nitrate, an ammonium salt or as an amino acid. Nitrogen is essential for the production of plant proteins, chlorophyll and nucleic acids.

2. *Phosphorus*. This is supplied as a soluble phosphate or hydrogen phosphate salt in the medium. Phosphorus is essential for the energy-generating systems in plant cells and is needed for ATP and NADP production and nucleic acid formation.

3. *Potassium*. This element is generally supplied as the cation accompanying other nutrients. Although potassium is not an essential component in any of the plant cell structures, it is required as a cation for enzyme function and thus plays an important role in many cellular processes.

Other important major elements are:

4. *Iron*. This is supplied as a ferrous sulphate salt. Iron is important in metabolic processes in the plant cell, such as photosynthesis and respiration.

5. *Calcium*. This element is supplied as a nitrate or chloride salt. Calcium is needed in a range of cellular processes, such as cell signalling, and in the formation of the pectates in the cell wall.

6. *Magnesium*. This element is supplied as a sulphate or chloride salt. Magnesium is essential for the formation of chlorophyll and as an activator of enzymic processes during photosynthesis.

7. *Sulphur*. This element is supplied as a sulphate salt. Sulphur is required in amino acids such as methionine and cysteine, and is essential in some of the components of the respiratory pathways.

Minor elements

Alongside the major elements, plants also require a range of elements in very small concentrations:

(a) *Boron*. This is supplied as a borate salt. Boron is required for a range of processes, such as cellular transport, nitrogen metabolism and cell division.

(b) *Chlorine*. This is supplied as a chloride salt. Chlorine is required in minute quantities for photosynthesis and

Table 18.1 Composition of three of the most frequently used plant tissue culture media.

Components	Murashige and Skoog basal salts (g/l)	Gamborg's B5 basal salts (g/l)	Chu's N6 basal salts (g/l)
Macronutrient salts			
Ammonium nitrate	1.65	n/a	n/a
Ammonium sulphate	n/a	0.13	0.46
Calcium chloride anhydrous	0.33	0.11	0.12
Magnesium sulphate anhydrous	0.18	0.12	0.09
Potassium nitrate	1.90	2.50	2.83
Potassium phosphate monobasic	0.17	n/a	0.40
Micronutrient salts			
Boric acid	0.0062	0.0030	0.0016
Cobalt chloride·$6H_2O$	0.000025	0.000025	n/a
Cupric sulphate anhydrous	0.000025	0.000025	n/a
Ferrous sulphate·$7H_2O$	0.0278	0.0278	0.0278
Manganese sulphate·H_2O	0.0169	0.0100	0.0033
Molybdic acid sodium salt·$2H_2O$	0.00025	0.00025	n/a
Na^2–EDTA·$2H_2O$	0.03726	0.03730	0.03725
Potassium iodide	0.00083	0.00075	0.00080
Zinc sulphate·$7H_2O$	0.00860	0.00200	0.00150
	Murashige and Skoog vitamins (g/l)	Gamborg's B5 vitamins (g/l)	Chu's N6 vitamins (g/l)
Glycine	0.002	n/a	0.002
Myo-inositol	0.100	0.100	n/a
Nicotinic acid	0.0005	0.001	0.0005
Pyridoxine–HCl	0.0005	0.001	0.0005
Thiamine–HCl	0.0001	0.010	0.001
	Murashige and Skoog complete medium w/agar (g/l)	Gamborg's B5 complete medium w/agar (g/l)	Chu's N6 complete medium w/agar (g/l)
Sucrose	30	20	20
Agar	8	8	8

plant growth. The chloride salts provided in tissue culture are usually well in excess of what is needed by the plant.

(c) *Cobalt*. This is supplied as the chloride salt. Cobalt is needed in vitamin B_{12} synthesis.

(d) *Copper*. This is supplied as the sulphate salt. Copper is required in the energy-conversion processes of the plant cell.

(e) *Iodine*. This is supplied as an iodide salt (KI). Iodine is included in tissue culture but its omission has no consistent side-effects.

(f) *Manganese*. This is supplied as a sulphate salt. Manganese is essential for the functioning and integrity of the chloroplast membrane.

(g) *Molybdenum*. This is supplied as a molybdate salt. Molybdenum is required for the conversion of nitrate to ammonium for amino acid synthesis.

(h) *Zinc*. This is supplied as a sulphate salt. Zinc is involved in the formation of chlorophyll and auxins.

Carbohydrates

Under normal circumstances plants are autotrophic and do not need to be supplied with carbohydrate. However, in tissue culture photosynthesis is greatly depressed, so in order to achieve high growth rates in culture, a carbohydrate supply is required. In a most instances sucrose is the best carbohydrate source. This is hardly surprising, because it is the main sugar made as a product of photosynthesis in most plants. However, there are examples where glucose, fructose or maltose has been used as a carbohydrate source. In some cases even insoluble carbohydrates such as starch have been used. Starch is broken down gradually by the root system of the plant and acts as a low but constant supply of carbohydrate to the growing plant (Figure 18.2).

Vitamins

Under normal circumstances, a plant can synthesize all of the requirements for growth, but for excised plant tissues the organs that synthesize vitamins may not be present, or the tissue may not be able to synthesize vitamins in the quantities required to sustain high growth rates. As a result, the addition of vitamins to the culture medium can increase growth rates of tissue-cultured plant material. In early tissue-culture protocols, yeast extract was used as a supplement [this contains vitamins B_1 (thiamine), B_3 (niacin) and B_6 (pyridoxine)]. However, yeast extract is of an undefined composition and is therefore unpredictable. The vitamins mentioned earlier are still the major vitamins used in plant tissue culture. However, some others are used, such as inositol (a carbohydrate used in the B vitamin complex) and the other B vitamins.

Plant growth regulators

The roles of plant growth regulators in plants have been discussed in earlier chapters, but they are also important additions to plant tissue culture media to maintain growth and direct organ development (see Figure 18.9). The plant tissues in culture do synthesize their own growth regulators, but these will be at too low a concentration to influence growth strongly. Furthermore, manipulation of plant growth regulators allows the fate of a tissue-cultured plant to be altered in the direction desired by the tissue culturist. Plant tissues are very variable and there is no consistent response by them to plant growth regulators in tissue culture. Thus, few generalizations can be made, but some of the roles of plant growth regulators in tissue culture are discussed below.

Auxins

These are probably the most important of the plant growth regulators in tissue culture. The auxins generally used in tissue culture are indole-3-acetic acid, indole-3-butyric acid, 1-napththaleneacetic acid (NAA) and 2,4-dichloro-

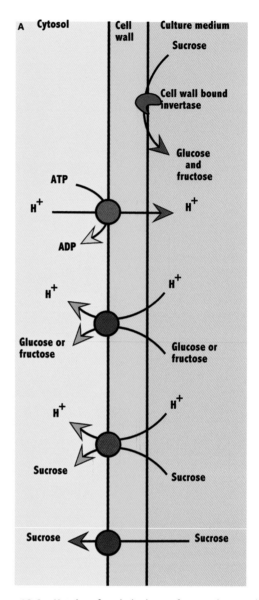

Figure 18.2 Uptake of carbohydrates from a tissue culture medium. The main carbohydrate used in plant tissue culture is sucrose (A). Sucrose transporters can either take up sucrose by free diffusion (no energy required), or H+/ sucrose transporter proteins can take up sucrose. The energy for this uptake is derived from the pumping of H+ ions across the plasma membrane at the expense of ATP. Alternatively in some plant tissues the sucrose is broken down into glucose and fructose prior to uptake by H+/ hexose transporters. Some of these transporters can transport both hexose sugars and others transport either glucose or fructose. This depends on the plant species in question. However, in a small number of examples of tissue culture starch is used as the carbohydrate source (B). The plant explants secrete amylase enzymes out into the agar medium. These then metabolize the starch to glucose or maltose and hese free sugars can then be taken up into the plant tissue.

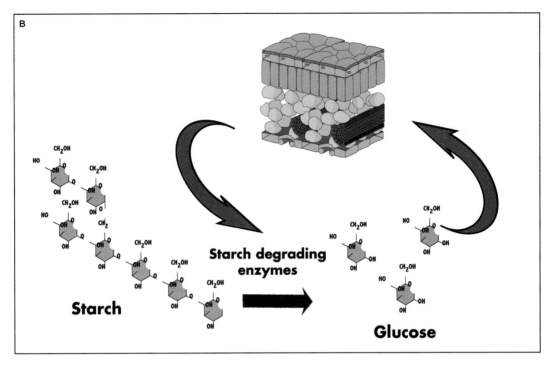

Figure 18.2 (*Continued*).

phenoxyacetic acid (2,4D). The former two auxins occur naturally and tend to have a weak effect on plant tissues. In addition, they are easily oxidized and degraded when illuminated. Thus, they do not persist in culture for long periods. The last two compounds are synthetic compounds that have an auxin-like effect. They are extremely potent and persist for long periods in the medium. In tissue culture, auxins stimulate cell division, which frequently results in callus formation and/or root initiation. When used in combination with cytokinins, they can promote the formation of shoots (Figure 18.3).

Cytokinins

These and auxins are the most frequently used growth regulators in tissue culture. Cytokinins stimulate cell division and they release the apical dominance effect of auxins, which can lead to the production of side-shoots (see Figure 18.8). They can also stimulate the proliferation of shoots from callus tissue. Commonly used cytokinins, in order of decreasing potency are zeatin riboside, 2-isopentyladenine (2iP), 6-benzylaminopurine (BAP) and kinetin.

Gibberellins

There are over 100 different types of gibberellin and they vary markedly in potency to influence development in different plant tissues. The two forms most frequently used are GA3 and GA7, since they are commercially available

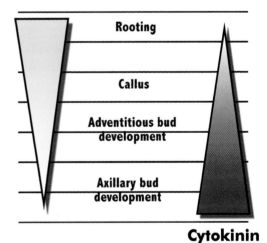

Figure 18.3 The presence of auxins and cytokinins in tissue culture media. The main plant growth regulators used in tissue culture are auxins and cytokinins; other plant growth regulators are only used under special circumstances. The figure shows the roles of the auxins and cytokinins in the medium. High auxin concentrations stimulate root formation, whereas lower auxin concentrations, with perhaps some cytokinins, favour the formation of callus. High cytokinins favour the formation of shoots (no roots usually form). Of course the two growth regulators can be mixed in any number of different combinations until the desired response is achieved.

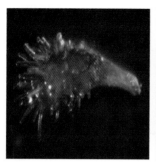
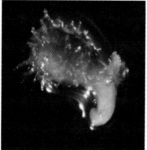
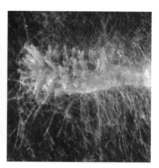

Figure 18.4   Seed sterilization – difficult seeds. To work in tissue culture, plants need to be sterile. This can be a hard result to achieve, since plant tissues are covered in fungal and bacterial spores that will readily grow on plant tissue culture media. In fact they will grow more rapidly than the plant and this invariably results in the death of the plant tissue. The easiest way to sterilize a plant tissue is to take a seed and place it in 10 % v/v bleach solution. After 20 minutes most spores will be dead. Then the bleach needs to be washed away with sterile water and the seed can then be placed on the medium. Here are shown some seeds of *Darlingtonia californica*, which proved to be very hairy and were almost always contaminated with fungi, even after bleach sterilization. The seeds were eventually sterilized by adding a 70 % ethanol wash into the procedure.

and most of the others are not. They are rarely used in tissue culture. Gibberellins are used for seed germination in tissue culture and to lengthen stems for root initiation and propagation. In rare cases they can be used to supplement auxins and cytokinins during cell division.

Abscisic acid

Abscisic acid (ABA) is normally seen as a stress signal in plants. However, it has recently found a use in the production of artificial seeds generated from tissue culture. Since ABA accumulates in seeds on maturation and signals the metabolic events in the seed to allow dormancy, this is hardly surprising. In addition, ABA has been used for the stimulation of precocious flowering in tissue-cultured plants. This only works in plants which rely on a period of drought to trigger flowering.

Tissue culture sterility

Actively dividing plant cells divide every 24–48 hours as the tissue grows. This rate of growth is slow compared

with that of bacteria or fungi (Figure 18.4). Therefore, if there are fungal and bacterial contaminations in tissue culture, they will tend to overgrow the plant tissue and in many instances will kill the plant tissue, through competition for nutrients, attack or the liberation of toxins into the medium. As a consequence, for tissue culture, plant material needs to be sterile. Once a tissue is sterile, it is a relatively easy job to maintain it in culture for long periods of time.

The easiest way to sterilize a plant tissue is through handling seeds. Seeds naturally possess a thick testa, which is resistant to harsh treatments and protects a sterile embryo within. Therefore, exposure of the seed to a compound such as bleach (10% v/v) for 20 minutes is usually sufficient to sterilize the seed. Other compounds, such as mercury-containing compounds, sulphuric acid or hydrogen peroxide, can be used for sterilization (Figures 18.4, 18.5). These compounds need to be washed from the plant tissue prior to placing it in culture. This is usually achieved through successive washes using sterile water. Seeds that possess hairs or projections in the testa can be troublesome, as they frequently harbour bacterial or fungal spores. These can be very resistant

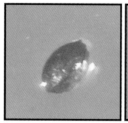
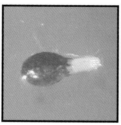

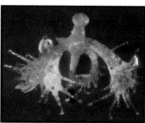

Figure 18.5   Seed sterilization — easy seeds. In contrast to the *Darlingtonia californica* seeds discussed in Figure 18.4, the seeds of *Drosera capensis* are smooth-coated, meaning that there is little opportunity for fungal spores to lodge in the seed coat. These seeds were therefore easily sterilized with a simple bleach wash.

Figure 18.6 Aerial tissue sterilization and culture. Aerial tissues can be treated in much the same way as seeds but, since they lack the resistant testa, they need to be treated more delicately during sterilization. Here tissue explants of *Drosera capensis* are shown. The leaf petioles in this plant spontaneously form meristems when on the tissue culture medium. This gives masses of little plantlets after several weeks. Not all plant tissues respond like this, but propagation of tissue is a major consideration if aerial tissues are to be introduced into tissue culture. Some plants need a meristem on the tissue explant in order to form a basis for propagation; other tissues will spontaneously form them.

to toxic chemicals, so an extra step of washing the seeds in 70% v/v ethanol for around a minute is required (Figure 18.6). A seed may not germinate very readily if added to tissue culture containing sugars and high concentrations of mineral ions, and therefore it may be necessary to place sterile seeds on sterile agar plates containing only water. Seeds can also be dormant and may need additional treatments, such as cold exposure or temperature shocks. Once germinated, provided the plant responds well to the ion components of the medium, the plant should grow if illuminated (Figures 18.5, 18.6).

Aerial tissues can also be sterilized and used as a means of introducing plants into tissue culture. This can be much more difficult. Aerial tissues are continually bombarded with fungi spores and these frequently lie on the leaf surface or in grooves on the stem. Careful cleaning can minimize infections, but even the smoothest of tissues may take several attempts before a sterile plant is obtained. Usually an axillary bud is required to allow growth of the plant, unless the whole tissue responds easily to growth regulators and forms new meristems easily.

Once sterile, all further manipulations on a plant tissue need to be performed in a sterile environment, such as a laminar flow bench, which filters all life forms from the air (Figure 18.7). The bench can be sterilized using ethanol, industrial methylated spirits or UV light to make it a clean work surface.

Types of plant tissue culture

There are numerous types of plant tissue culture. Plants can be simply grown in culture and propagated through cuttings. In culture, the plants revert to a juvenile form and produce small leaves and thin shoots. Thus, the plants can be grown in miniature. This form of propagation can be more rapid than traditional propagation methods, but really uses little of the true power of tissue culture. Through the addition of a cytokinin and an auxin, the cutting can be manipulated to produce many side-shoots. This can vastly increase the productivity of tissue culture. For some tissues, the addition of cytokinin can force embryogenesis from the

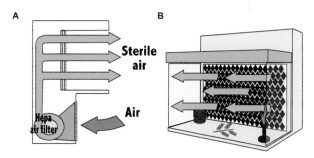

Figure 18.7   The laminar flow bench. The growth of plant cells in tissue culture is slow and sterility therefore needs to be maintained throughout the process. One way to achieve this is through the use of a laminar flow bench. Air is sucked in beneath the bench (A), where it is filtered to remove all dust and spores. The air is then blown back over the bench surface (B). Whilst working on the bench, the sterile air flow ensures that no contamination by fungal or bacterial spores will occur in the tissue culture.

culture. If embryos form from the sexual organs of the plant, then this is termed embryogenesis; however, if they form from other tissues it is called somatic embryogenesis.

Some tissues exhibit little response to tissue culture and need high auxin levels to begin cell division. Once division is initiated, a mass of cells forms, known as a callus (plural, calluses). These cells are disorganized and bear no discernible plant organs (Figure 18.8). A callus can have two fates:

it can continuously divide and generate more calluses, or it can be triggered to produce shoots or roots. Continuously dividing calluses can eventually lose the capacity to generate shoots. There is a second form of tissue culture, known as a suspension culture, which is the equivalent to a callus culture but in a liquid medium. The medium is constantly swirled around to aerate it and this leads to the callus cells being disrupted into small cell clumps.

Applications of plant tissue culture

There are seven major applications for plant tissue culture:

1. Micropropagation.

2. Cloning.

3. Plant genetic manipulation.

4. Hybridization.

5. Dihaploid plants.

6. Removal of viruses from plant tissues.

7. Generation of somaclonal variation.

Figure 18.8   Callus culture of pea (*Pisum sativum*). The cells have been stimulated to form callus through the application of auxins in the culture medium. (A) The callus, which is a disorganized mass of plant cells. By changing the composition of the medium, the callus cells begin to form organized meristems (B) and ultimately shoots (C).

Figure 18.9 Tissue culture of Venus flytrap (*Dionoea muscipula*). Some plants will propagate without going through a callus stage. By adding a cytokinin to the medium, the leaf explants spontaneously form shoots and propagate (A–C). Using this method, a single leaf of this plant can be used to produce over 10 more plants. It is therefore possible to mass-produce this plant in millions very easily, using tissue culture. In addition, the plants grow 10 times faster than they would in soil. A plant in soil will reach maturity and flower after 5 years, but in tissue culture the same plant can reach the same stage in around 6 months.

Micropropagation

Tissue culture allows the replication of plants in miniature. The methods for achieving this have been discussed in the types of tissue culture section, but this can result in large-scale production of plants. For instance, if a Venus flytrap is grown from seed, it will take 5 years to achieve a mature flowering plant (Figure 18.9). At this size the plant can then be propagated asexually, yielding around one extra plant per year. Thus, there is a considerable investment in time and money in growing flytraps in this way. In North and South Carolina this slow growth placed considerable stress on the wild stocks of plants, as it is far easier to dig a plant up and sell it than wait 5 years for a seedling to mature. In tissue culture, a germinated seed will achieve around 3 years of growth in 3–4 months. At this point, each leaf can be cut off to produce a further two or three plants within a period of 2 months. This, then, offers a dramatic increase in production capacity, as quite literally hundreds of thousands of mature plants can be generated in a year. When this is combined with a five- to ten-fold increase in growth rate, then masses of mature plants can be grown in a fraction of the usual time. It is not just flytraps that can be propagated this way, as many species of orchid can be grown similarly. A single *Oncidium* orchid can be micro-propagated to 1 000 000 plants in a single year! Micro-

propagation is one of the greatest assets of plant tissue culture (Figure 18.10). The only problem with orchid tissue culture is that it is not possible to grow the plants to a large enough size in culture vessels to permit flowering within 12 months of coming out of culture. For most *Cymbidium* orchids, it takes around 5 years of soil growth for the plant to attain flowering size.

Cloning

All micropropagated plants are clonal and thus a plant with a particularly desirable phenotype can be cloned and multiplied, rather then relying on intensive breeding strategies to perpetuate the phenotype. In many respects, sexual reproduction is a poor means of replicating plants with a valuable trait, since a phenotype is frequently lost in the next generation. Tissue culture allows controlled large-scale asexual reproduction to be forced.

Genetic manipulation

In order to generate a transgenic plant, a particular gene needs to be introduced into every cell of a plant, which is probably impossible to do to a mature pant. However,

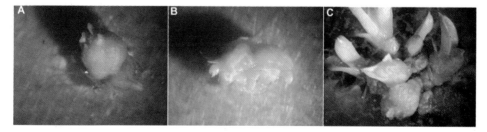

Figure 18.10 Tissue culture of orchids (*Oncidium* sp.). Tropical orchids were some of the first plants ever to be used in tissue culture and plants such as this *Oncidium* are routinely propagated in tissue culture (A–C). Great numbers of these plants can be produced in a short time frame, which has helped make orchids the popular pot plants they are today.

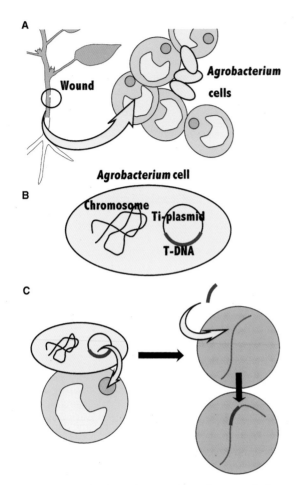

Figure 18.11 The bacterium *Agrobacterium tumefaciens*, a natural genetic engineer. In the wild it attacks the stems of plants that have been injured (A). The bacterium senses the tissue damage and releases DNA (called tDNA) from its cells (B). This tDNA enters the plant cells and integrates with the chromosomes of the plant. Thereafter, as the plant cells divide, so does the T-DNA (C). The DNA encodes some genes that stimulate the production of plant growth regulators, and this causes the plant to form a gall or tumour. The bacterium than feeds in the tumour and multiplies. *Agrobacterium tumefaciens* is quite a common pest of horticulturally grown plants.

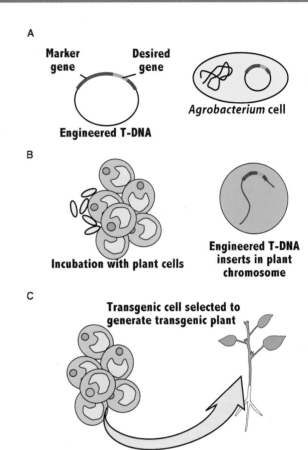

Figure 18.12 *Agrobacterium tumefaciens* used as a tool. *A. tumefaciens* is a natural genetic engineer but this ability has been used by plant biologists to engineer plants genetically. The genes encoded by the tDNA have been removed (A) and genes that a plant biologist wishes the plant to express have been added. Therefore, when an engineered *A. tumefaciens* bacterium infects a plant, this time it will produce a transgenic plant. Using special marker genes, the engineered plants can be selected and, using tissue culture, whole plants can be made from single engineered cells (B, C). *A. tumefaciens* has therefore made the genetic engineering of all plants susceptible to its infection relatively easy.

because plants are totipotent, it is not necessary to introduce a gene into every cell of a plant; only one cell needs to gain the gene, and then that cell needs to be instructed to build a whole plant. This is precisely the role of tissue culture in the production of transgenic plants.

In most instances, for the production of transgenic plants, a step involving tissue culture is required. In a few cases there are examples of flowers soaked with *Agrobacterium tumefaciens* (see below) having been used to derive transgenic plants, but these really are in the minority. Tissue culture is so useful because one of the desired traits

of a plant that is genetically modified is that all of the cells of the plant contain the same genetic construct of the transgene. Thus, the transgenic plant needs to be derived from a single plant cell. Tissue culture offers the possibility to achieve this, because altering the plant growth regulators in the tissue culture medium has profound effects on the fate of the cultured cells and whole plants can be regenerated from single cells. In many respects, the lack of a system for regeneration of plants from cultured cells is the major obstacle to the application of transgenic technologies to a wider range of plant species. Hence, there is great interest in devising new protocols to achieve this.

For dicotyledons, many of the obstacles to the genetic transformation of plants was overcome by the discovery of *Agrobacterium tumefaciens*, a natural pest of plants which has evolved a method of introducing genes into a plant host. The bacterium contains a plasmid known as a Ti-plasmid (Figure 18.11). One piece of this DNA, known as the T-DNA, bears genes that encode proteins which manipulate carbohydrate and plant growth regulator metabolism. It is this DNA that is transferred from the bacterium into the host plant cells and finally integrates into the chromosomes. In the natural state, these genes cause the production of plant growth regulators and novel carbohydrates in the infected tissue. This leads to a tumour forming on the plant, which is an ideal environment for the bacterium to live in. Through simple gene manipulation, the genes borne by the T-DNA have been replaced by other genes which are intended to be inserted into plants (Figure 18.12).

For most T-DNA plasmid constructs, two genes are present. One gene acts as a selectable marker gene that, once expressed in the plant, confers the resistance to some particular treatment. A list of genes is shown in Table 18.2. Through tissue culture, the infected cells can be treated, e.g. with an antibiotic that kills all non-transgenic cells but allows genetically modified ones to survive. Thus, the surviving plant cells can then be stimulated to form a whole transgenic plant. Occasionally the plants formed are chimeric in the cell types present and certain cells are transgenic whilst others are not. For this reason, the potential transgenic plants are grown to maturity and then selfed. Seed from these plants can be germinated in the presence of the antibiotic selection used in tissue culture, and those that germinate and grow will be composed of solely transgenic cells, since the embryo forms from a single cell during development.

One drawback in the use of *Agrobacterium* for plant transformation is that it requires the presence of acetosyringone as a signal for infection and DNA transfer. This is released by dicotyledons after damage to the cell wall (which is when *Agrobacterium* tends to infect plants). However, acetosyringone is not released by monocotyledons after cell damage. In addition, it has proved to be very difficult to add acetosyringone during *Agrobacterium* infection of monocotyledons to allow DNA transfer. Moreover, many of the most important crop plants are monocotyledons. Hence, it was very important that new methods were designed for monocotyledons.

The most successful method for monocotyledons has been the biolistic approach. With biolistics, minute particles of gold are coated with the target DNA. These are given an electrical charge and then, using an electric field, they are projected at the target plant tissue (Figure 18.13). Through careful optimization of the process, it is possible to judge the charge required to lodge the gold particles in the nuclei of the cells. In some cases there is stable integration of the DNA coating the particles in the chromosomes of the target tissue. These cells are then stimulated to divide and form plants. Although this technique is effective for monocotyledons, it is far less efficient than the *Agrobacterium*-mediated method described earlier.

Hybridization

Most plants can hybridize readily with closely related species; however, hybridization becomes more and more

Table 18.2 Selectable markers used in plant transformation.

Marker gene	Selection
NPTII	Kanamycin resistance
Bar	Allows detoxification of gluphosinate (herbicide)
Csr-1	Makes plants resistant to sulphonylurea
PMI	Permits mannose metabolism in some plants
GUS	Glucuronidase gene allows metabolism of X-glucoronide substrate to blue colour
Luciferase	Gene taken from fireflies. When luciferin substrate is added, the transgenic plant tissue glows
Green fluorescent protein (GFP)	Gene taken from jellyfish. Under specific wavelengths of light, the transgenic plants fluoresce
Xylose isomerase	Permits xylose metabolism

The actual uptake of the transgene into individual plant cells is a rare event, so a marker gene is linked to the transgene which is wanted in the plant. Then, only cells which have taken up the desired gene and the marker gene will grow and divide when the transgenic plants are subjected to the selection treatment.

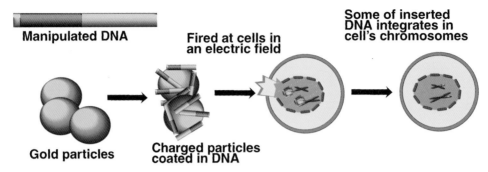

Figure 18.13   The use of biolistics to produce transgenic plants. *Agrobacterium tumefaciens* can infect most dicotyledonous plants but not monocotyledons. Since many of our crop plants, such as rice, wheat and barley, are monocotyledons, new methods have been designed to introduce foreign genes into them. One method is to use biolistics. Microscopic gold particles are coated in the gene construct to be inserted. Then they are given a charge and shot at plant protoplasts (plant cells without a cell wall). Occasionally the DNA enters the nucleus and is introduced into the chromosomes. Using tissue culture, whole transgenic plants can then be obtained from these protoplasts.

difficult as the species become more distantly related. Yet hybridization of different species is a valuable tool in plant breeding. Tissue culture has offered a way around these barriers to hybridization. Two techniques have facilitated plant hybridization, embryo rescue and protoplast fusion.

Many hybrids fail during the formation of the seed. Fertilization occurs and the embryo begins to develop and then, for as-yet unknown reasons, the exchange of nutrients between the parent plant and the developing seed fails and the immature seed dies. One method around this problem is through embryo rescue. The developing seed pod is taken from the parent plant and then sterilized and placed onto a culture medium. If the period of development is optimized, it is possible in some cases to rescue the embryos before they die and support the development of a mature plant (Taji *et al.*, 2002a).

A second method for producing hybrids is protoplast fusion (Figure 18.14). Through the digestion of the cell wall, the protoplast of a cell can be released. Then chemical agents or an electrical field are used to fuse the protoplasts of two different plant species. In many instances this fusion fails, but in a small number of cases the two sets of chromosomes are mixed or rearranged and result in novel hybrid plants.

Dihaploid plants

Using immature pollen culture, haploid plants can be formed. These are almost always sterile and are of little use. However, using chemical treatment with colchicine, the number of chromosomes can be doubled, resulting in a

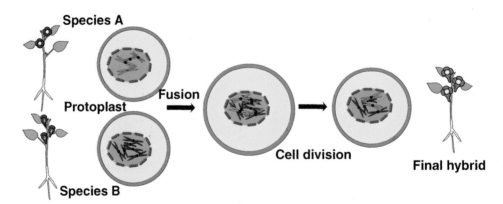

Figure 18.14   Protoplast fusion. Protoplasts have another use in tissue culture and that is for protoplast fusion. Placing them in an electric field can fuse two protoplasts from different genera of plants. The large hybrid cells can then be grown into a full plant, which allows the production of plant hybrids that were previously impossible to obtain.

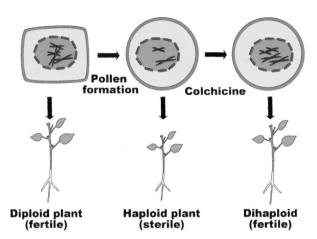

Figure 18.15 Dihaploid plants. Plant tissue culture can be used to produce dihaploid plants as a breeding tool. If a plant has a particularly desirable characteristic, such as resistance to an insect pest, this resistance can be quickly bred into mainstream cultivars of a crop. By culturing haploid pollen, the cells can be triggered to spontaneously double their chromosomes. This leads to dihaploid plants, which can then be used to cross-breed with high-yield cultivars to introduce the trait very quickly. The use of dihaploid plants usually halves the time required to breed new traits into a line of plants.

dihaploid plant (Figure 18.15). Such plants have huge potential in breeding programmes, because through conventional breeding it can take several generations to achieve a true breeding line. Tissue culture offers the potential of producing such a line very rapidly indeed and this then allows controlled breeding of specific plant characteristics.

Removal of viruses from plant tissues

The presence of viruses can be damaging to plant growth. One phenomenon noticed early on with the discovery of plant tissue culture was that, on cell division, viruses did not replicate as rapidly as the plant cells (Morel, 1960). Thus, if a plant tissue is grown and propagated several times in tissue culture, a viral infection can be gradually reduced and ultimately removed. This is particularly important in agriculture and horticulture. One nice example is the salad blue potato. This unusual variety of potato was originally developed as a Victorian oddity (Figure 18.16). The tubers formed contain a purple anthocyanin throughout the flesh. The variety was rediscovered recently in the Lost Gardens of Heligan, but the variety was old and had become riddled with viruses. The

Salad Blue   **Burgundy Red**

Figure 18.16 Tissue culture used to remove plant viruses, which divide more slowly than plant cells. If the plant cells can be stimulated to divide very rapidly, as in tissue culture, regular division of the plant can result in the complete elimination of the virus. These two potato varieties, salad blue and highland burgundy red, were recovered from old Victorian stock using this technique.

presence of viruses in potatoes severely reduces the yield and makes the variety useless for cultivation. However, the variety was sterilized, introduced into tissue culture and propagated frequently until the virus particles had been eliminated. Now this unusual potato variety is on the market again and is being sold as a Heritage Potato variety. A similar problem occurred with the horticultural Surfinia varieties. In the mid-1990s, *Petunia* Surfinia varieties were sold containing viral infections. In order to remove these infections, tissue culture was used and then the plants could again be mass-produced for horticultural sale.

Generation of somaclonal variation

During tissue culture, plant cells divide rapidly and during these divisions mutations can arise. If selective conditions are applied in the culture medium to favour a particular phenotype, it is then possible that a heritable beneficial phenotype can be generated. This variation was given the name 'somaclonal variation' (Larkin and Snowcroft, 1981). It was believed that useful genetic variation in plant tissues could be caused by the tissue being cultured. There are two main theories to explain the source of this observed variation. First, the variation could be a result of heterogeneity of the tissue initially used for tissue culture. That is to say, a tissue such as a leaf is made up of many different cell types which could be in various states of cell division or ploidy. There is certainly good evidence for this, as certain leaves can contain haploid, diploid and tetraploid cells. Second, there is good evidence that much of the variation in tissue culture is a direct result of genetic rearrangement in the tissue-culture process. In some instances there is evidence of duplication of chromosomes, increasing the ploidy of the cells. In other examples, there is evidence of movement of transposable elements and of chromosome damage (Taji *et al.*, 2002b). Regardless of the source of the variation, this can be very useful for gaining new variability in plants which can then be used as a breeding stock. In tissue culture, plants can be selected for various phenotypes, such as disease resistance, and then used as a breeding stock to introduce the trait into high-yield breeding lines.

Plant tissue culture has undoubtedly had a major impact on plant breeding over the past 50 years and has been responsible for much of the increase in yield observed over this time. It has proved to be a powerful tool for agriculture, horticulture and plant conservation. However, will it probably have its greatest impact through supporting plant genetic manipulation.

References

Larkin PJ and Snowcroft WR (1981) Somaclonal variation. A novel source of variability from cell cultures for plant improvement. *Theoretical Applications of Genetics*, **60**, 197–214.

Morel G (1960) Producing virus-free *Cymbidium*. *American Orchid Society Bulletin*, **29**, 495–497.

Morel G (1964) Clonal multiplication of orchids. In *Orchids: Scientific Studies* (ed. Withner CL) Wiley, New York.

Murashige T and Skoog F (1962) A revised medium for the rapid growth and bioassays with tobacco tissue cultures. *Physiologia Plantarum*, **15**, 473–497.

Taji A, Kumar P and Lakshmanan P (2002a) *In Vitro Plant Breeding*, Food Products Press, New York, pp. 57–67.

Taji A, Kumar P and Lakshmanan P (2002a) *In Vitro Plant Breeding*, Food Products Press, New York, pp. 101–109.

van Overbeek J, Conklin ME and Blakeslee AF (1941) Factors in coconut milk essential for growth and development of very young *Datura* embryo. *Science*, **94**, 350–351.

19

Remarkable plants

Most plants are remarkable in some way or other, as this whole book has demonstrated, with a wide range of examples from many of the plant families. But some species particularly stand out from the rest as being truly noteworthy.

When talking to people about plants, certain species always instantly create interest and these are what this chapter is about. Carnivorous plants are one group that never fails to attract attention from an audience, probably because the thought that plants can turn the tables on animals captures people's imagination. Orchids and resurrection plants also generate a great deal of interest and it is these groups that are examined in this chapter.

Insect mimicry in the orchidaceae

Of all of the plant families, only the Orchidaceae use insect mimicry as a means of attracting and getting insects to transfer pollen from the anthers to the stigmatic surface. The particular example of the European bee orchids (*Ophrys* genera), which has already been discussed in Chapter 15, is the most remarkable example of the relationship between orchids and bees (Figure 19.1). However, there are many other examples of mimicry in the orchid family. The Australian hinged dragon orchid (*Drakonorchis drakeoides*) mimics the female thynnid wasp in appearance and scent. The female wasp is wingless and climbs on grasses and other plants when it is sexually active. From this position it releases pheromones that attract the male wasp. The male wasp uses these scents to approach the female and then uses visual stimuli to locate and land on her. The male picks up the female and mates with her in mid-air. The hinged dragon orchid takes advantage of this sequence of events and flowers just prior to the emergence of the female wasps. The flower releases sex pheromones similar to those produced by the female wasp. Furthermore, the flower of the orchid bears an appendage that resembles the female wasp in shape and colour. The male is attracted to the flower by these stimuli and attempts to land on the mimic wasp and fly off with her. Unfortunately for the wasp, the appendage is well connected to the flower and pivots and swings the male wasp into the anthers. Here the male is coated in orchid pollen and, after repeated attempts to abduct the mimic wasp, the male gives up and tries his luck elsewhere. This cycle is repeated several times until the wasp gets wise to the mimic, but by then the orchid should have been pollinated.

An alternative form of mimicry is observed in the *Paphiopidelum* orchids. These orchids bear labellum pouches to the flowers, which have spotted translucent bottoms. This speckling mimics a flower with an infestation of aphids. Certain species of flies which lay their eggs amongst aphids are attracted to the flowers and enter the pouches. Here they become trapped and the only route out of the chamber is via a tunnel leading past the stigma and anthers. The flies, on release, are coated in pollen and on repeat of the whole episode, the orchid is fertilized.

The *Oncidium* orchid is a tropical species from Central and South America, frequently sold as a houseplant. The genus contains between 400–700 different species. The flowers are yellow and brown, which is consistent with them being mainly pollinated by bees. Two strategies seem to be adopted by *Oncidium* species for pollination. First, some *Oncidium* flowers are thought to mimic the flowers of other species which produce an oil to reward visiting bees. On visiting an *Oncidium* flower, the bees obtain no oil and then fly off to visit a second flower, pollinating it in the process. Second, some bees gather around flowers as a breeding territory. The flowers of orchids such as *Oncidium hyphaematicum* look sufficiently like a bee that male bees attack the flowers in an attempt to remove a trespassing male bee. During this struggle, the male bees frequently get pollinia deposited on them and then, after further conflicts, end up pollinating the flowers. This is

Physiology and Behaviour of Plants   Peter Scott
© 2008 John Wiley & Sons, Ltd

Figure 19.1 Bee orchid (*Ophrys*), of which there are over 50 recognized species in Europe. These plants use a very sophisticated form of sexual mimicry to fool certain species of bees to pollinate the flowers. A series of euophrys and pseudophrys flowers are shown. Euophrys flowers (A–C, F–H) orientate the visiting bee such that the head of the insect abuts the pollinia, whereas pseudophrys flowers orientate the insect the other way round. The yellow bee orchid (D) and the sombre bee orchid (E) are pseudophrys. (A) Late spider orchid (*O. holoserica*); (B) Fly orchid (*O. insectifera*); (C) Bee orchid (*O. apifera*); (D) Yellow bee orchid (*O. lutea*); (E) Sombre bee orchid (*O. fusca*); (F) *O. moresii*; (G) Woodcock orchid (*O. scolopax*); (H) mirror orchid (*O. speculum*).

known as pseudoantagonism, and is yet another example of the diverse range of orchid mimicry.

This range of mimicry techniques has probably led to the great diversity of the orchid family, which contains over 25 000 different species. Although orchids are renowned by most people for their beautiful and showy flowers, they are certainly more remarkable as the undisputed masters of mimicry in the plant world.

Venus flytrap

No plant evokes more interest and wonder than the Venus flytrap (*Dionaea muscipula*). The Venus flytrap is certainly not known for its success as a species. It is only found in a limited coastal region in North and South Carolina and even in these states it is restricted to pine savannahs (Schnell, 2002). Even as a carnivorous plant, the Venus flytrap is not the most successful insect catcher; that accolade probably belongs to the North American pitcher plants. The flytrap's novelty lies in the fact that it can move quickly and is one of the fastest moving plants known. The plant bears around five leaves, each extending up to 7 cm. At the end of

each leaf there is a two-lobed trap, hinged in the middle, resembling a man-trap. Appendages on the edges of the lobes form tines which act to trap prey, once the trapping mechanism is activated.

The traps work by first attracting prey. This is achieved by the lobes bearing a red coloration (resembling a flower), and there are nectar glands all around the rim of the lobes which secrete a sticky sucrose solution. When insects visit the trap, they begin feeding on the sucrose but will invariably wander around the leaf in search of more sugar. This brings them into contact with one of six trigger hairs (three on each lobe). Touching a hair once will not trigger the trap, but if the same hair or another one is touched within about 10 seconds the leaf lobes rapidly move together (Figures 19.2–19.4). This is one of the few instances where a plant exhibits a memory, as the plant can recall how recently it has been touched. As the leaf lobes move, the tines interlock and hold the insect between the lobes. If the insect is small, it will be able to escape through the gaps between the tines; however, if it is large and trapped, it will begin to scramble around the trap, trying to find a way out. This puts the insect into contact with the hairs again and activates the second phase of the

Figure 19.2 Structure of the trap of the Venus flytrap (*Dionaea muscipula*), whose leaf ends in a two-lobed trap. Hairs known as tines fringe the trap and in the centre of the lobes there are three or more trigger hairs. If these trigger hairs are touched twice within 20 seconds, an electrical impulse passes across the leaf surface, during which the cells on the inner section of the leaf surface lose much of their turgor and collapse. This flips the leaf over very rapidly, throwing the interlocking tines together and thus capturing the prey and closing the trap.

closure mechanism. The two lobes become pressed together over about half an hour, and the trap seals. At this point, digestive enzymes, including proteases and phosphatases, are released into the chamber and digestion begins. Digestion takes several days to complete, depending on the size

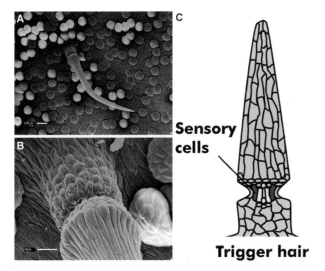

Figure 19.3 Structure of the trigger hair of the Venus flytrap (*Dionaea muscipula*), whose role was discussed earlier. An electron micrograph of the hair is shown (A) , with close-up detail (B) of the constricted region at its base. A schematic diagram (C) shows the sensory cells in this region. These cells sense the pressures applied to the trigger hair and initiate the electrical impulse that closes the trap.

of the prey (the larger it is, the longer digestion takes). The trap then opens again and the indigestible chitinous shell of the insect is left. This shell is washed out by rainfall and the trap can be used a second time; however, it is usually less effective as a trap as a result of distortions of the leaf structure.

The actual mechanism for leaf movement is still not well understood. If the tip of the trigger hairs is touched, it bends at a point where the hair narrows at the base. It is this point which generates the signal for the trap to close. A signal then passes across the leaf surface very rapidly. This signal may be in the form of an electrical impulse across the leaf. Certainly an electrical current has been measured, but it is not clear whether the current is a signal or a result of ion movements associated with the closure. The lobe that is stimulated begins to close first and within one-15th of a second the signal has arrived at the other lobe and closure is initiated. Closure involves a collapse of turgor in a layer of cells between the upper and lower epidermis of the leaf. This leads to a 30% shrinkage of the leaf thickness. The most likely explanation for the rapid movement of the lobes is that they are what are known as prestressed structures. If the leaf of a Venus flytrap is dissected and the upper and lower epidermis are parted, both sides curl outwards. The leaf is in a stressed state and its natural state is in a closed position with the tines interlocked. However, they are held open by the turgor pressure within the layer of cells between the upper and lower epidermis. On stimulation, an electrical signal passes through the cell

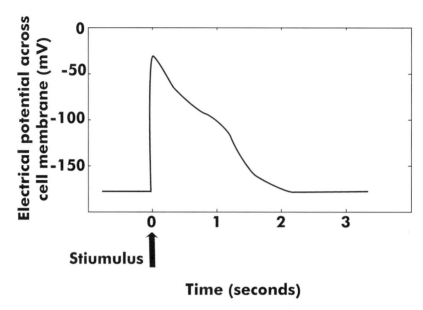

Figure 19.4   The electrical impulse across the Venus flytrap (*Dionaea muscipula*) leaf, which is very similar to that observed in a human nerve cell. A large movement of ions occurs as the impulse moves across the leaf surface, causing the collapse of turgor in the cells in the trap lobes.

layer and this triggers the opening of channels in the cell membrane, causing a rapid leakage of water from the cells. Once the leaf lobes begin to move, this leakage is probably exacerbated by pressure on the cells caused by the closing trap. As a consequence, the trap mechanism is rather like a spring which is released by a touch. Once a cage is formed and an insect is trapped, a second phase to the trapping mechanism begins. This involves the lengthening of the cells on the outside of trap lobes. This expansion is slow and takes between 30 minutes and 2 hours to complete. This expansion is also permanent growth and is most likely a result of acidification of the apoplastic space, causing loosening of the cell wall and subsequent cell expansion. The expansion drives the lobes together and they meet at the base of the tines. The sugary secretion in this region helps to gum the leaf lobes together. The trap lobes are visibly at their thinnest just prior to the leaf margin where they widen out again. This thinning causes the leaf margin to bend outwards as the lobes are pressed together, and at this point expansion ceases. After digestion of the prey, the cells on the inner side of the trap begin to expand; this resets the whole trap and rainwater or winds flush out the undigested remains of the prey. Only prey such as some slugs that lack a chitinous shell are completely absorbed by the plant, leaving no remains. The trap has undergone permanent growth during capture and digestion and therefore is now slightly bigger than it was initially. It is also usually slightly buckled and the trigger hairs will have sustained some damage when the lobes were pressed

together. These lead to the trap being poor for capture of a second prey item.

This rapid movement and the gruesome nature of the trapping mechanism make the Venus flytrap one of the world's most remarkable plants. It was mentioned earlier that in the wild, the flytrap is a relatively poor fly catcher, is rare, and is restricted to a small habitat, but as a result of its remarkable abilities the plant has been extraordinarily successful. Millions of plants are sold commercially each year and it has been planted out in many new habitats across the world, thereby becoming spectacularly successful.

The oldest and biggest plants in the world

It is surprisingly difficult to separate these two records in the plant world. If a plant keeps on growing, then inevitably one would expect the largest plant to be the oldest plant, but this does not strictly follow.

The oldest complete tree is the bristlecone pine (*Pinus longaeva*), which grows in the mountains of eastern California (Figure 19.5). One particular specimen, the Methuselah tree, is thought to be 4789 years old, which makes it around 1000 years older than the next oldest tree of another species. This means that this particular specimen would be considered old even when the pyramids were being built. The oldest trees are present in the White-Inyo Mountains in California and the oldest plants are unmarked to protect

Figure 19.5 Bristlecone pine (*Pinus longaevia*), individuals of which have been dated as being the oldest living plants. This one is from the Cedar Breaks State Park in Utah.

them from damage from visitors. The bristlecone pine grows slowly and the tree is only around 4 m tall, therefore its size does not betray its great age. The age of a living tree specimen is determined by driving an augur through the trunk to remove a core. The annual rings can then be counted and the age determined. This is not an easy task for so old a plant in such a cold habitat. Most of the tree is dead and only some branches continue to grow. But this is still not the oldest specimen ever known. Another bristlecone

pine (called Prometheus) once grew in the same area and whilst taking an augur core sample from the tree in 1965, the augur got stuck. In order to remove the augur, permission was sought to fell the tree and the National Parks Board granted this. The tree was felled and only after this was it realized what a treasure had been lost. The Prometheus tree was a staggering 4900 years old; after all those years it had been felled in such a careless accident. This led to a tightening of conservation in the National Parks and since then the identity of many special plants, such as the oldest tree and the tallest tree, is now kept secret.

The identity of the largest tree in the world has never been kept a secret even though it is hidden amongst many other enormous trees. The General Sherman tree is in the Giant Forest region of the Sequoia National Park in California (Figure 19.6). It is a *Sequoiadendron giganteum* or giant sequoia, and the National Park holds four of the five largest sequoias in the world. The General Sherman tree is not the tallest tree in the world; it measures only a mere 83.8 m and many trees are taller than that (see later for details of the world's tallest plants). With a base circumference of 31.3 m it does not have the biggest trunk either, as there is a cypress tree in Mexico which has a circumference of 49.4 m (these data were obtained from the Sequoia National Park website). But for volume of wood the tree has no equal. The trunk of the tree alone is thought to weigh 1400 tons and to have a volume of 1486.6 m^3. As a consequence, it is thought to be the largest living organism and weighs more than 15 adult blue whales. If the whole plant was considered, it would probably weigh almost 6000 tons. Once a branch fell from

Figure 19.6 Examples of two giant sequoias (*Sequoiadendron giganteum*) from the Sequoia National Park, California, USA. The General Sherman tree (not shown) is generally taken to be the largest living organism on Earth.

Figure 19.7 The honey fungus (*Armillaria ostoyae*). There are suggestions that this woodland fungus is the largest living organism on the planet. However, this is very difficult to verify and in fact the clonal stands of many tree species may in fact be larger.

the General Sherman tree that was so large it was bigger than any tree on the eastern side of North America!

One *S. giganteum* that was also famous was a tree in the Miraposa Grove in Yosemite National Park, California, which had a large piece cut out of it that was so large, a road was built through it. A famous picture exists of a Volkswagen Beetle car being driven through this tree bridge which allows the viewer to appreciate truly the enormity of these plants. Sadly, lightning in the 1970s destroyed this tree. The National Parks authorities deemed the trees too precious to ever cut a second tree in this way, so it is a spectacle never to be repeated.

The cases for the Methuselah tree and the General Sherman tree being the oldest and largest living organisms on earth would seem clear-cut. It could be argued that most of the General Sherman tree is deadwood and therefore not a living organism, but that aside, the evidence is compelling. However, there have recently arisen some competing organisms for the largest living thing: fungi. Since most of the body of a fungus lies underground, it is very difficult to judge its actual size. Estimates can be made as a function of the area covered by the fungus. However, to establish that, the identity of the fungus needs to be authenticated through analysis of its genetic structure. Using these methods, a honey fungus, *Armillaria ostoyae*, has been found to cover over 600 hectares on Mount Adams, Washington State, and 890 hectares in the Malheur National Forest in Oregon (Figure 19.7). In addition, the fungal clones have been estimated to be between 2000 and 7000 years old and therefore could threaten the old age record of the bristlecone pines. No estimate has been made of the weight of the fungal clones, but they could dwarf the General Sherman tree. However, before fungi take the honour of being the oldest and largest living organisms, the fact is that these are *clones* of the fungus, and when we consider clonal plants, more evidence needs to be considered.

There are several species of plant that form large stands of clonal individuals, e.g. the creosote bush (*Larrea tridentata*), the box huckleberry (*Gaylussacia brachycera*)

and the quaking aspen (*Populus tremuloides*) (Madson, 1996). Through genetic analysis it can be established that there are substantial clonal populations of these plants across North America. The quaking aspen is a tree that reaches about 30 m in height and with a trunk size of 70 cm diameter it would appear to be unremarkable, but it has a tremendous capacity to reproduce asexually (Little, 1980). The root system forms suckers that, in turn, form new trees (Figure 19.8). The root system of a mature plant can produce a maximum of a million new shoots per acre (Mitton and Grant, 1996). The shoots also grow at the rate of 1 m/year. A combination of forest fires, logging and avalanches helps to produce the clearings needed for these shoots to establish themselves, and clonal stands of quaking aspen are rapidly established. One giant clone in Utah, known as the Pando clone, covers over 43 hectares and includes over 47 000 trees (Mitton and Grant, 1996). An estimate of the weight of this clone is 6000 tons, not

Figure 19.8 Stands of clonal trees (*Populus tremuloides*). Clonal tree stands that could still be connected through their root systems could in fact be the largest organisms in the world. It is possible to date their origin by the size and growth rate of the stand of trees, which could also make clonal plants some of the oldest organisms on Earth.

including the root system. This would mean this clone is larger than the General Sherman tree, but whether it is larger than the *Armillaria* fungus is not known.

By following the rate of progress of growth of the stands of clonal plants, it is possible to predict the age of the plants. The Pando clone of quaking aspens is probably around 8000 years old, which would mean the original plant establishing itself soon after the last glacial period, almost 10 000 years ago. Moreover, creosote and huckleberry stands have been estimated to be 11 000 and 13 000 years old, respectively (Wherry, 1972; Madson, 1996). This would bring the record for the oldest living organism well and truly back to the plant kingdom. However, a recent discovery in Tasmania has pushed the record for the oldest plant even further back than other clones. The endangered shrub species King's holly (*Lomatia tasmanica*) is restricted to a small habitat in south-western Tasmania. It is a triploid plant and therefore is severely restricted in its ability to reproduce sexually, and hence uses asexual reproduction by means of rhizomes (Brown and Gray, 1985). The plant has never been observed to set fruit and individuals of the plant are all thought to be clonal. There is only one population known to exist on Tasmania. A study by Lynch *et al.* 1998 set out to investigate the diversity in the population of King's holly and found it to be completely clonal. Fossil King's holly leaves identical to those of the living plants have been discovered near to the site and dated at around 43 000 years old (Jordan *et al.*, 1991). The plant has a slow growth rate (0.26 mm/year) in the cold climate of Tasmania, which would make the 1.2 km population very old. It has been suggested that, judging by the proximity of the fossil find, the limited distribution of the plant and the similarity of the plant structure to the fossil, the clonal plant may be as old as 45 000 years. Even though, measuring by dendrochronological methods, the average age of an individual stem is around 240 years old (Brown and Gray, 1985), this would make King's holly the oldest plant known.

Tallest plants

The tallest plants in the world are not the oldest or the largest. The current tallest tree is called the Mendocino tree, a specimen of *Sequoia sempervirens* (coast redwood) which is in the Montgomery State Reserve, California, and stands at 112 m tall (five storeys taller than the Statue of Liberty). Thus, the tree exceeds the height of the General Sherman tree by over 28 m. It is a mystery how a plant manages to transport water up to such a height. It has been suggested that the forest in which it grows is subject to frequent mists which may provide valuable water needed by the leaves and upper branches. The needles of the plant have been reported to act as a surface for condensation of the mists and can lead to a considerable volume of water being dripped onto the forest floor. The tree is only 1000 years old and has a 3 m diameter, making it very long and thin, and thus its mass does not approach that of the giant sequoias. However, the Mendocino tree is not the tallest tree ever reported. A *Eucalyptus regnans* specimen and an unidentified *Eucalyptus* in Victoria, Australia, measured 143 m and 150 m, respectively. Both of these trees were only measured after loggers had felled them. Given humankind's history for poor treatment of remarkable tree specimens, the Mendocino tree is not labelled in the Montgomery State Reserve, to protect it from unwanted attention.

Cycads

Cycads are remarkable for several reasons. They could be referred to as living fossils, as this group of plants was at its height during the Mesozoic Period (around 200 million years ago). Fossils from that period and earlier still resemble the species around today. They are still found in tropical regions across the globe. Cycads are gymnosperms and there are three known families of cycads: Cycadaceae, Stangeriaceae and Zamiaceae. The plants are woody and they form lignin (Giddy, 1984). They resemble palm trees but are not related to them. They were very abundant during the Triassic and Jurassic periods and therefore would have been around when dinosaurs roamed the earth. In many respects they are the crocodiles of the plant world and have not changed for millions of years. However, they are markedly in decline now and one of them is what could be termed the world's rarest plant (Figure 19.9). Cycads are dioecious and they exist as separate male and female individuals. In Kirstenbosch Botanic Gardens, Cape Town South Africa, there is a male specimen of *Encephalartos woodii*. It is the only plant of this species that has ever been found. Unable to reproduce, there only ever will be one specimen of this plant. Cycads live for thousands of years so it is unlikely to die soon, and the specimen has been cloned but all of the plants will be genetically identical.

Welwitschia mirabilis

Welwitschia mirabilis is endemic to Namibia in southern Africa. The plant is perfectly adapted to life in the Namib desert. It forms a long trunk, most of which is underground, and a long tap root (Figure 19.10). The plant produces two strap-like leaves that grow continually at their base. Due to

Figure 19.9 Wood's cycad (*Encephalartos woodii*). Four individuals of Wood's cycad were discovered in 1895 in Ngoye forest, KwaZulu-Natal, South Africa. The specimens were all clonal male plants and since that time no female plant has ever been discovered. All the wild specimens are now dead, leaving a small number of cloned individuals in botanical gardens in South Africa. This specimen was photographed in Kirstenbosch Botanic Gardens.

weathering, the leaves split and become broken but they continue to extend as a result of the presence of an intercalary meristem. The shoot tip dies whilst the plant is still young and it rarely exceeds 50 cm in height. Once the plant has stopped growing upwards, it grows outwards, forming a trunk from which the leaves grow. The plants are dioecious and use CAM to survive in the desert. The desert area where they live receives very little rain, but the leaves can effectively absorb water from sea mists that blow across them periodically. The desert temperatures can rise as high as 65°C at times. It is remarkable that *Welwitschia* can survive for over 1500 years in such a harsh climate with so little rainfall. In addition, since the leaves they bear

Figure 19.10 *Welwitschia mirabilis*, the only member of the family Welwitschiaceae. This desert plant bears several similarities to both the Gymnosperms and the Angiosperms and is thought to be a missing link between these two divisions. This individual was photographed in the Stellenbosch Botanic Gardens, South Africa.

grow continually, they are the longest living leaves of any plant; the leaves of a mature plant can be around 3 m long but have the capacity over the plant's lifetime to be 100 m long. This length is never realized, as winds and the harsh environment tend to kill off the leaf tips. Unconfirmed reports suggest that some specimens of *Welwitschia* have leaf areas of over 55 m^2 and they possess around 20 000 stomata/cm^2. For most desert plants, the presence of high densities of stomata and large leaf areas are positive disadvantages for surviving in arid environments. *Welwitschias* have evolved another strategy; the habitat in which the plants live is subjected to frequent sea mists. These plants function in reverse compared to other plants. The leaves absorb condensing water from the mists through the stomata and this mechanism is capable of supplying the water needs of the whole plant. Therefore, although the plant possesses a deep root system, which is known to exceed 30 m in depth, the main way of obtaining water is thought to be through the leaves.

Both *Welwitschia mirabilis* and *Ephedra viridis* are members of Gnetophyta. Gnetophyta are a group of plants which are gymnosperms but they possess xylem vessels instead of tracheids which the phylum ordinarily contains. In addition, species of *Ephedra* use double fertilization (see Chapter 1) which is not used by any other Gymnosperm. *Welwitschia* and *Ephedra* are regarded as missing links between the Gymnosperms and the Angiosperms (the flowering plants).

The castor bean plant

The castor bean plant (*Ricinus communis*) is a common garden plant and an invasive pest of subtropical regions of

Figure 19.11 The castor bean plant (*Ricinus communis*), which contains the lectin ricin, the most toxic product identified in plants. This is a common weed in wasteland across the world and is sometimes even grown as an ornamental plant in gardens. The mature plant (A), seed pod (B) and seeds (C) of the plant are shown.

the world (Figure 19.11). It is used commercially for the extraction of castor oil and in the garden setting it is grown for its interesting coloured leaves, seeds and seed pods. However, there is a major problem with this plant; it is the most toxic plant known. To recover the oil from the seeds, hot pressing is used to minimize the possibility of toxin contamination of the oil.

The castor bean plant produces the lectin *Ricinus* agglutinin. One component of this lectin is a protein called ricin. This lectin is exceedingly dangerous to mammals and is one of the most toxic substances known to be produced by plants. An amount the size of a pin-head (0.5 mg) is enough to kill an adult human. Ricin is present in the castor bean seeds and makes them extremely dangerous to eat without proper treatment first.

This lectin has been in the news several times. Two Bulgarian dissidents close to Communist President Todor Shivkov in Bulgaria defected to France and the UK. These dissidents wrote about their experiences in Bulgaria and the President ordered their execution. The first attempted assassination was of Vladimir Kostov in Paris in August 1978; this failed and the victim was only ill for a short while. Then later, in 1978, Georgi Markov (the second dissident) was assassinated in London by being skewered by an umbrella containing ricin particles. Markov took several days to die but within that time he reported the umbrella incident. In his leg was found a lead ball impregnated with the toxin ricin. Later it was found that Kostov had had a similar ball in his back, but it had not fully penetrated his skin. It was later uncovered that the Russian KGB had been perfected a method of assassination using ricin and had sold this on to the Bulgarian secret service. This was the first example of a plant toxin being weaponized. Ricin has continued to be in the news more recently, with growing concern that it could be used as a weapon of mass destruction. Since the amount of ricin required to kill someone is very small, there is a real threat that the toxin could be used as an aerosol bomb or used to pollute foodstuffs and water supplies.

Ricin is one of the most toxic substances known and it is relatively easy to grow and mass-produce. Once poisoned with it, there is no known antidote and death occurs within 8 days of exposure. Ricin acts by affecting ribosome binding and prevents protein synthesis in cells. Needless to say, castor beans are more than adequately protected from predators!

Garlic

Garlic belongs to the family Alliaceae (onion family) and has been used for millennia for the flavouring of foods (Figure 19.12). There are around 600 different varieties of garlic in cultivation, but they are all from two main subspecies, *Allium sativum ophioscorodon* (hard-necked garlic) and *Allium sativum sativum* (soft-necked garlic). The soft-necked varieties have arisen as a result of centuries of breeding. There are other plants sold as garlic, such as elephant garlic (*Allium gigantum*), which is in fact more closely related to a leek than garlic itself. However, it does still bear a recognizable garlic flavour. There are several remarkable features of garlic, the first being its unique taste. It is one of the strongest of food flavourings. Moreover, there is a great deal of evidence that garlic possesses many medicinal properties. Garlic has been shown to cause a 12 % drop in blood cholesterol levels within 4 weeks of regularly consuming it (Silagy and Haw, 1994). It also contains a compound, alliin, which is converted to allicin on crushing the cells. Alliin makes up around 0.25 % of the weight of a garlic bulb. There is good evidence that alliin is an antithrombotic factor and reduces blood clotting. This is thought to give health benefits with respect to heart conditions. Alliin also acts as an antifungal agent. This compound probably acts to protect the bulb when in the soil.

Thus, garlic is one of those rare things in the world that taste nice and are beneficial to health!

Figure 19.12   Garlic (*Allium sativum*), thought to have originated in Central Asia. Its use seems to have rapidly spread around the world and it is recorded as being used in Ancient Egypt in 1600 BC. It is reputed to possess many medicinal properties. Here the bulbs are shown on sale in a French market. The active ingredients are shown (right). Alliin in the bulbs is converted to allicin on crushing the cells. Allicin has been demonstrated to act as an antithrombotic factor and can reduce clotting in blood.

Theobroma cacao

There are several plant genera which produce the flavour we associate with garlic, but one could not mention special plants without a brief mention of a plant that produces a flavour no other species can – chocolate. *Theobroma* is a tropical understorey evergreen from South America. There are around 20 species in the genus and they belong to the family Sterculiaceae. The main species used commercially is *Theobroma cacao*. The Aztecs first cultivated *Theobroma* and the beans were converted into a sacred drink called *xhocolatl* (which has since been corrupted to 'chocolate'). The beans were also highly prized and were used as a form of currency by the Aztecs. The beans were exported to Europe in the sixteenth century, but the first chocolate bars were not made until the mid-nineteenth century.

When the tree is around 3 years old, it produces small flowers along the stem. The flowers are self-incompatible and need to be fertilized with pollen from another plant in order to set seed. Small midges pollinate the flowers. After 6 months, a large pod develops, containing 40 seeds. All commercially grown plants are grown in the shade, as the plant cannot be grown in the open sunlight. The pods are harvested, the beans are extracted and then fermented for 1 week in order to convert all sugars into alcohol and then organic acids. It is at this point that the chocolate flavours begin to develop. The beans are then dried and roasted and further flavours develop as a result of the Maillard reaction. The shell of the seed is then removed and the cotyledons are mashed to give a paste known as the nibs or cacao mass

(Figure 19.13). Around 60 % of the nibs is made up of a complex fat known as cocoa butter. It is this paste which is blended to make chocolate products. If the cocoa butter is extracted, cocoa powder is left. Chocolate also contains two alkaloids, theobromine and a small amount of caffeine, which both act as stimulants and make up the bitter taste that cocoa possesses.

The first cocoa products were drinks with cacao beans added to water and drunk by the Mayas (Zackowitz, 2004). Dominican friars in 1544 introduced a drink combining cacao beans, wine, peppers and sugar into Europe and the drink became popular. Then in 1847, the first chocolate bars appeared in the UK and the first confectionery form of chocolate began. Most chocolate is grown in West Africa now, but relatively little chocolate is consumed by people in warmer countries. Cocoa butter melts at about 34°C and therefore chocolate does not store well in warmer climates and costly refrigeration is required. However, this low melting point is crucial for the flavour of chocolate as it melts in the mouth, allowing full appreciation of the plant products. The major consumers are Switzerland, Germany, Belgium and the UK, with these countries consuming around 9–10 kg/person/year (Zackowitz, 2004).

Wheat and agriculture

Wheat as a cultivated plant is believed to have originated in the Fertile Crescent around 10 000 B.C. (Neolithic–stone age; Zohary and Hopf, 1993). This area was very rich in nutrients and was well irrigated by the Tigris and

Figure 19.13 *Theobroma cacao*. There are around 20 different species of *Theobroma* plant (Sterculiaceae family) that are used in the production of chocolate. The genus is native to the tropical regions in Central and northern South America. The Aztecs used the beans of the plant for currency and as a drink. The Spanish introduced this drink to Europe and plantations were established in the tropical regions of central Africa. Linnaeus gave the name Theobroma to the plant, meaning food of the Gods. The term chocolate was derived from the Axtec word xhocolatl meaning bitter water. A) A mature *Theobroma* pod is shown. These pods contain a number beans (B). If roasted the beans can be separated into cocoa butter (C) and cocoa mass (D). These can be recombined to make chocolate (E). The presence of the alkaloid theobromine gives chocolate its bitter taste. To some organisms, such as dogs, the alkaloid is very toxic, but to most humans the taste is very appealing.

Euphrates rivers. History does not tell us what happened precisely with wheat and why it became a plant of interest to hunter-gatherer communities, but two of the most likely possibilities are, first, that wheat gave a high yield of nutritious seeds, and second, that some plants exhibited an unusual property for a member of the Poaceae – the seeds were non-shattering. Shattering is a general property of all seeds on maturity (Figure 19.14). An abscission zone forms between the mature seed and the plant. Winds and rain easily dislodge the seeds and self-seeding occurs.

Under natural conditions, non-shattering is not an advantageous trait, as it prevents effective seed dispersal. However, for the purposes of gathering seeds, it allows rapid harvesting with little crop loss. Through selection of wheat with high yield and little shatter, the crop has evolved to what we have today.

The first corn to be domesticated is thought to be wild einkorn, which is made up of several separate species, *Triticum aegilopoides*, *T. thaoudar* and *T. urartu*. The variety of einkorn cultivated now is *T. monococcum*

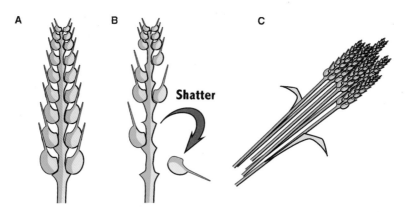

Figure 19.14   Shatter in grasses. As most grass seeds (A)mature, they undergo a process called 'shatter'(B), in which the seed breaks away from the old inflorescence very easily. It is thought that one of the crucial features of the development of the crop plant wheat was failure of this shatter process. If shatter fails, (C) the seeds will fail to spread effectively – not an evolutionary advantage to a plant. However, the failure of shatter also makes it possible to harvest a large crop very easily from a plant. Early human civilizations are thought to have taken advantage of this and cultivated the plants, giving rise to the wheat plants we have today.

which, like all of the wild einkorn, has the genome AA (Figure 19.15). The wild varieties exhibit shatter much more than does the cultivated einkorn. Einkorn continued to be the major crop in Europe until the end of the Bronze Age (5000–4000 B.C.), when emmer wheat began to take over. However, einkorn is still grown to this day in isolated regions of the world (Perrino and Hammer, 1984). Einkorn is particularly successful in dry, cool conditions and when not grown under intensive farming (Vallega, 1992).

Emmer wheat also originated in the Fertile Crescent region and there is evidence of wild emmer wheat (*Triticum turgidum*, group *dicoccoides*) being present in

Figure 19.15   The evolution of wheat as a crop plant. Wheat is the most important food grain consumed by humankind. A series of photographs are shown of different varieties of wheat that have been involved in the evolution of this important food crop.

Figure 19.16 Wheat is one of the most successful plants in the world, covering vast areas of the world's surface. Here a picture of straw-coloured wheat fields in northern England are shown to emphasize the dominance of this plant in terms of growing area.

very early civilized sites at the same time as einkorn (Zohary and Hopf, 1993). The use of cultivated emmer wheat (*Triticum turgidum*, group *dicoccum*) became much more widespread than einkorn and displaced it. The emmer genome is tetraploid and has the make-up AABB, with the BB genome possibly coming from crosses with *T. speltoides*, *T. searsii* or *T. tripsacoides* (Kimber and Sears, 1987). The addition of the BB genome broadened the range of habitats in which wheat could be grown, since emmer can grow more successfully than einkorn when exposed to higher temperatures. The durum wheat (*Triticum turgidum*, group *durum*), used in pasta making, is derived from emmer wheat.

Spelt wheat is hexaploid and has the genome AABBDD. Spelt wheat has its origins in Iran in the Neolithic period (5000–6000 B.C.; Zohary and Hopf, 1993). Emmer wheat must have been dispersed in the near vicinity to *T. tauschii* (a wild grass in Iran). A hexaploid hybrid was formed where the grass DD genome was incorporated into the nucleus to give spelt wheat. During the Bronze Age (4000–1000 B.C.) the use of cultivated spelt wheat spread across Europe and North Africa. Spelt wheat ultimately gave rise to the bread wheats (*T. aestivum*), which is the major form of wheat grown today, and there are currently around 200 000 different varieties. The major protein in the bread wheats is gluten (11–15 % of weight of grain) and it is the presence of this protein that is essential for the trapping of gas bubbles during the rising of dough. Wheat varieties containing lower protein quantities are used for baking and cake making.

The addition of the DD genome increased the yield of the plant but at a great cost, as now the seeds are so large they cannot be released from the stem without assistance. Thus, through the history of wheat an alliance has been struck between humankind and wheat (Figure 19.16). We have used it as a food supply and eaten a large proportion of its seeds, but always enough has been planted again to broaden its range. Vast areas of land have been cleared for its growth. The remarkable feature of wheat is no longer its yield or its ability not to shatter, but the sheer area of land devoted to its growth. By this measure, wheat has become the most successful plant on Earth. Wheat has been so successful that many plant species have taken advantage of its success. By possessing seeds of a very similar shape and size to wheat, certain plant species have taken advantage of a niche habitat, occupying fields as weeds (Figure 19.16).

This Chapter has covered just a handful of a huge number of remarkable plants whose presence on the planet makes the world a richer place. In reality all plants are remarkable in some way. The last example, wheat, highlights how humankind has influenced and used plants to its advantage.

References

Lynch AJJ, Barnes RW, Cambecedes J and Vaillancourt RE (1998) Genetic evidence that *Lomatia tasmanica* (Proteaceae) is an ancient clone. *Australian Journal of Botany*, **46**, 25–33.

Brown MJ and Gray AM (1985) *Lomatia tasmanica* – a rare endemic plant from Tasmania's south-west. *Tasmanian Naturalist*, **83**, 1–3.

Giddy C (1984) *Cycads of South Africa*, 2nd edn (revised) C. Struik, Cape Town, South Africa.

Jordan GJ, Carpenter RJ and Hill RS (1991) Late Pleistocene vegetation and climate near Melaleuca Inlet, southwestern Tasmania. *Australian Journal of Botany*, **39**, 315–333.

Kimber G and Sears ER (1987) Evolution in the genus *Triticum* and the origin of cultivated wheat. In *Wheat and Wheat Improvement*, 2nd edn Heyne E.G. (ed.), Agronomical Monographs 13. American Society of Agronomists, Madison, WI.

Little EL (1980) *The Audobon Society Field Guide to North American Trees*, Alfred A. Knopf, New York.

Madson, C. (1996) Trees born of fire and ice. *National Wildlife*, **34** (6), 28–35.

Mitton JB and Grant MC (1996) Genetic variation and the natural history of quaking. *BioScience*, **46** (1), 25–31.

Perrino P and Hammer K (1984) The Farro: further information on its cultivation in Italy, utilization, and conservation (1). *Genetics Agriculture*, **38**, 303–311.

Schnell D (2002) *Carnivorous Plants of the United States and Canada*, Portland, OR.

Silagy CS and Haw N (1994) Garlic as a lipid lowering agent – a meta-analysis. *Journal of the Royal College of Physicians*, **28**, 39–45.

Vallega V (1992) Agronomic performance and breeding value of selected strains of diploid wheat. *Triticum monococcum. Euphytica*, **61**, 13–23.

Wherry ET (1972) Box-huckleberry as the oldest living protoplasm. *Castanea*, **37**, 94–95.

Zackowitz MG (2004) One sweet world: on the trail of chocolate. *National Geographic*, April.

Zohary D and Hopf M (1993) *Domestication of Plants in the Old World: The Origin and Spread of Cultivated Plants in West Asia, Europe, and the Nile Valley*, 2nd edn. Oxford University Press, New York.

Glossary

Abscisic acid (ABA) A plant growth regulator or hormone involved in maintaining dormancy in seeds and buds.

Allelopathy The excretion, by plants, of organic compounds into the soil that inhibit the growth of plants of other species.

Androdioecious Plants that have individuals that form only hermaphrodite or all-male flowers.

Angiosperms Classification division of flowering plants, also known as Magnoliophyta.

Annuals Plants that carry out their life cycle over a single year (see also Perennial and Biennial).

Anoxia Complete deprivation of a plant of oxygen (see Hypoxia).

Antennae complexes Chlorophyll/protein complexes bound to the light-harvesting complexes on the thylakoid membrane.

Anther Four-lobed organ in a flower that forms and releases pollen.

Apical meristem The growing point of a plant at the end of a shoot or root.

Apomixis Ability of some plants to form fertile seeds without any need for sexual reproduction.

Apoplastic Space inside a plant but outside of living cells (see Symplast).

Assimilation The incorporation of a nutrient or compound into the metabolism of a plant.

Autotrophy Ability of plants to make their own food (see Heterotrophy).

Biennial Plant that carries out its life cycle over a 2 year period (see also Perennial and Annual).

Bryophytes A phylum of simple plants including mosses, liverworts and hornworts.

C3 plants Plants that form a three-carbon intermediate as the first product of photosynthesis.

C4 plants Plants that form a four-carbon intermediate as the first product of photosynthesis.

Calvin cycle The metabolic cycle that fixes carbon dioxide into an organic form in plants.

Carboxylation The incorporation of carbon dioxide into an organic form in plants during photosynthesis.

Carotenoids Lipid-soluble red or yellow pigment in plants.

Cavitation Obstruction of the xylem by air bubbles.

Cell wall Rigid outer layer on most plant cells made up of cellulose fibres.

Cellulose A complex carbohydrate made up of b1,4-linked glucose molecules.

Chlorophyll A green pigment used to capture light energy in plants.

Chloroplasts A double membrane-bound organelle in plants that contains chlorophyll bound to proteins on the thylakoid membranes.

Circadian rhythm A daily rhythmic activity based on a 24 hour period.

Circumnutation The regular circular movement of the growing tip of a plant.

Companion cells Part of the phloem tissue associated with the loading of carbohydrate into the phloem sieve cells.

Crassulacean acid metabolism (CAM) A group of plants that accumulate organic acids during the night and metabolize them during the day to liberate carbon dioxide for photosynthesis.

Cryptochromes Light-sensitive protein in plants that absorbs blue light and allows the plant to respond to changes in blue light levels.

Cuticle Tough covering around the aerial structures of a plant, usually made up of secreted waxes.

Cyanobacteria Aquatic photosynthetic bacteria.

Cytokinins Class of plant growth regulators (hormones) that promote cell division and shoot formation.

Dehydrins A group of proteins believed to play a role in the response of plants to drought stress.

Dicotyledons   Plant of the class Magnoliopsida, the larger of the two great plant groups.

Dioecious   Plant species that has male or female individuals

Disaccharide   A sugar composed of two monosaccharides.

Dormancy   The ability of seeds not to germinate despite the conditions being favourable for the process.

Ethylene   Gaseous hydrocarbon (formula C_2H_2)

Filament   Structure in a flower supporting the anther.

Fructans   Carbohydrate made up of fructose molecules bound on to the fructose moiety of a sucrose molecule.

Gametophytic incompatibility   Incompatibility mechanism dictated by the gametophyte.

Gibberellic acid   Class of plant growth regulators (hormones) that promote elongation and relieve dormancy in seeds and buds.

Gluconeogenesis   Metabolic pathway that is the reverse of glycolysis.

Granal stacks   Groups of appressed thylakoid membranes in the chloroplast, associated with the activity of photosystem II.

Guard cells   Reinforced cells in the leaf epidermis that can flex to allow pores called stomata (singular: stoma) to open in the leaf surface for gaseous exchange to occur.

Gymnosperms   Classification division of cone-bearing plants, also known as Pinophyta.

Gynodioecious   Plants that have individuals that form only hermaphrodite or all-female flowers.

Hemiparasites   A parasitic plant that can grow independently of a host.

Heterotrophy   Ability of an organism to use existing organic molecules for food (see Autotrophy).

Holoparasites   A parasitic plant that needs a host to be able to grow.

Hypoxia   Deprivation of a plant of enough oxygen to respire efficiently using the Krebs cycle (see Anoxia).

Intercalary meristem   Growing point at the base of a leaf in grasses.

Internodes   Space between the nodes on a stem.

Jasmonic acid   This compound and its methyl esters are natural plant growth regulators that play an important role in wounding and plant resistance to pests.

Krebs cycle   Metabolic cycle responsible for producing most of the energy generated in respiration.

Leaf primordial   The initial formation of a leaf from a meristem.

Lignin   Complex polyphenolic compound used to support the structure of a plant.

Meristem   Dividing group of cells that generate further plant material, such as shoots, leaves and roots.

Microbodies   A group of membrane-bound cellular components in plants, also called peroxisomes and glyoxysomes, which have particular metabolic functions in the cell.

Micropyle   Region on the embryo sac where the pollen tube enters in order for fertilization to occur.

Microspore   Immature pollen grain.

Mitochondria   (singular: mitochondrion) Double membrane-bound cellular organelles in which respiration occurs.

Monocotyledons   Plant of the class Liliopsida, the smaller of the two great plant groups

Monoecious   Plant species that has male and female flowers on the same plant.

Mycorrhizal associations   A link between a plant and a fungus through the root system, in which the plant alone or both the plant and the fungus gain an advantage from the association.

Mycotrophic   Literally means 'fungus-eating' – a term used to describe the feeding strategy of saprophytic plants.

Nastic movements   Movements of a plant in response to a non-directional stimulus, e.g. temperature or humidity.

Node   Leaf attachment point on a stem.

Non-cyclical photophosphorylation   Phosphorylation of ADP using light energy, using the photosystems I and II.

Organelles   A differentiated structure within a plant cell, such as a chloroplast, mitochondrion or vacuole.

Osmosis   The diffusion of water molecules through a semipermeable membrane.

Papillae cells   Finger-shaped cells that line the stigma of a flower.

Perennial   A plant that carries out its life cycle over several years (see also Biennial and Annual).

Phloem   Living section of the vascular tissue in plants that transports primarily sugars and amino acids.

Photoperiodism   Ability of plants to sense and respond to changes in day length.

Photorespiration   Metabolic pathway used to recover carbon from the production of 2-phosphoglycollate during photosynthesis.

Photosystem II   Protein complex on the thylakoid membranes that shuttles electrons during photosynthesis, resulting in the breakdown of water to form oxygen.

Phyllotaxis   The arrangement of leaves around the stem of a plant.

Phylogenetic tree   A diagrammatic representation of the evolution of plants, depicting the interrelationship between the different phyla.

Phytochromes   Pigments in plants bound to proteins that allow the plant to sense red and far-red light.

Phytomer Repeated pattern of leaf formation or root formation in plants.

Plant genetic manipulation The ability to insert new genes into plants.

Plasmodesmata Channels that link the symplast of one cell with that of another.

Pollen tube The formation of a long, thin, tube-like cellular extension in pollen grains during pollination.

Pollination The movement of pollen grains from the anthers to the stigmatic surface.

Protochlorophyllide A precursor molecule for the synthesis of chlorophyll.

Pteridophtyes A phylum of ferns

Q-cycle Part of the electron transport complex in the chloroplast that shuttles H^+ ions from the stroma to the thylakoid lumen

Raffinose series oligosaccharide A carbohydrate molecule with galactose molecules added to a sucrose molecule.

Recalcitrant seeds Seeds that cannot survive drying below around 50% relative water content.

Rhizoid A root-like structure from which mosses, liverworts and ferns gain nutrients and attachment to a surface. In orchids, 'rhizoids' is a term used to describe the formation of hair-like filaments from the protocorm.

Root cap A layer of reinforced cells that overlie the dividing apical meristem in a root. They serve to protect these cells during root growth through the soil. The root cap is constantly replaced.

RUBISCO A shortened name for the enzyme ribulose bisphosphate carboxylase/oxygenase, used to fix carbon dioxide in the Calvin cycle.

Saprophytic nutrition The ability to break down non-living organic components to support growth.

Scarification Practice of scratching a seed in order to get it to germinate.

Seismonasty Response of a plant to vibrational movement.

Self-incompatibility The ability of a plant to recognize pollen grains that have originated from itself or a very closely-related individual plant.

Sink tissues Plant tissues that have a demand for a particular product, such as sucrose or amino acids. These are obtained from a linked source tissue.

Sporophytic incompatibility Incompatibility mechanism dictated by the sporophyte.

Starch Polymeric carbohydrate formed from the monomer of glucose.

Stigma The surface of the carpels that received pollen grains in plant sexual reproduction.

Stoma (plural: stomata) A pore in the leaf surface formed by movement of two guard cells.

Stratification The storage of seeds for a period under cold conditions in order to achieve germination.

Stroma The contents of the chloroplast outside of the thylakoid membranes.

Style Structure between the stigma and the ovaries in a flower.

Symplast Space inside a plant but inside living cells (see Apoplast).

Thylakoid membrane A membrane within the chloroplast that bears the chlorophyll and photosystems.

Thylakoid lumen Space bound by the thylakoid membrane, where H^+ ions are pumped during photosynthesis.

Tonoplast Alternative name for the vacuole.

Tracheids The vessels in which water is moved in the xylem. These are found mainly in gymnosperms and angiosperms.

Transpiration Evaporation of water from the leaves of a plant and its replacement from the xylem tissues.

Tropism Movements of a plant in response to a directional stimulus, e.g. touch or light.

Vacuole Membrane-bound compartment in a plant cell that can take up around 90% of the cell volume in some cells. In plants the vacuole mainly has a storage function.

Vascular cambium Layer of dividing cells around the stem of a plant that forms the phloem and xylem tissue as the plant grows.

Xylem Dead section of the vascular tissue in plants that transports primarily water from the roots to the leaves.

Xylem vessels The vessels in which water is moved in the xylem. These are found mainly in flowering plants.

Index

Abscisic acid (ABA)
  in drought stress 176–**177**, 207
  made in roots 63
  in seeds 217–218, 220
  in tissue culture 274
Abscission 188, **206**
AIDS 263–264, 267
Alkaloids
  in parasitism 106
  in defence **251–252**
  in medicine 257, 259, **261**, **263**, **265–268**
Allelopathy **252–253**
Amino acids
  non-protein **249**
  photorespiration 31–32
  synthesis 17, 56, 59, **80–82**, 87
  transport 43–44, 50, 103
  in nectar secretions 118
Amyloplast
  in gravity sensing 167–**168**
  in starch storage **56**–57
  import into **56**–57
Androdioecious 142–**143**
Angiosperms **2**–7, 290
Annuals 66, 107, 112, 200
Anoxia **186**–187
Antennae complexes
  structure **20**–21, **23**
  association with other proteins 26
Anther
  in reproduction **133–136**, 141, 145, 229, 235, 283
  structure 146
Antimalarial drugs
  quinine 251, **261–262**
Apical meristem
  shoot 9, 11, **151–154**, 157
  root 11, 154–155, **157**
Apomixis 132–133
Apoplastic space
  phloem transport 35, 39, 43–53, 57
  xylem transport 67–69, 73, 184, 189, 171

  auxin movement 163–164
  plant movement 171
Apple
  Cox 149
  climacteric 205–206
  clones 132
  cyanogenic glycosides 249–**250**
  incompatibility groups 144
Arabidopsis
  circadian clock 198
  cold stress 189
  ethylene sensitivity 206
  fertilization 140, 144
  hydrotropism 167
  light perception 165, 194, 198
  mutants 158–159
  photoperiodism 199–200
  thigmotropism 170, 202
  water transport 184
Araceae 234–235
Asexual reproduction 131–133, 161, 211
Assimilation
  of iron 83–85
  of phosphorus 78–80
  of nitrogen 80–81
  of sulphur 85–86
Autotrophy 18–41, 104
Auxin
  abscission 206
  lateral roots 80, 155
  plant movement 127–128, 159, 161–174
  tissue culture 269, 271–276

Banana
  ripening 56–57, 205–206
  tissue culture 269
  tannins 247
Basipetal 161, **163**, 165–166
Biennial 63
Biomes 7–9
Bladderwort 116–117, 120–124, 126

Physiology and Behaviour of Plants Peter Scott
© 2008 John Wiley & Sons, Ltd